Advanced
Welding

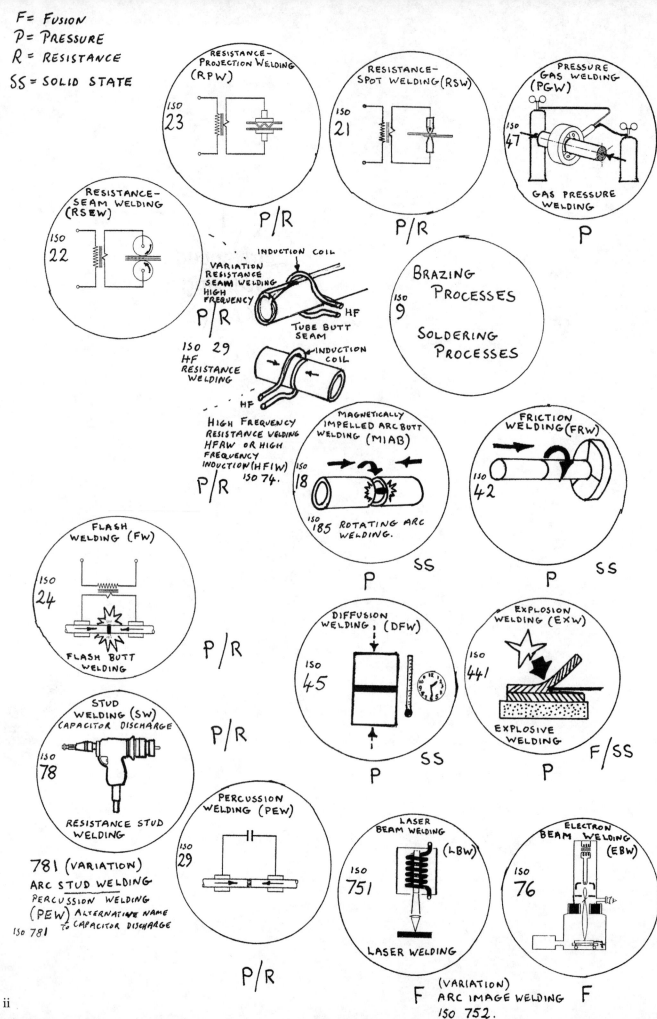

F = FUSION
P = PRESSURE
R = RESISTANCE
SS = SOLID STATE

RESISTANCE-PROJECTION WELDING (RPW)
ISO 23
P/R

RESISTANCE-SPOT WELDING (RSW)
ISO 21
P/R

PRESSURE GAS WELDING (PGW)
ISO 47
GAS PRESSURE WELDING
P

RESISTANCE-SEAM WELDING (RSEW)
ISO 22
P/R

VARIATION RESISTANCE SEAM WELDING HIGH FREQUENCY
INDUCTION COIL
HF
TUBE BUTT SEAM
ISO 29
HF RESISTANCE WELDING
INDUCTION COIL
HF
HIGH FREQUENCY RESISTANCE WELDING HFRW OR HIGH FREQUENCY INDUCTION (HFIW) ISO 74.
P/R

BRAZING PROCESSES
SOLDERING PROCESSES
ISO 9

MAGNETICALLY IMPELLED ARC BUTT WELDING (MIAB)
ISO 18
ISO 185 ROTATING ARC WELDING.
P   SS

FRICTION WELDING (FRW)
ISO 42
P   SS

FLASH WELDING (FW)
ISO 24
FLASH BUTT WELDING
P/R

DIFFUSION WELDING (DFW)
ISO 45
P   SS

EXPLOSION WELDING (EXW)
ISO 441
EXPLOSIVE WELDING
F/SS
P

STUD WELDING (SW) CAPACITOR DISCHARGE
ISO 78
RESISTANCE STUD WELDING
P/R

781 (VARIATION) ARC STUD WELDING
PERCUSSION WELDING (PEW) ALTERNATIVE NAME TO CAPACITOR DISCHARGE
ISO 781

PERCUSSION WELDING (PEW)
ISO 29
P/R

LASER BEAM WELDING (LBW)
ISO 751
LASER WELDING
F
(VARIATION) ARC IMAGE WELDING ISO 752.

ELECTRON BEAM WELDING (EBW)
ISO 76
F

ii

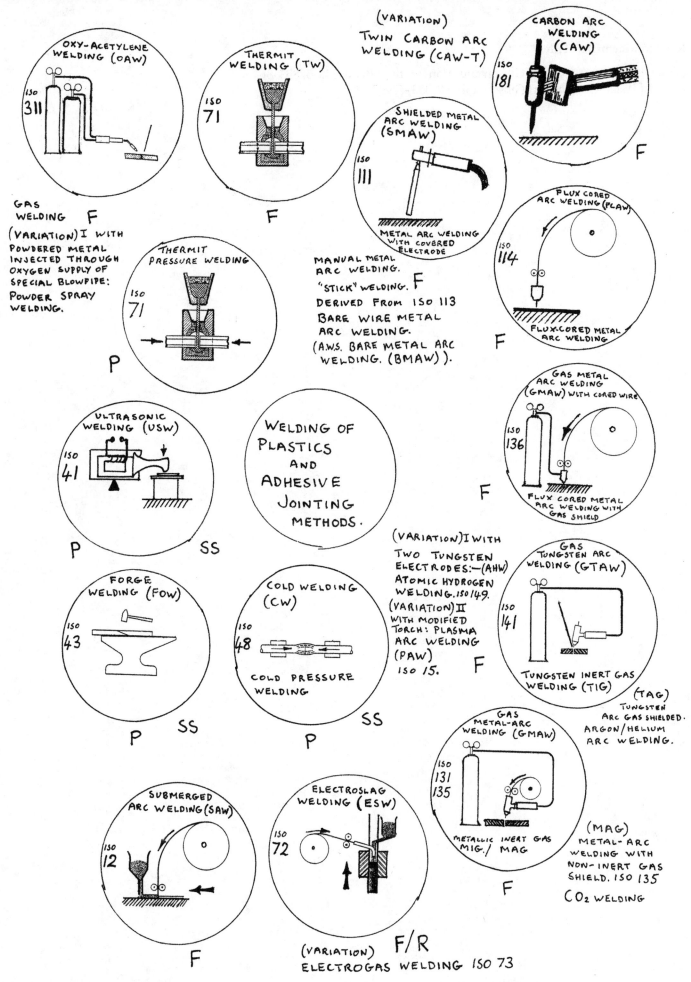

OXY-ACETYLENE WELDING (OAW)

ISO 311

GAS WELDING F

(VARIATION) I WITH POWDERED METAL INJECTED THROUGH OXYGEN SUPPLY OF SPECIAL BLOWPIPE: POWDER SPRAY WELDING.

THERMIT WELDING (TW)

ISO 71

F

THERMIT PRESSURE WELDING

ISO 71

P

(VARIATION) TWIN CARBON ARC WELDING (CAW-T)

SHIELDED METAL ARC WELDING (SMAW)

ISO 111

METAL ARC WELDING WITH COVERED ELECTRODE

MANUAL METAL ARC WELDING. "STICK" WELDING. F DERIVED FROM ISO 113 BARE WIRE METAL ARC WELDING. (A.W.S. BARE METAL ARC WELDING. (BMAW)).

CARBON ARC WELDING (CAW)

ISO 181

F

FLUX CORED ARC WELDING (FCAW)

ISO 114

FLUX-CORED METAL ARC WELDING

F

GAS METAL ARC WELDING (GMAW) WITH CORED WIRE

ISO 136

FLUX CORED METAL ARC WELDING WITH GAS SHIELD

F

ULTRASONIC WELDING (USW)

ISO 41

P   SS

WELDING OF PLASTICS AND ADHESIVE JOINTING METHODS.

FORGE WELDING (FOW)

ISO 43

P   SS

COLD WELDING (CW)

ISO 48

COLD PRESSURE WELDING

P   SS

(VARIATION) I WITH TWO TUNGSTEN ELECTRODES:—(AHW) ATOMIC HYDROGEN WELDING. ISO 149. (VARIATION) II WITH MODIFIED TORCH: PLASMA ARC WELDING (PAW) ISO 15.

F

GAS TUNGSTEN ARC WELDING (GTAW)

ISO 141

TUNGSTEN INERT GAS WELDING (TIG)

(TAG) TUNGSTEN ARC GAS SHIELDED. ARGON/HELIUM ARC WELDING.

GAS METAL-ARC WELDING (GMAW)

ISO 131 135

METALLIC INERT GAS MIG./ MAG

(MAG) METAL-ARC WELDING WITH NON-INERT GAS SHIELD. ISO 135

CO₂ WELDING

SUBMERGED ARC WELDING (SAW)

ISO 12

F

ELECTROSLAG WELDING (ESW)

ISO 72

(VARIATION) F/R ELECTROGAS WELDING ISO 73

F

Related Macmillan Titles

Basic Welding Stuart Gibson and Alan Smith [ISBN 0–333–57853–8]
Practical Welding Stuart Gibson [ISBN 0–333–60957–3]

Advanced Welding

Stuart W. Gibson

MSc, DME, Cert Ed, C&G (FTC), Sen M Weld I, MAWS, MAWTE,
Registered Welding Engineer

Former Lecturer in Charge of Welding, Hopwood Hall College

MACMILLAN

To the late MICK KERR, whose patience,
sense of humour and enthusiasm have
helped many along the welding path

First published 1997 by
MACMILLAN PRESS LTD
Houndmills, Basingstoke, Hampshire RG21 6XS
and London
Companies and representatives
throughout the world

ISBN 0–333–65384–X

A catalogue record for this book is available
from the British Library.

This book is printed on paper suitable for recycling and
made from fully managed and sustained forest sources.

10   9   8   7   6   5   4   3   2   1
06   05   04   03   02   01   00   99   98   97

Designed and typeset by 🇦 Tek-Art, Croydon, Surrey

Printed in Hong Kong

Every effort has been made to ensure that the information
contained in this book is accurate at the time of going to
press. However, the author and publisher do not assume
responsibility or liability for any applications produced
from the information. In welding, the application can vary
considerably from job to job and it is the responsibility
of the user to carry out work to statutory requirements.

Contents

Preface

Advanced Welding has been developed as a direct response to requests for a book to carry on from where *Basic Welding* left off. *Basic Welding* (ISBN 0–333–57853–8) is intended to act as an introduction to welding and, as such, covers the basic principles. To reduce repetition and save time, space and cost, *Advanced Welding* refers the reader back to *Basic Welding* where appropriate. For this reason, it would be advisable either to have read *Basic Welding* before studying the present book, or to have access to a copy to refer back to.

The term 'advanced' is, of course, relative. Hopefully, its use will be considered acceptable in this instance, as the text advances from *Basic Welding* and introduces the reader to advanced welding processes and their applications. It also covers the conventional welding processes in greater depth, while introducing some background history and metallurgy.

Advanced Welding is intended to act as a reference book for engineers and as a text-book for students following the higher levels of welding and general engineering courses. By covering work in some areas at slightly higher levels, it is envisaged that *Advanced Welding* would also be of help as an introduction to elements of the European / International Welding Engineer syllabus.

Grateful thanks are due to the following individuals and organizations for their help: My wife (for great patience!!), John Winckler – Macmillan Press Ltd, Dr Graham Wylde, Fred Delany and Margaret Slepyan – TWI, Dr Frank G. DeLaurier and Jeff Weber – AWS, John Hicks – IIW, Dr Steve Wood – BOC Gases, Europe, Dr Ralph Yeo – Lincoln Arc Welding Foundation, David Keats – Hydromech, Terry Bell – Dresser-Rand U.K., Ken Richmond – Precision Beam Technologies Ltd, Lawrence Darby – Thompson Friction Welding Ltd, Roy Bradshaw, Ian Carter – Imperial War Museum, UTP (UK) and UTC Engineering, Nigeria and Ghana, Leica U.K. Ltd, Murex Welding Products Ltd, Edward Nock – Westbrook Welding Ltd, Farrell Engineering Ltd, and Kuka GmbH.

Macmillan Press Ltd (Macmillan Education) are thanked for permission to use certain illustrations from the author's *Practical Welding* book (ISBN 0–333–60957–3).

Terminology

AWS (American Welding Society) terms for processes are used throughout, in accordance with the chart at the front of the book. For example, GMAW (gas metal-arc welding) is used in most instances with MIG (metal inert gas) and MAGS (metal-arc gas shielded) in brackets. Also GTAW (gas tungsten arc welding) is employed with TIG (tungsten inert gas) and TAGS (tungsten arc gas shielded) in brackets. For although AWS terms are preferred and used internationally, these other terms appear in welding literature and various schemes and syllabuses. Other well-used terms, which may or may not be standard, are included in brackets where appropriate.

The metric system is used throughout with imperial measurements in brackets, temperature being quoted in °C with °F in brackets in most cases.

Temperature conversions can be found on pages 282 and 283.

STUART GIBSON
Manchester, 1997

History and background

– And Zillah, she also bare Tubal-Cain, an instructor of every artificer in brass and iron

Genesis 4:22 (around 3950 years BC)

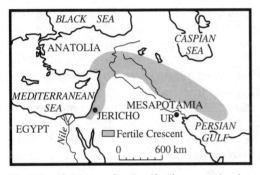

Fig. 1.1 *Sketch map showing 'fertile crescent' region about 10 000 years ago, when the area had good rainfall and fertile soil. The hills were also well wooded*

Fig. 1.2 *Sketch from an Egyptian wall painting from an 18th Dynasty tomb (about 1500 BC) showing the smelting of bronze in a furnace. The bellows are being operated by the metalworkers' feet*

Forge welding, the process in which heated metals are hammered together until they fuse, was practised by the ancients, as was brazing and soldering.

It is not clear when forge welding was first used, as primitive peoples in different parts of the world appear to have developed methods of metalworking independently over the centuries, and these principles could have been lost and rediscovered many times.

Around 6000 to 7000 BC, items were being made from copper in Anatolia (see Figure 1.1) and it is believed that copper was readily available in this region. Evidence of the early use of copper has also been found in Spain and Italy as well as in the British Isles. It is quite possible that its usefulness was discovered independently in some of these regions.

Initially copper was worked without heat, simply by hammering, then it was discovered that it could be melted and poured into a mould to cast a shape. Some of the first moulds were made from stone.

Eventually it was discovered that impure ore could be refined by further heating and that alloys could be made by blending metals together. The first of these alloys to be discovered appears to have been a mixture of approximately one part of tin with ten parts of copper to form bronze.

It is possible to put approximate dates to the use of bronze, as in some regions it coincided with the use of written documents and wall paintings (Figure 1.2).

Bronze was made and used in the region of Mesopotamia (Figure 1.1) earlier than 3000 BC and in China around 2000 BC. Examples of bronze objects from Thailand, however, have been dated around 3600 BC.

Bronze could be cast much more easily than copper and would maintain a much better cutting edge, making it suitable for tools and weapons.

Gold was another metal used by the ancients, mostly for decorative purposes. The discovery and use of gold were probably helped by the fact of it being in deposits nearer the surface in ancient times.

Simple mouth-blown blowpipes were used for soldering and forming metals by enhancing the heat of a fire (Figure 1.3).

Examples of these blowpipes can be seen in wall paintings of the tomb of Ti at Saqqara in Egypt, which is thought to date from around 2500 BC. There is, however, no proof that such blowpipes were ever used for the welding of metals by fusion in these times (Figure 1.2).

Gold beads discovered in the royal graves at Ur show evidence of soldering, the Ur empire being based in Mesopotamia around 2112 to 2004 BC (Figure 1.1).

Iron did not make an appearance until well after the first civilizations were established and its history at this stage is vague, as iron artefacts from early history have tended to be lost by oxidation. It is generally accepted, however, that the first traces of ironworking can be found in Asia Minor, although the use of iron did not become widespread until around 1000 BC. Welding in these early times was employed not only in making an end product, but also within the ironmaking process itself.

Fig. 1.3 *An Ancient Egyptian smith using a blowpipe to raise the temperature in a furnace*

The method used in smelting the iron in ancient and medieval times required a mixture of ore and charcoal to be heated within a blast of air in order to reduce the ore to iron without melting. The small pieces of iron were then taken from the furnace, cleaned and reheated in order that they could be hammered together to form larger pieces or **blooms** of iron. These small blooms were then, in turn, welded together to form pieces of useful size. Special welding furnaces were built specifically for this purpose.

The Romans copied the ancients in using these methods and remains of a Roman welding furnace can be found near Corbridge on Tyne in Northumberland UK,[1] at the site of an old military outpost south of Hadrian's Wall.

This direct reduction method gave a carbon content between 0.02 and 0.10 per cent, which meant that it could not be suitably hardened in order to make swords and agricultural or domestic cutting tools. Steel was costly and in short supply during Roman times and the Middle Ages, and it was quite common, therefore, to weld a thin strip to the edge of a tool and then to harden and temper this for use as the cutting edge. This method is known as **steeling**.

Viking smiths made swords by carburizing thin strips of iron and then forge welding them together lengthwise.

The term 'weld' is derived from the Old English 'weallan', to boil or 'well-up', and from the German 'wallen'.

The barbarism of the Fifth Century, and in particular Attila's Huns of around AD 447, destroyed the culture and economy of the Roman Empire throughout Europe and the trail of destruction left very few technological artefacts. Grave finds from this period have, however, indicated that swordmaking was well developed, based on earlier Roman tuition in the Rhineland.[2] Some types of chainmail were also manufactured by welding rings together. Swordmaking and ironworking also appear to have been taking place in other parts of the world, such as China (Figure 1.4), quite independently.

It is interesting to look at the medieval legends surrounding the mastersmith Wayland, who, among other things, is credited with having made the three swords – Naegling, Ekkisax and Mimming in *Thidrik's Saga*, together with the giant Beowulf's 'splendid mailcoat' mentioned in the Eighth Century Anglo-Saxon poem.

English tradition places Wayland's or another ancient smith's forge in a cave near the White Horse in Berkshire. Evidence makes it clear, however, that the story has origins in Jutland, based at the court of King Nithhad, where Wayland is lamed in order to keep him captive so that the king can have exclusive access to his swordmaking skills.

The Franks' Casket at the British Museum is dated from the Eighth Century and depicts Wayland's revenge in scenes of carved whalebone. In one, Wayland can be clearly seen to be holding the head of what is said to be one of King Nithhad's sons over an anvil.

Thidrik's Saga says that Wayland had laboured for seven days to produce a sword that could cut a felt sheet floating in the river, but Wayland sought further improvement.[3] With a file he reduced the sword to dust, then mixed the dust with meal to make cakes which were fed to some starving poultry. The bird droppings were then put in the forge and all the soft parts of the iron removed. A further sword was then made from this metal, but to the king's surprise, Mimming was created from metal that had undergone the entire process again.

Analysis of Wayland's methods tends to give credibility to this part of the saga, as prior to being subjected to the high temperature of the blast furnace, steel manufacture could be greatly enhanced by fine division, as pre-filing would allow the carbon diffusion to be accelerated and the poultry feeding would yield slag-free iron strips more suitable for forge welding together. It

Fig. 1.4 *A Chinese blacksmith forging a swordblade. The Chinese developed very efficient bellows, allowing them to smelt iron earlier than was possible in other parts of the world. In the 5th Century BC they were making cast iron cauldrons, spades and parts for chariots*

Fig. 1.5 *Sketch based on a 14th Century print of an armourer at work on a knight's helm*

Fig. 1.6 *Early iron cannons could not be cast in single pieces at the time. They were made like barrels, with iron staves bound together with iron rings, (hence the term 'gun barrel'). The strips of metal were forge welded together. Some designs completely covered the welded strips with welded rings in an attempt to give extra strength*

is also feasible that, after passing through the birds' stomachs and being pickled by high acid levels, the residual phosphorus content would also be greatly reduced.

In the Middle Ages many knights' helmets were forge welded (Figure 1.5). The helmet belonging to Edward, the Black Prince (1330–76), was made from three pieces of metal, welded in such a way that the joints could not be seen.[4]

Around the year 1330, the Italians used gunpowder to fire missiles from new weapons called 'thunder tubes', the word 'cannon' being derived from the Latin 'canna', meaning tube. The large iron guns of this time could not then be cast in one piece. They were made by welding iron staves together and then strengthened with welded iron rings, in the same way that a barrel was made, giving the name 'gun barrel'. One example in France in 1375 was the building of a cannon employing thirteen smiths, three forges and 1040 kilograms (2300 pounds) of iron.

Although metal rings were welded round the gun barrels in an attempt to stop them bursting, many of the early welds were weak causing guns to burst in action. King James I of Scotland was killed in such an incident at the test firing of a cannon in 1437, when the barrel exploded. Figure 1.6 shows sketches of early welded cannons.

The method of welding small lumps of iron together by heating in a charcoal furnace and then hammering, continued through the Middle Ages until about the Fifteenth Century, when water power was harnessed to operate bellows and larger furnaces were constructed. The large, mechanically operated bellows provided increased temperatures which could for the first time melt the iron into a liquid state. This new type of iron, or cast iron, was probably first discovered by the accidental overheating of a charcoal bloomery furnace. At first, there was no obvious use to which it could be put, as it was hard and brittle and could not be shaped by hammering. It did not take long, however, before it was realized that cast iron could provide a short cut to the production of large quantities of wrought iron and the iron smelting industry, based on the charcoal-fuelled blast furnace, came into being. The precise origin of the blast furnace is uncertain, but it is possible that it was developed somewhere in the region that is now Belgium, before AD 1400.

A major consequence of the development and spread of blast furnace technology was the availability of an inexpensive and efficient casting material. Iron castings were soon being used to make many products that had previously been made from welded wrought iron. Probably one of the most important examples was cannon barrels, which were very successfully cast by the foundries of that era.

As casting technology improved, the use of welding diminished, with forge welding still remaining the main process until towards the end of the Nineteenth Century when water-gas welding came into existence. Lavoisier produced water gas in 1793, by passing steam, unmixed with air, through glowing coke. Water gas consists essentially of hydrogen, carbon monoxide and nitrogen. Its application to welding was pioneered by Fitzner[5] in Germany around 1878, when he developed water-gas burners and anvils for end-to-end, lap and wedge welding (Figure 1.7) [see also Chapter 2].

Another welding process employed by some foundries of this period used molten metal to repair cracks or blowholes in castings. A mould was made around the area to be repaired, molten iron poured in until it was fluid all around the edge of the repair and then the runner was closed, leaving the iron to cool down in the mould. This process is known as 'flow or cast welding', and it was the forerunner of modern fusion welding, because it also melted the edges being joined as well as providing molten 'filler metal'.

It is possible that Professor G. Lichtenberg of Germany may have joined metals by 'electric fusion' in the laboratory as early as 1782, however, most

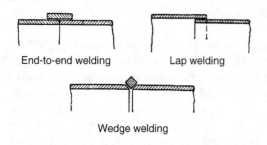

Fig. 1.7(a) *Welding preparations used by F. Fitzner around 1878*

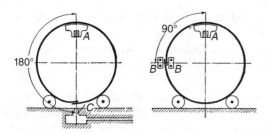

Fig. 1.7(b) *Showing the advantages of welding a seam by water gas over the use of a coke fire.*

On the left the joint can only be heated from one side using the coke fire 'C'. The seam then has to be rotated through 180° to be hammered on the anvil 'A'.

On the right the joint can be heated to a more uniform temperature by water gas burners 'BB' on each side. Also the joint has only to be turned through 90° to the anvil 'A'

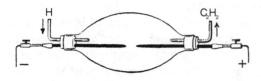

Fig. 1.8 *Berthelot's synthesis of acetylene using a carbon arc burning in hydrogen*

sources trace the history of electric welding to 1801, when Sir Humphrey Davy (1778–1829) discovered that an arc could be created by bringing the two terminals of a high-voltage electric circuit near to each other.

Davy demonstrated the arc to the Royal Institute in 1808, however it was not used practically until many years later, Davy only using the term 'arc' for his discovery some 20 years later.

The first person known intentionally to join metals by electric welding was an Englishman named Wilde.[6] About the year 1860, he melted small pieces of iron together. In 1865 he was granted a patent on his process, being the first patent relating to electric welding.

In 1836, Edmund Davy (a cousin of Humphrey Davy), obtained a hydrocarbon by the action of water on potassium carbide, the product at the time being called klumene.[7] Berthelot rediscovered the substance in 1860 and called it acetylene. Berthelot showed that it can be formed by passing ethylene or alcohol vapour through a red-hot tube, and by direct synthesis from its elements when an electric arc burns between carbon poles in an atmosphere of hydrogen (Figure 1.8). Small quantities of methane and ethane are also formed. In 1862, Wöhler demonstrated that acetylene could be produced by the action of water on calcium carbide but it was still some years before other developments allowed the birth of gas welding using acetylene. Meanwhile, De Meritens used a carbon arc process to join the plates of a storage battery. (A diagram of the process as shown in the French patent of 1881 is given on page 37 of *Basic Welding*.)

Two scientists, Nikolas de Bernardos and Stanislav Olszewski, experimented with the De Meritens' process and improved upon it, and in 1885 they were issued with a British patent for a welding process employing carbon electrodes. Bernardos also filed a Russian patent for a process in which the carbon electrode was connected to the positive terminal of a DC circuit. The carbon was not fixed as in the De Meritens' process, but was fitted with an insulated handle in order that it could be manipulated easily by hand and used for welding cast iron, wrought iron and steel. This process was patented in 1887 and, therefore, Bernardos is generally credited as being given the first patent for arc welding.

During the 1880s, many scientists were carrying out experiments with electricity and it is thought that Joule made a resistance weld in his laboratory in England in 1857, although this was never developed, and it was Elihu Thompson in the USA who, quite independently, built a machine for this purpose. Encouraged by a certain amount of success, he developed a larger machine with water-cooled jaws and in 1886 took out the first patent for welding by electric resistance.

Professor Thompson's discovery of this important process came about by chance as he was working in the laboratory of the Franklin Institute, Philadelphia, discharging a Leyden battery through the fine wire winding of an induction coil. The fine wire became a high potential primary, the ends of the coarse wire winding being brought into light contact and welded together. From this experiment the 'Thompson process' was developed. No electric arc is employed, the heat which effects the welding being due to the resistance of the parts of the metal pieces at the point of contact where they are to be welded together. The process is fully described in Chapter 9. Figure 1.9 shows a print of an early automatic chain welding machine employing the 'Thompson resistance welding process'. The machine shown was made in Germany around 1906.

In 1888 Bernardos took out a patent for spot welding using carbon electrodes while, around this time, Henry Howard brought the process of carbon-arc welding to the UK from Russia; his firm later became Messrs Stewarts & Lloyds Ltd, and by the 1890s they were producing welded steel piping of

Fig. 1.9 *Print showing an early German resistance welding machine for welding large chains.*

Such machines used the 'Thompson process' and were made in many different designs for welding particular components.

(The machine shown was made by the firm of Hugo Helberger GmbH about the year 1906)

various sizes "for use as conduits for water, sewage, gas or air, and for any working pressure up to 500 lbs per square inch". In the United States, the Baldwin Locomotive Works established a shop in 1892, using carbon-arc welding for locomotive maintenance. Acceptance of the carbon-arc welding process was slow, however, mainly because the procedures employed at the time introduced particles of carbon into the weld metal, transferred from the carbon electrode through the arc. These particles of carbon could make the welded joint hard and brittle.

A major step forward for welding technology was made just two years after the Bernardos' patent for carbon-arc welding was granted. It was another Russian, N.G. Slavianoff, who announced a process in which he had replaced the carbon electrode by a metal rod electrode. The arc from the rod not only melted the work at the joint to be welded, but the rod also gradually melted, adding metal to the weld. The rod therefore had to be fed downwards, to maintain the arc length, as well as being moved along the joint at the required speed. In this same year, 1889, and completely unaware of Slavianoff's process, Charles Coffin was granted a US patent for a similar metal-arc welding process. Charles Coffin later became the president of the General Electric Company. Commercial applications of the metal-arc process during the following years were dependent upon improvements to the electrode, the earliest being made from bare wire obtained from Norwegian or Swedish iron, which produced weak and brittle welds. The arc often overheated the weld metal and there was no protection against embrittling reactions with the air. In attempts to overcome these problems, researchers produced electrodes that had light coatings of various minerals and organic compositions. In the year 1907, Oscar Kjellborg of Sweden took out such a patent and is credited as being one of the original developers of the flux-covered electrode.

These early coverings mainly helped to stabilize the arc but did little to purify the weld metal or to shield it from the atmosphere. It was only when Strohmenger, in the UK, produced a heavily coated electrode in 1909[8] and took out the US patent in 1912, that the welding industry had a means of producing weld metal with good mechanical properties. This was the famous quasi-arc electrode which used a wrapping of asbestos string. Again, chance appears to have played a hand in the progress of welding technology, for

Arthur Strohmenger was a consultant chemist and two of his clients were the Cape Asbestos Company and the Tudor Accumulator company. Strohmenger became interested in the experiments on electric arc welding being carried out by an engineer named Slaughter who worked for Tudor Accumulator. He analysed some of the electrode coatings in use and then suggested making an electrode by wrapping a welding wire with some blue asbestos from the Cape Asbestos Company, a product which happened to be on his bench at the time. The electrode was a success and would also run on AC, in contrast to the wash-coated wire electrodes then being used. Strohmenger showed this new type of electrode to his friend Professor Sylvanus P. Thomson, who stated that the arc produced was not a normal metallic arc and so named it a 'quasi-arc'.*

The Quasi-Arc company which was formed to produce the asbestos electrode enjoyed much success, but later, as the demand for asbestos electrodes declined, it gradually went out of business.

A major increase in the use of welding occurred during World War I, when there was a sudden demand for a large number of transport ships, which could not be met by the existing production methods employing the relatively slow process of riveting. Government officials in the USA realized that faster methods of manufacture were required and Professor Comfort Avery Adams, Dean of Harvard Engineering School, was asked to appoint a committee to investigate the problem, the first meeting being held in 1917. Many members of the committee were of the opinion that the Thomson resistance method was the way forward, however, after visiting England, where they witnessed metal-arc welding with bare and covered electrodes being used on shipbuilding, bomb, mine and torpedo manufacture, they returned to the USA convinced that some form of arc welding was the way forward. The increase in arc welding in the UK at the time was due to gas shortages restricting the use of oxy-acetylene welding. Another major factor that promoted arc welding at this period was its use in the fast repair of scuttled German ships in New York Harbour. Welding experts from two railroad companies were called in and, with the help of arc welding, the ships were rapidly repaired and used to help the Allied war effort. In 1918. A.O. Smith of the USA devised a cellulosic electrode by wrapping paper soaked in sodium silicate round a steel wire to replace the British Quasi-Arc electrode, which was then unobtainable because of the war. The cellulosic electrode would only operate on DC and therefore required a motor-generator welding machine. This promoted a preference for DC sets which had existed in the USA for many years.

On 27 March 1919, Professor Adams became the first president and founder of the American Welding Society.

Alongside the events taking place in electric welding, oxy-fuel gas welding was also progressing gradually. The acetylene industry first developed in order to supply a source of artificial light and this was also true, of course, of the electric arc industry with the early carbon-arc lamps. The first types of blowpipes, however, used oxygen and hydrogen. Robert Hare Jr (1782–1858) demonstrated an oxy-hydrogen blowpipe to the Chemical Society of Philadelphia in December 1801.[9] In the 1850s Henri Sainte-Claire Deville (1818–81), then professor of chemistry at the École Normale Supérieure in Paris, made a number of oxy-hydrogen blowpipes for his researches on the metallurgy of platinum (Figure 1.10).

In 1838 Eugène Desbassayns de Richemont (1800–59), the son of a count from what is now Réunion Island in the Indian Ocean, patented the process of fusion welding. The following year he was awarded a gold medal at the Paris Exposition Publique des Produits de l'Industrie Française for the fusion welding of lead with his 'airhydric' blowpipe. He named his process soudure autogène, translated into English as 'autogenous welding', which was not

Fig. 1.10 *An oxy-hydrogen blowpipe devised by Henri Sainte-Claire Deville in the late 1850s for his platinum researches. The blowpipe is about $1\frac{1}{3}$ metres long and it was used for cupellation*
(From H. Sainte-Claire Deville and H. Debray, 'Du Platine et des Métaux Qui l'Accompagnent,' Annales des Mines, Mémoires, Vol. 16, 1859, Plate 1, Fig. 3)

*This was at the time when any possible health risks in the use of asbestos were not fully appreciated.

strictly correct but remained popular for many years until the term 'fusion welding' more accurately described a liquid-state welding process.

In 1888 Thomas Fletcher of Warrington, England, demonstrated his oxygas cutting torch to the Liverpool Section of the Society of the Chemical Industry. The torch cut by melting (fusion), not with an oxygen stream, the oxygen being for the purpose of increasing the flame temperature. The demonstration showed how a large hole could be cut through a $\frac{1}{4}$ inch (6.4 mm) thick wrought iron plate in a matter of seconds. This concerned bankers, who visited Mr. Fletcher's works. Banks were worried further when, on 22 December 1890, a Mr. Brown attempted to rob the Lower Saxony Bank in Hannover. He had successfully cut through one steel door when his oxygen supply ran out, forcing him to leave on a 'business trip' from which he did not return.

The Société L'Oxhydrique Internationale was founded by Félix Jottrand (1863–1907) in 1896 in Brussels for the manufacture of oxygen and hydrogen by the electrolysis of water. The company patented their oxy-hydrogen blowpipe and system of oxy-hydrogen welding in 1901. Dr L.A. Groth writes in 1906: "The blowpipe introduced by L'Oxhydrique under the name of 'Pyrox', and patented in almost every civilised country, has during its existence of five years proved to be entirely satisfactory in its action, no accident whatever having occurred during the operations with same, which speaks greatly in favour of its safety and ability of keeping a homogeneous composition of the flame; besides, it is light in weight and easy to handle."

Some 56 years after Edmund Davy had first discovered acetylene, Willson, a Canadian electrical engineer working in the USA and also at Moisson in France, discovered how to make calcium carbide on a commercial scale.

Efforts to store and transport acetylene compressed in steel cylinders foundered initially, since pure acetylene becomes a dangerous explosive in the absence of oxygen or air, when compressed to between one and two atmospheres. During 1896, George Claude (1870–1960), who was working as an electrical engineer for the Compagnie Française Thomson-Houston, had the idea that it might be possible to store acetylene dissolved in a liquid such as carbon dioxide in mineral walter (seltzer). The company liked the idea and set him up in a laboratory to work with Albert Hess (who died on the *Titanic* in 1912).

They discovered that 25 times as much acetylene could be stored, for a given pressure, when the acetylene was dissolved in acetone. Also, the acetone stabilized the acetylene and allowed higher pressures to be maintained without the danger of explosion. A patent was taken out in 1896 and in February 1897 the Compagnie Française de l'Acétylène Dissous (CFAD), (French Dissolved Acetylene Company) was founded. It was here that the oxy-acetylene torch was born, four years later.

A crucial factor towards the development of the oxy-acetylene torch was a discovery in 1895, by Henry-Louis Le Châtelier, professor of industrial chemistry at the École des Mines in Paris. He had found that a higher flame temperature could be obtained by burning together oxygen and acetylene rather than by burning oxygen and hydrogen.

Le Châtelier's brother was a director of CFAD, and put him in contact with Charles Picard, an engineer working at the CFAD laboratory.

Le Châtelier discussed the subject of porous material for filling dissolved acetylene cylinders in order to adsorb the liquid solvent for the acetylene. He also gave Picard the idea of making an oxy-acetylene torch. The first attempt in 1900 failed and then in 1901 Picard designed a nozzle to give an exit velocity higher than the flame propagation speed of over 200 metres per second with the oxygen-acetylene mixture.

In order to lower this flame propagation speed, Picard saturated the acetylene with petroleum spirit.

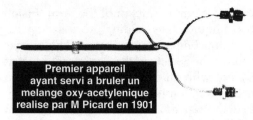

Fig. 1.11 *Charles Picard's oxy-acetylene blowpipe of 1901 at the Institut de Soudure in Paris*

On 21 March 1901 Picard lit the torch (which was basically a mixing tube fed by two smaller-diameter connecting tubes, see Figure 1.11). Taking no chances, he stood behind a partly opened door and used a long pole with a taper attached to the end. It was a success and the small bluish flame burned quietly.

Edmond Fouché, a director of CFAD, who had encouraged Picard's work, was telephoned and came to see it working. Straight away, the torch was used successfully to repair a cast iron component from an acetylene pump.

Fouché announced this success at a meeting of the Société des Anciens Elèves des Arts et Métiers on 3 November 1901, and to the Société Française de Physique on 6 December. The oxy-acetylene torch had been born!

By 1902 a torch had been developed that would work satisfactorily without additives to slow down the propagation speed. Figure 1.12 shows a drawing of the CFAD oxy-acetylene torch that was marketed in 1904, complete with internal flashback arrestor. Figure 1.13 shows this particular type of torch being used to weld an iron vessel.[10] The CFAD was experiencing many difficulties at this time and was forced to let Fouché go as they could no longer afford his salary. Fouché had been working on an injector system for

Fig. 1.12 *Torch marketed by the French Dissolved Actylene Company in 1904 and intended for use with high-pressure acetylene also marketed by the firm. The actylene flows through a small drum of 'porous material' (probably aluminium shavings, as these were used later in torches) to prevent flashbacks and then into a perforated tube to allow the acetylene to mix with the oxygen*
(From G. Chalmarès, 'Le Chalumeau Oxy-Acétylènique, Soudure Autogène et Lampe de Projecteur,' La Nature, Vol. 32, No. 1613, 23 April 1904, p.328, Fig. 1)

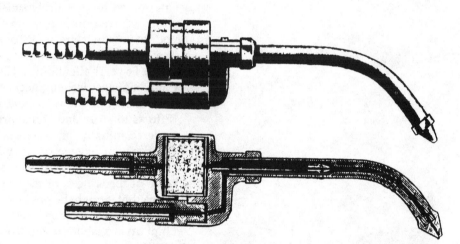

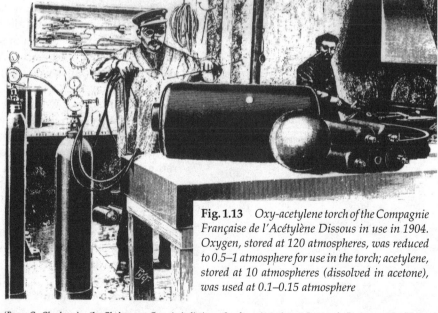

Fig. 1.13 *Oxy-acetylene torch of the Compagnie Française de l'Acétylène Dissous in use in 1904. Oxygen, stored at 120 atmospheres, was reduced to 0.5–1 atmosphere for use in the torch; acetylene, stored at 10 atmospheres (dissolved in acetone), was used at 0.1–0.15 atmosphere*

(From G. Charlmarès, 'Le Chalumeau Oxy-Acétylènique, Soudure Autogène et Lampe de Projecteur,' La Nature, Vol. 32, No. 1613, 23 April 1904)

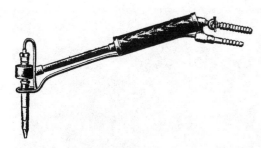

Fig. 1.14 *Fouché Cyklop Blowpipe*

a low-pressure torch and when he was taken on by the Société des Applications de l'Acétylène he patented the device (Figure 1.14). Fouché's torches were classed as being among the first trustworthy, well-constructed and safe blowpipes available. Two were sent to the United States and were used to repair a machine by welding.

In 1902, Carl Linde had discovered a method of liquefying air and producing nitrogen and oxygen, and it was only when this method became commercially available that oxy-acetylene welding was free to develop on a wide scale. The first oxygen-producing plant using the Linde process was opened in 1906 at Buffalo in New York State.

Another important welding process – Thermit welding, owes its discovery to C. Vautin, who, in 1894, found that finely powdered aluminium had an affinity for metallic oxides, and if such mixtures were ignited a very high temperature of around 3000°C is obtained. Dr Goldschmidt applied the process commercially by showing that iron bars could be welded within a mould, by using the molten iron produced by the reaction. The first patent was granted in 1897.

In 1903 Bouchayer produced a commercial spot welding machine but it still employed carbon electrodes. Harmatta made a machine in 1912, substituting the carbon electrodes for ones made from copper.

In 1920 the first vessel with an all-welded hull, *The Fullager* was launched by Cammell Laird of Birkenhead, and Henry Ford successfully defended himself against an action from the Thomson Company for the infringement of resistance welding patents, thus allowing possibly faster progress in resistance welder application. A similar lawsuit arose in 1929, when A.O. Smith sued J.F. Lincoln over extruded electrode coating. Lincoln's success in winning the case is said to have opened up opportunities for further progress.

In 1923 the Institution of Welding Engineers (later to be the Institute of Welding and subsequently The Welding Institute) was formed under its first president, Sir Peter Rylands, Bart.

In 1935 the submerged-arc welding process was developed by Linde in the USA.

Many discoveries and developments took place in the next few years as part of the war effort and in the early post-war years, as a direct result of war research. Some of the major events are mentioned here, in this brief history, together with the invention dates or first commercial usage of new processes. The processes themselves are discussed in detail in subsequent chapters.

Submerged-arc welding was developed by Linde USA in 1935, and in 1938 Germany used electric arc welding extensively in order to save weight in the construction of 'pocket' battleships. In 1939 arc stud welding (which had previously been used by the Royal Navy in 1919) was used in the USA on a large scale in the manufacture of aircraft carriers and other structures.

In 1940 the construction of 'Liberty' ships gave a major boost to the allied war effort and some spectacular failures drew attention to areas of quality control and the investigation of fracture problems in welded structures.

In 1941 inert-gas tungsten-arc welding was announced by the US aircraft industry, while 1944 saw the introduction of continuous covered electrode automatic welding with the UK Fusarc machine and high metal recovery iron powder electrodes from Philips in The Netherlands.

A novel application of flash-butt welding during the war was the 'PLUTO' project (PipeLine Under The Ocean), in which nearly 1000 miles (1600 km) of 3 inch (76 mm) diameter pipe was welded together using 200 000 welds. Lengths of the pipe were wound onto giant bobbins 27.5 m × 15 m (90 ft × 50 ft), floated out to sea and then unwound across the Channel after D-Day, allowing 1 000 000 gallons of petrol per day to be pumped to French depots (Figures 1.15 and 1.16).

Fig. 1.15 *Pipe being stored for the PLUTO project – the photograph shows nearly 200 miles of pipe*
(Photograph by permission of the Imperial War Museum, London)

Fig. 1.16 *The tug* Britannic *joining up the 27 mile pipeline from England to France at Boulogne in 1944. This was the seventeenth pipeline between England and the European mainland. The pipes for PLUTO were flash welded from basic 40-foot lengths. The concept for the pipelines was developed by Captain John Hutchings, OBE DSC RN*
(Photograph by permission of the Imperial War Museum, London)

1946 saw the introduction of HF-stabilized AC tungsten-arc welding of aluminium. Cold pressure welding was also invented in that year by the General Electric Co. Ltd. (U.K.). In 1948, inert-gas metal-arc welding (MIG) was introduced by the Air Reduction Company in the USA (gas metal-arc welding) (GMAW). In 1951 Electroslag welding was first used on a production line in Russia and Stiegerwald in Germany developed an electron-beam machine. 1953 saw the first use of CO_2 shielding on automatic machines for spray-transfer type welding in the USA and on semi-automatics in Russia. The Russians also announced friction welding in 1956 and electro-gas welding in 1957. The theory of laser light was also first suggested in 1957 by Charles Townes and Arthur Schawlow in the USA, but the first laser was not built until 1960, by another US scientist, Theodore Maiman. Also in 1960, vacuum diffusion welding was developed in Russia and ultrasonic welding was further developed.

Explosive welding was discovered in the USA after Bahrani and Crossland had accidentally joined components to formers when explosive forming.[11]

Magnetically impelled arc butt welding became commercially available in the early 1970s, after initial work in Germany. Since 1977, the Welding Institute has carried out a great amount of research work and developed new machines.[12]

Over the last few years, the emphasis has been on the improvement and 'fine-tuning' of many existing processes, in order to reduce costs and increase quality. The further miniaturization of electronic components and such innovations such as pulsed arc MIG and TIG with inverter power sources (GMAW and GTAW) are but a few of the major factors influencing welding trends.

Automatic welding machines have continued to be introduced for repetitive applications and there has been an increase in the use of robotics and computer control systems throughout the 1980s and 1990s.

The role of the manual welder comes to the fore in very skilled applications of welding, such as pipewelding or where there is much variation in the type of work.

There is still a great need for the highly skilled welder, who has a sound knowledge of most of the welding processes and the welding characteristics of the ferrous and non-ferrous materials likely to be encountered.

These factors are important in all companies that undertake welding but have particular importance in a small concern, which might employ just one welder, because of new quality requirements (European and International) that demand a 'named' individual to be responsible for the welding activity of the company.

The illustration at the front of the book shows the main material joining processes. In order to present the information in the space available, brazing and soldering processes have been reduced to one unit on the chart, as have the welding of plastics and adhesive joining methods. These topics and welding processes that are variations of the main ones on the chart are covered in subsequent chapters.

A more comprehensive list of processes with variations developed from different heat sources can be found in *Basic Welding* on page 129, or by reference to International Standards. At the end of this book you will find The American Welding Society's Master Chart of Welding and Allied Processes.

How to use the master chart

Each major welding process is portrayed as a sketch, or mentioned as a variation of the sketched process. Where a welding process is known by more than one name, the American Welding Society name and letter abbreviation is given above the sketch and the ISO (International Standards Organization) name below it. The ISO numerical indication of process is given to the left of each sketch.

If a process is also known by a common name or names other than those within the circle, then this is given outside the circle.

The letter 'F' next to a process indicates a fusion welding process. 'SS' indicates a solid-state process. All processes are either one or the other, with the exception of an explosive weld, which contains areas of fusion and solid-state welding.

A letter 'P' signifies that the process requires pressure to complete a weld and an 'R' shows that heating due to electrical resistance is employed.

References/further reading

1. Eric N. Simons, *Welding*. Frederick Muller Ltd
2. P. Dixon, *Barbaric Europe*. Elsevier-Phaidon

3. *Historical Metallurgy,* Vol. 10, p 84
4. Stuart Gibson and Alan Smith, *Basic Welding,* Macmillan
5. L.A. Groth, *Welding and Cutting Metals.* Constable, 1908
6. *The Procedure Handbook of Arc Welding.* The Lincoln Electric Company
7. J.R. Partington, *General and Inorganic Chemistry.* Macmillan
8. Peter Houldcroft, Steps in welding innovation and achievement. In *Metal Construction,* Vol. 5, No. 12, December 1973
9. A.C. Nunes, Jr., Gas welding origins. In *AWS Welding Journal,* Vol. 56, No. 6, June 1977
10. G. Chalmarès, Le chalumeau oxy-acétylènique, soudure autogène et lampe de projecteur. In *La Nature,* Vol. 32, No. 1613, April 1904
11. A.S. Bahrani and B. Crossland, Explosive welding and cladding. In *Inst. Mech. Eng. Proc.,* Vol. 179, Pt 1, No. 7, pp 264–81
12. K.I. Johnson, A.W. Carter, W.O. Dinsdale, P.L. Threadgill and J.A. Wright, The MIAB welding of mild steel tubing. The Welding Institute Research Bulletin, Vol. 19, October 1978

Forge welding and oxy-fuel gas welding

Forge welding (FOW; ISO 43) also known as 'fire welding', is the method that was practised by the blacksmith for many generations (see previous chapter).

Some workshops still have a coke forge, where a controlled forced draught supplied by an electric blower regulates the temperature of the fire. These are convenient for heating ferrous components that require forging (shaping by hammering or pressing), as well as forge welding.

In many instances, the oxy-acetylene blowpipe can replace the forge and is often employed as a 'portable forge', in that the heat source can be taken to the work, instead of the other way round.

Many of the blacksmith's tools, such as tongs (Figure 2.1) and the anvil and hammers (Figure 2.2), are invaluable in the modern workshop, particularly for carrying out artwork and decorative metalwork.

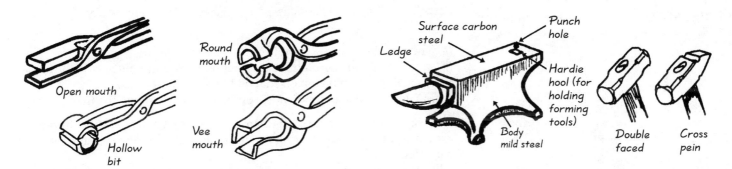

Fig. 2.1 *Types of tongs*

Fig. 2.2 *The anvil and sledge hammers*

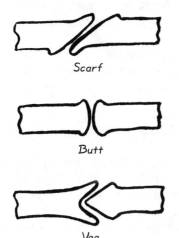

Fig. 2.3 *Types of forge weld end preparations*

In forge welding, the ends of the parts to be joined are first upset to the required shape (Figure 2.3). This is to ensure that the mating surfaces are convex, so that any scale will be squeezed out during hammering, allowing metal-to-metal contact. Such shapes also provide extra metal, avoiding a reduction in cross-sectional area when the hammering of the welding operation is carried out.

Wrought iron usually contains layers of readily fused slag and is relatively easy to forge weld. Carbon steel requires the use of a flux to make the oxide more fluid. Clean silica sand, fluorspar or borax should be sprinkled on the weld faces prior to hammering together.

Other variations of forge welding roll, draw or squeeze the components together instead of hammering them.

In order to obtain the ends of the metal in a plastic state for welding, they must be heated to a white incandescent heat (over 1000°C). This plastic state ceases and oxides begin to form on the surface after about 10 to 15 seconds from being taken out of the fire (depending on the size of the components being joined) and therefore, if the weld is unfinished after this amount of time, the ends must be reheated.

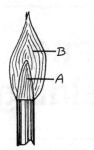

Fig. 2.4 *The structure of a hydrogen or carbon monoxide flame with two cones*

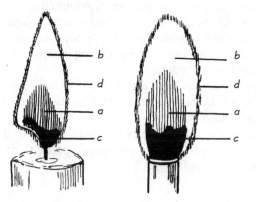

Fig. 2.5 *The structure of hydrocarbon flames. (After Berzelius)*

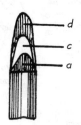

Fig. 2.6 *The small hydrocarbon flame with continuous blue region C*

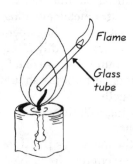

Fig. 2.7 *Experiment using match suspended in the tube of a bunsen burner*

Fig. 2.8 *Burning the gas from the inner cone of a candle flame*

Oxy-fuel gas welding (OFGW) and oxy-acetylene welding (OAW), ISO 311

This heading covers and includes any welding process or operation that employs the heat of combustion for welding, where the fuel gas combines with oxygen producing heat, light and flame.

The heat produced by the gas flame or flames, (usually a single flame produced at the tip or nozzle of a specially designed welding torch/blowpipe) is used to melt the edges of the base/parent metals to be joined and also the filler metal, if used, which is usually in the form of a rod, or wire.

Molten metal from the edges and filler rod intermix, forming a molten pool. As the blowpipe is moved along the joint, a new molten pool is formed as the previous pool solidifies. The perimeters of successive solidified pools form the ripples on the surface of a completed weld.

A flame is a zone in which chemical reaction between gases is occurring (combustion), accompanied by the evolution of heat and light; in brief, it is a glowing gas (first discovered by Van Helmont, in 1648).

A hydrogen or carbon monoxide flame burning in air or oxygen (Figure 2.4) has two cones, an inner A, of unburnt gas and an outer B, in which a single overall reaction

$$2H_2 + O_2 \rightarrow 2H_2O \text{ or } 2CO + O_2 \rightarrow 2CO_2$$

occurs. A flame of pure hydrogen burned from a metal jet or nozzle in dust-free air would not emit a visible light.

The fact that hydrocarbon flames contain four regions was first shown by Berzelius in *The Use of the Blowpipe*, published in 1822 (Figure 2.5). This figure shows:

(a) the dark inner cone of unburnt gas or wax vapour;
(b) a yellowish-white, brightly luminous region, occupying most of the flame;
(c) a small bright blue region shaped like a cup at the base;
(d) a faintly visible outer mantle, completely surrounding the flame.

If the supply of gas is reduced, the flame will become smaller, with the luminous area (b) gradually disappearing and the region (c) forming a continuous inner cone (Figure 2.6).

That the inner cone (a) is cold inside and contains unburnt gas can be demonstrated by two simple experiments:

1. Suspend a match head upwards in the tube of a Bunsen burner, by carefully sticking a pin through the stalk of the match, then light the Bunsen flame. The match head will not ignite for some considerable time (Figure 2.7).
2. Insert one end of a glass tube into the inner cone of a candle or Bunsen flame. The unburnt gas will travel up the tube and can be ignited at the upper end (Figure 2.8).

The use of oxy-hydrogen and oxy-acetylene

If oxygen and hydrogen are supplied separately to a blowpipe jet (Figure 2.9) and the mixture is ignited, a blue, pointed, intensely hot (approximately 2700 to 2820°C) flame is produced. The oxygen-hydrogen ratio is very difficult to adjust, since there is no visible cone in the flame. The flame also has a relatively low heat content and is therefore limited to the welding of low melting point metals such as lead or thin sections and small components in other materials.

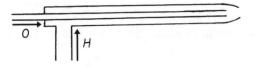

Fig. 2.9 *Schematic diagram of oxy-hydrogen blowpipe (after Partington)*

The flame produced by burning acetylene in air is large, bushy and yellow in colour. It gives off large amounts of sooty smoke, owing to incomplete combustion of the carbon, and produces very little heat.

By using a special welding blowpipe, of either the high- or low-pressure variety, the acetylene can be mixed with oxygen. The oxy-acetylene flame produced is hotter and has a temperature within the range of 3100–3250°C. The steam formed is almost completely dissociated:

$$C_2H_2 + O_2 \rightarrow 2CO + H_2$$

The fact that the flame is strongly reducing, together with its high temperature, makes it highly suitable for welding materials (Figure 2.10).

For cutting iron or steel, an additional central oxygen tube is employed; this usually corresponds with a central hole in the cutting nozzle, with the pre-heating flame or flames surrounding it. The metal to be cut is heated to a bright red heat (ignition temperature) and then the inner oxygen jet is turned on. The iron or steel rapidly oxidizes away directly beneath this central oxygen stream, burning brilliantly and emitting sparks.

Combustion of the oxy-acetylene blowpipe flame

The following procedure is recommended for obtaining neutral welding flame. The acetylene valve on the blowpipe is opened slightly and the acetylene gas lit at the nozzle tip with the aid of a spark lighter or pilot light. The valve is adjusted until there is no smoke being given off and there is no gap between the end of the nozzle and the flame. The oxygen valve is then opened slowly, providing the extra oxygen needed for complete combustion. As this extra cylinder oxygen is introduced, the yellow sooty flame gradually changes in appearance to a long white plume. A small blue-white inner cone is formed as the volume of oxygen and acetylene become equal at the blowpipe tip. This small inner cone is surrounded by a larger flame, completing combustion with atmospheric oxygen.

When the volumes of the two gases leaving the blowpipe tip are equal, the oxy-acetylene flame burns, yielding carbon monoxide and hydrogen. The reaction also results in the very high temperature (approximately 3250°C) at a point just in front of the small blue-white inner cone (Figure 2.10). This is the first stage of combustion and takes place in the brilliant, narrow combustion zone surrounding the inner cone of unburnt gas.

The hydrogen and carbon monoxide formed from this first reaction are also combustible and these burn in the second stage, taking the required oxygen from the atmosphere (Figure 2.10). The hydrogen reaction is

Hydrogen + Oxygen = Water vapour
$$H_2 \qquad O \rightarrow H_2O$$

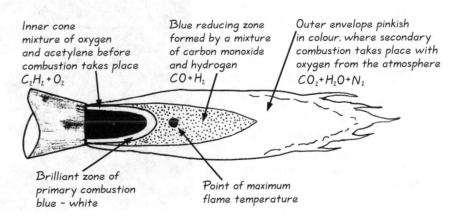

Inner cone mixture of oxygen and acetylene before combustion takes place $C_2H_2 + O_2$

Blue reducing zone formed by a mixture of carbon monoxide and hydrogen $CO + H_2$

Outer envelope pinkish in colour, where secondary combustion takes place with oxygen from the atmosphere $CO_2 + H_2O + N_2$

Brilliant zone of primary combustion blue – white

Point of maximum flame temperature

Fig. 2.10 *The structure of the neutral oxy-acetylene flame*

The carbon monoxide reaction is

Carbon monoxide + Oxygen = Carbon dioxide
$$CO \qquad\qquad O \quad \rightarrow \quad CO_2$$

Therefore, the hydrogen and carbon monoxide formed in the first stage of combustion are burned completely in the second stage.

Theoretically, to burn one volume of acetylene, two-and-a-half volumes of oxygen are required. Therefore, to complete the second stage of combustion, more oxygen comes from the surrounding atmosphere than the amount required from the cylinder in order to form the blue-white cone of the first stage.

This type of flame shows complete combustion with equal amount of oxygen and acetylene from the cylinders and is known as a **neutral** flame (Figure 2.11, part ②). Because there is no excess of carbon or oxygen in the flame, the reducing gases, carbon monoxide and hydrogen, protect the weld from the atmosphere. The *neutral* flame is therefore desirable for many welding applications, including the welding of low carbon steel.

A flame burning with an excess of acetylene is known as a **carburizing** flame and shows a white feather around the blue-white inner cone (Figure 2.11, part ①). The feather is caused by the excess of carbon in the flame. This carbon is given off by the flame and if a piece of steel were heated to red-heat by such a flame, some of this carbon would be absorbed into the surface layer, reducing its melting point. One welding technique, known as Linde welding, was designed to take advantage of this reduced melting point for increased welding speeds, the extra carbon being compensated for by the use of low carbon content filler rods. The carburizing flame is also employed for special applications, such as hardsurfacing.

A flame containing an excess of oxygen is known as an **oxidizing** flame. This extra oxygen is obtained by adjusting the blowpipe valve to give above the amount required for the first stage of combustion and will mean that the reducing zone becomes contaminated with oxygen. The oxidizing flame has an inner cone which is shorter and more pointed than that of the neutral flame and the flame is harsher (Figure 2.11, part ③). If such a flame is used for

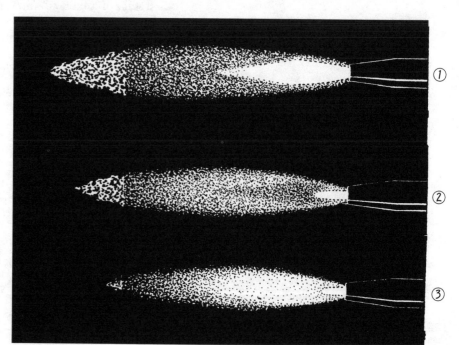

Fig. 2.11 *The three oxy-acetylene flames:*

① *Carburizing, with excess acetylene;*
② *Neutral, with equal amounts of oxygen and acetylene;*
③ *Oxidizing, with excess oxygen*

16

welding steel, oxides and possible porosity will result. The oxidizing flame can, however, be used when torch brazing or when fusion welding brass. In these instances, the inner cone should be reduced to about half its normal neutral length to reduce zinc oxide fumes.

The part of the flame used for welding is the hottest part, just in front of the inner cone, as indicated in Figure 2.10. This point is also well within the reducing zone. Because the second stage of combustion takes oxygen from the surrounding air, it follows that there is also a high nitrogen level, as each volume of oxygen in air is accompanied by 3.78 volumes of nitrogen. The very outer parts of the flame, farthest from the nozzle, tend to entrain air, giving a pinkish coloured envelope which is oxidizing and has a high nitrogen content (Figure 2.10).

Flame and cone shapes

Figure 2.11 shows the standard shape of flames produced with medium-sized nozzles and a laminar gas flow. With the standard blowpipe flame, because the velocity at the centre of the gas stream will be highest, the flame will be longer at the centre. Frictional losses reduce the velocity at the wall of the nozzle, making this outer portion of the flame shorter.

The size of the nozzle (bore) or tip will determine the shape of the inner cone. Smaller-bored tips generally produce pointed cones or semi-pointed cones, with the cone becoming more rounded to blunt in shape, as the diameter of the bore increases.

The high- and low-pressure oxy-acetylene welding systems

Acetylene is generated from calcium carbide and water in special generators of either the water-to-carbide or carbide-to-water type (Figures 2.12 and 2.13). When a generator of either the stationary or portable type is employed to supply acetylene through a piping system and hydraulic back-pressure valve directly to the welding operation, this is known as the low- or medium-pressure system. The use of such equipment involves special safety precautions and regulations, together with the use of a blowpipe that has an injector system to enable the lower-pressure acetylene to be sucked through the blowpipe by the higher-pressure oxygen.

Mainly for convenience and increased control over rate of supply, dissolved acetylene has superseded the use of generators in many instances. Acetylene is transported and used dissolved in acetone solution held in seamless steel cylinders with this system.

Additional safety is obtained by packing the cylinder with an inert absorptive solid material; kapok (the seed-hairs in the pods of a tree, *Eriodendron anfractuosum*, which grows mostly in India and other hot climates) was one of the first materials to be used for the purpose, but mixtures of kieselguhr and charcoal can also be used (kieselguhr is a diatomaceous earth, being a mass of hydrated silica (SiO_2) formed from the skeletons of minute plants known as diatoms – it is very porous and absorbent). Synthetic porous packing of an inert nature can also be used.

These safety measures are necessary because free acetylene gas, under certain temperature and pressure conditions, can dissociate explosively.

By dissolving the acetylene and dividing the interior of the cylinder into a mass of small cells, a safe method of containing acetylene is obtained.

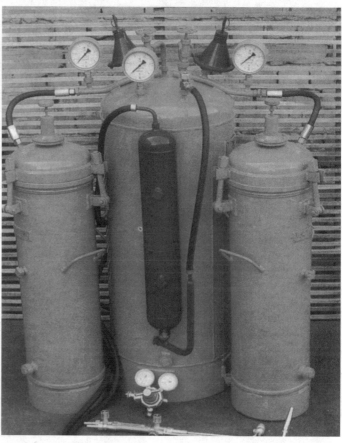

Fig. 2.12 *Examples of movable acetylene generators*
(Courtesy of Russian Scientific–Technical Welding Society, 1 Shelaputinsky Lane, Moscow 109004)

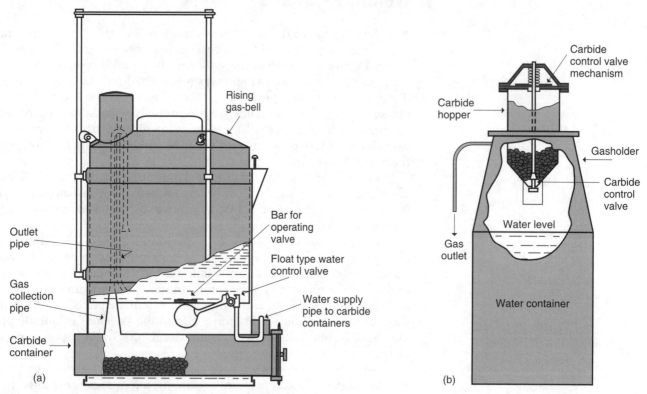

Fig. 2.13 *Acetylene generators: (a) water-to-carbide; (b) carbide-to-water*
(Courtesy of TWI, Cambridge, UK)

The solvent acetone is capable of absorbing 25 times its own volume of acetylene per atmosphere of pressure.

Acetylene cylinders are fitted with a fusible safety plug in the base; this is made from a metal that melts at around 100°C, thus allowing the gas to escape should the cylinder be subjected to excessive heat.

Further details of the acetylene cylinder and a cross-sectional sketch are given in *Basic Welding*, Chapter 3.

Production of acetylene gas

Acetylene can be produced by the action of an electric arc on other hydrocarbons, however the calcium carbide route, as mentioned above, is the one most used when producing acetylene for welding purposes in an 'on-site' or portable generator.

Calcium carbide is produced by mixing limestone, which is a sedimentary rock consisting mainly of calcium carbonate, $CaCO_3$, with carbon (usually in the form of coke), in a thermoelectric furnace. The most common type of calcium carbide furnace employs up to three continuous, self-baking electrodes. These are made using a thin mild steel casing of circular section, about $1\frac{1}{2}$ metres in diameter and open at both ends. When positioned in the furnace, this cylindrical casing is filled with a mixture of anthracite and pitch as a binder. Such electrodes are known as Söderberg electrodes, after their Norwegian inventor.

As the electrodes, supplied with electrical power, descend into a crucible containing the limestone and coke, the heat produced bakes the paste in the casing into solid, amorphous carbon, which of course is a good conductor of electricity. In this intensely hot reaction zone, the steel casing will melt and appear as an impurity (ferro-silicon) in the calcium carbide.

Since these electrodes are consumed at the lower end, fresh sections of steel casing are welded to the upper end and filled with electrode paste. The filling and welding operations are carried out from a special floor above the level of the top of the furnace.

By means of these electrodes, electrical energy is applied to bring about a high-temperature reaction between carbon and lime. Conversion of this electrical energy to thermal energy is obtained in part from the resistance of the coke–lime mixture to the passage of the current and in part from a large quantity of very small electric arcs between the coke particles.

Acetylene generators

There are two main types of generator (Figure 2.13):

1. Water-to-carbide and
2. carbide-to-water.

In both types, the chemical reaction that takes place is the same. The carbon from the calcium carbide combines with the hydrogen of the water, forming acetylene gas, while the calcium combines with the oxygen and hydrogen to form a calcium hydroxide residue:

$$CaC_2 + 2H_2O \rightarrow C_2H_2 + Ca(OH)_2$$

In most countries, a special licence is required in order to install and operate an acetylene generator, and the generator must be certified to comply with the regulations of the country concerned. The calcium carbide is usually crushed to form small pieces (no bigger than 50 mm square) which are packed into air-tight containers. It is usual for each container to hold around 45 kg (100 lb) of carbide for transportation and storage. Again, in most countries, a special licence is required from the authorities in order to store calcium carbide.

In the simplest type of water-to-carbide generator, water drips on the carbide which is in a tray (carbide container) at the base of the generator. The acetylene is collected in a water-sealed gasholder or stored by the displacement of water in the generator. The amount of water fed to the carbide, and therefore the amount of acetylene produced, is controlled by the height of the gasholder bell. When the gas level is high, the bell will rise allowing the ball-type water control valve to shut off. As the level of gas falls, the bell will sink, pressing on the float and opening the water control valve, allowing acetylene to be produced once more. Because the gasholder part of the generator will rise and fall steadily during operation, this type of acetylene generator is often known as 'the rising bell' type (Figure 2.13a). The water-to-carbide tends to be the preferred type in Europe, while in the USA, the carbide-to-water method tends to be favoured.

This latter type of generator (Figures 2.12 and 2.13b) discharges small lumps of carbide from a hopper through a discharge control valve mechanism into a relatively large body of water. The large body of water helps to dissipate the heat produced from the reaction. The carbide control valve is usually spring loaded and can be set to a pre-determined pressure, so that when the acetylene level builds up the valve closes, reducing and gradually stopping the production of acetylene until the pressure falls again.

Obviously, details of manufacture and operation will vary with different makes of generator and it is therefore very important that the manufacturer's instructions are fully read and understood before attempting to operate this type of equipment.

Most generating equipment will operate automatically once the pressures have been set, requiring only standard servicing and sludge removal (*sludge* is the term used for the spent calcium carbide, which has to be removed using 'spark-free' tools and special procedures).

Stationary acetylene-generating plant should be installed in the open air (with weather protection) or in a well-ventilated building, away from the main workshops (100 metres minimum is stated as a guide; specific details should be checked with the regulations in force in the country of installation).

Internal and external walls of generator houses should be made from fireproof material such as bricks or concrete blocks. Floors should be either of wood, or concrete or asphalt. It is common to employ rubber matting or surfacing to prevent accidental sparking.

The generator house must be properly ventilated to prevent the risk of an explosive or toxic atmosphere being formed. (Acetylene forms a potentially explosive mixture with air within the range of 2–82 per cent of acetylene).

Separate rooms should be provided for the generator and carbide storage, and these should be well illuminated. Adequate lighting can be obtained by well-designed windows giving natural lighting in daylight hours and allowing external electric lights to shine through them in hours of darkness, as there must be no internal electric wiring or switches in the generator house, again to eliminate the danger of sparks or minute arcing which could cause an explosion. Heating, when required, should be by hot-water radiators, connected to an external boiler system.

Smoking, flames, welding plant and flammable materials are not allowed in the generator house or near an open-air generator.

Many of these precautions also apply to portable generators. These should only be used, cleaned or recharged in the open air or in a well-ventilated workshop, well away from any sources of ignition such as smoking or welding and cutting operations.

Provision must be made for the removal of sludge and if this is not required for other chemical processes on the same site, then one method is to have a sludge pit from which the sludge is removed and transported by road tanker on a regular basis. Sludge pits must be located outside the generator house. They should have tight-fitting lids and be equipped with vent pipes that rise above the roof levels of surrounding buildings.

Generator houses should be clearly labelled as such and have the appropriate hazard warning signs positioned at the entrance: they must also be in the protection zone of a lightning conductor.

Hydraulic back-pressure valve and purification filter

Because acetylene straight from the generator is liable to have a high moisture content and other impurities detrimental to the welding process, a purifying unit is usually installed in the supply line from the generator.

Such units generally consist of a cartridge, made from three or four different substances, that the acetylene gas has to pass through. A typical cartridge would have a layer of filter wool, in order to collect particles of lime, a layer of ferric iron in a soluble material such as powdered pumice or good-quality kieselguhr, to act as an active oxidizing agent, and a layer of pure pumice to extract water vapour. The purifying unit has a removable lid, to allow the whole cartridge to be changed at regular intervals.

A properly designed back-pressure valve must also be fitted to the supply line between the generator and each blow-pipe, to prevent a back-fire or reverse flow of gas reaching the generator. The valves should be regularly inspected, and the water level should be checked every day. Figure 2.14 shows a section through a typical design of hydraulic back-pressure valve.

As mentioned earlier, only blowpipes of the injector type, designed for low-pressure acetylene use, should be employed. Use of a high-pressure blowpipe, consisting simply of a mixing chamber with regulating valves, could cause an explosion on this type of system.

The simple cross-sectional diagrams in Figure 2.15 show the difference between the high- and low-pressure blowpipes. Descriptions and more detailed sketches are given in *Basic Welding*, pages 24 and 25.

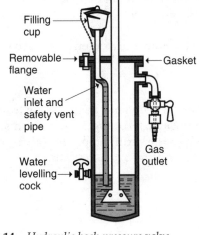

Fig. 2.14 *Hydraulic back-pressure valve*

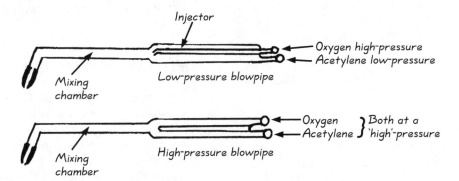

Fig. 2.15 *Simple cross-sectional diagrams showing difference between high- and low-pressure blowpipes*

Other fuel gases

MPS fuel gas mixtures

Methylacetylene propadiene stabilized fuel gas, known as MPS, is a liquefied, commercially prepared fuel which can be handled and stored in a similar manner to liquid propane (that is, it is sold and transported in steel cylinders which usually contain up to 45 kg (100 lb) of liquefied gas, by road-tanker for large consumers).

MPS is produced by mixing together several hydrocarbons – methylacetylene, propane, propadiene, butane etc. The other compounds dilute the unstable methylacetylene sufficiently to allow safe handling.

Although MPS has very similar characteristics to acetylene (AWS gives an approximate flame temperature of 2927°C when burning in oxygen, it requires about twice the volume of oxygen per volume of fuel gas. The oxygen cost will therefore be higher if MPS gas is used instead of acetylene. Heat distribution within the flame is, however, more even than with acetylene, allowing for increased control of heat input in certain applications.

Natural gas

Although, of course, the exact composition of a natural gas will vary depending on the location of the gas field or well, the main constituent is methane (CH_4). Methane is also known as marsh gas, being formed from decaying organic matter, and it also occurs in coal gas.

One volume of methane requires two volumes of oxygen to obtain complete combustion. The chemical reaction is

$$CH_4 + 2O_2 \rightarrow CO_2 + 2H_2O$$

The flame temperature is lower than for acetylene, being within the range of 2600 to 2770°C (depending on composition). For this reason, the main use of natural gas is for pre-heating when oxygen cutting and for other heating operations, where extremely high temperatures are not required.

Propylene

Propylene (or propene, C_3H_6) is a colourless gas, being an oil-refinery product that has characteristics very similar to the MPS types of gas. The average temperature when burning in oxygen is around 2900°C and although not suitable for welding or hardsurfacing, it is employed for oxygen cutting, torch brazing and flame-spraying operations in some countries.

Propane and butane

Propane (C_3H_8) and butane (C_4H_{10}) can be obtained from crude oil, as a fraction of natural gas or via the oil-refining process. They are used mainly for pre-heating during oxygen cutting. Because they both have high critical temperatures and low critical pressures, they can be easily liquefied at relatively low pressures, enabling these gases to be used and transported in welded steel cylinders, with each cylinder containing around 45 kg (100 lb) of liquefied gas. If demand is high, large storage tanks can be built on site and bulk delivery made by road tanker. These gases are also sometimes known as liquefied petroleum gas (LPG). Flame temperatures can vary depending on the composition of the gas – in some countries mixtures of propane and butane are used. Approximate temperatures for oxy-butane flame are up to 2815°C and for oxy-propane up to 2810°C.

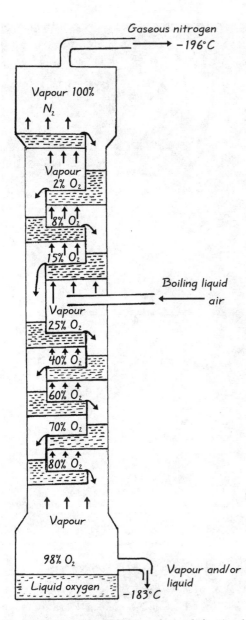

Fig. 2.16 *The essential features of a simple fractional distillation column, where the lower-boiling-point nitrogen is separated from the oxygen in liquid air. In this system, aluminium or brass trays which contain many small holes are positioned at regular intervals inside a single air-separation column. The liquid is held on these trays while vapour bubbles upwards through the many holes. The vapour has sufficient velocity to prevent the liquid dripping back. On each tray there are one or more points through which the liquid can run down pipes on to the tray beneath. As vapour passes up through the column it becomes richer in nitrogen. As the liquid spills down the column, it becomes richer in oxygen*

Hydrogen

As mentioned earlier, the oxy-hydrogen flame has a lower temperature (2700°C to 2820°C) and a relatively low heat content, compared with the oxy-acetylene flame. The oxy-hydrogen flame is also difficult to adjust, as the cone is not visible. The use of this gas is therefore restricted to certain flame brazing operations and the welding of low-temperature metals such as aluminium and lead. It is also used for underwater cutting operations with oxygen cutting equipment.

Oxygen

Oxygen gas is colourless, odourless and non-flammable, but will support and increase the combustion of flammable materials. The Earth's atmosphere contains approximately 20 per cent oxygen by volume, which is sufficient to support fuel gas combustion. However, the use of pure oxygen supplied to blowpipes gives a faster burning reaction and higher flame temperatures.

It is very important that a working atmosphere does not become enriched with oxygen (from a leaking hose or pipe, for example), as the slightest spark under these conditions can cause a 'flash' fire in which everything flammable in the area is burnt in a matter of seconds.

Oxygen can also ignite oil, so it is also very important that the workshop and working clothes are clean and free from any oily deposits.

Oxygen for welding purposes is mostly manufactured by extraction from the atmosphere, using a process known as **fractional distillation**.

Fractional distillation is based on the fact that when a mixture of liquids of different boiling points is heated, the vapour given off will be richer in the component with the lower-boiling point, while the remaining liquid will contain more of the higher-boiling component.

In the manufacture of oxygen, air is compressed, cooled and filtered to remove any dust, carbon dioxide and water; the air then passes through coils (Figure 2.16) and is allowed to expand to a low pressure. The air will be cooled during the expansion, and is then passed over the coils of incoming air, which will begin to liquefy. (As an example of the drop of temperature that can be obtained in this way, if air at a pressure of 100 bar and a temperature of 20°C was expanded to 1 bar in an efficient machine, the gas would start to liquefy at –191°C.)

Boiling liquid air is sprayed into a fractional distillation tower (Figure 2.16). The vapour travels up through the column, passing through the liquid at each level. As the vapour rises it becomes richer in the lower boiling nitrogen. Since the liquid spills over while going down the column, it will become richer in oxygen right up to the final stage, when it is around 98 per cent pure and at –183°C. Gaseous nitrogen can be collected from the top of the tower.

There are many variations of this process. One such variation is the Linde double column system, which operates in two stages, where the fractions of oxygen and nitrogen are taken to a second stage, with the nitrogen being fed to the top as a reflux. The second column produces oxygen as either a liquid or a vapour at up to 99.7 per cent purity and a top vapour product containing 97–99 per cent nitrogen. In plants designed to produce liquid oxygen, this can be stored or transported by road tanker for bulk use, while gaseous oxygen is withdrawn for compression into standard-sized steel cylinders of solid drawn construction. These cylinders are usually charged to a pressure of 173.5 bar (2500 lbf/in²).

Fig. 2.17 *Distillation column, liquid oxygen storage tank and road tanker for bulk deliveries* (*Courtesy of BOC Gases*)

Liquid oxygen tanks

If consumption is very high, oxygen and other gases such as nitrogen, argon and liquefied petroleum gases can be supplied by specially designed vacuum-insulated road tankers for storage and usage from vacuum-insulated evaporator units on site.

Different types of tank for storing liquid oxygen are shown in Figures 2.17 and 2.18. The tank illustrated in Figure 2.18 consists of a thin-walled brass sphere suspended on chains, inside a steel shell. At the top of the sphere is an inspection manhole, which can be sealed. A sump is located outside the sphere. The space between the shell and the sphere is filled with thermal insulating material (this could be magnesium carbonate, mypore or aerogel).

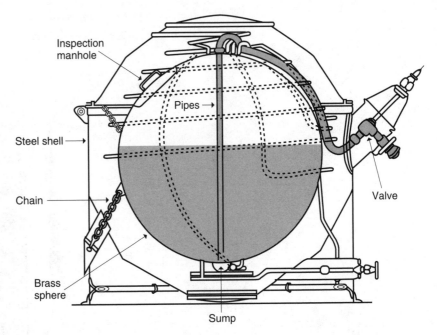

Inspection
manhole

Pipes →

Steel shell →

Chain →

Brass
sphere

Valve

Sump

Fig. 2.18 *One type of storage tank for liquefied oxygen (after D. Glizmanenko and G. Tevseyev)*

The tank is filled and emptied through a pipe fitted with a valve, via a detachable metal flexible hose. Liquid oxygen enters the hose from another reservoir (the road tanker) under pressure. Storage tanks are fitted with pressure gauges. Figure 2.17 shows a fractional distillation column and a cylindrical liquid oxygen storage tank.

Losses of liquid oxygen in this type of storage tank, as a result of evaporation, can range between 0.4 and 0.7 per cent per hour of the oxygen held in the tank. The percentage loss increases as the tank is emptied.

Liquid oxygen can also be stored in vacuum-insulated vessels. The air from the space between the walls of such vessels is evacuated to a residual pressure of between 10^{-5} and 10^{-6} mm of mercury.

Still another variety of storage includes the use of vessels with vacuum-powder insulation. This consists of powdered magnesia, aerogel or silica gel filling the space between the walls, which have been evacuated to a residual pressure of between 10^{-1} and 10^{-2} mm of mercury.

Vacuum-insulated and vacuum-powder-insulated tanks weigh less than those with conventional insulation, and losses of oxygen due to evaporation from them are between 0.1 and 0.15 per cent per hour for small and medium tanks, while large tanks, storing over 7000 litres (1500 gal) of liquid oxygen, may lose as little as 0.1–0.2 per cent of the contents per day.

Equipment for oxy-fuel gas welding

Gases for welding can be distributed and used from single cylinders (Figures 2.19 and 2.20), or from generators and by pipelines fed from bulk supply systems. Pipes for transporting acetylene must **not** be made out of copper, as this would cause an explosive compound known as copper acetylide to be formed. It is usual to use steel pipes for the acetylene supply.

The rate at which acetylene is withdrawn from a cylinder must not exceed 20 per cent of the contents per hour, otherwise the acetylene will contain quantities of acetone vapour, giving reduced flame temperatures.

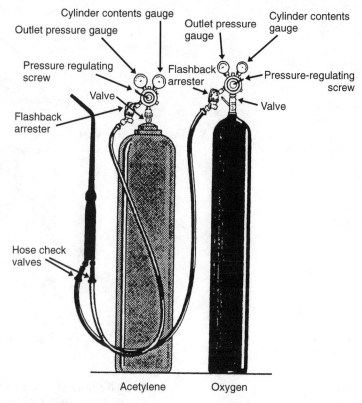

Fig. 2.19   *Typical high-pressure welding outfit*

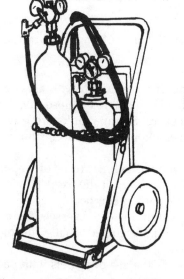

Fig. 2.20   *A cylinder trolley can be either purchased or fabricated from angle iron and tubing; this greatly increases the portability of the equipment*

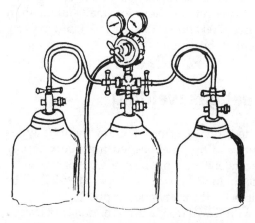

Fig. 2.21   *Method of manifolding three cylinders using 'pigtail' connector pipes and a three-way adaptor*

The withdrawal rate for liquefied fuel gases from cylinders will be determined by the temperature of the cylinder, the amount of fuel in the cylinder and the operating pressure being employed.

If the workshop demand for gas is greater than can be provided by a single cylinder, but not sufficient to require on-site generation or a bulk liquid system, then a manifold system in which two or more cylinders are coupled together may be the answer. Note that, from 1995, the maximum filled pressure of the larger oxygen cylinders was increased to 230 bar at 15°C. This means that all fittings should now be rated for this pressure even though the standard cylinder remains at 175 bar at 15°C.

Figure 2.21 shows how three cylinders can be coupled together using 'pigtail' pipes and a special three-way adaptor when an increased rate of discharge is required.

Specialist suppliers can install large manifold systems which will allow the coupling of large numbers of cylinders to meet heavy gas demands. These instructions must be in a separate, well-ventilated room fitted with safety lighting.

Such systems incorporate flash-back arrestor systems (see Figure 2.24 later in this chapter) and it is usual to have two banks of cylinders controlled by stop valves. This is to allow half to be in use while the other half can be changed. In this way, acetylene and oxygen can be piped from one manifold room to a number of welding points, and empty cylinders can be replaced without interrupting welding.

In most countries, this type of large manifold will require a certificate or licence in order to operate, and the system will require inspecting by an independent authority every twelve months.

The minimum equipment necessary to carry out manual oxy-fuel gas welding is shown in Figure 2.19. This comprises a cylinder of fuel gas, a cylinder of oxygen, blowpipe, connecting hoses and regulators. Such equipment is relatively inexpensive and self-contained. By either purchasing or making a cylinder trolley, as shown in Figure 2.20, it can also be made portable.

Various sizes of nozzle or tip can be fitted to the blowpipe and the regulators can be adjusted to provide the recommended gas pressure for a specific nozzle. In this way, various sizes of flame are available to match the thickness of plate and type of joint being welded. Welding filler rods of different sizes are also available.

Eyes must be protected by properly fitting goggles with the correct grade of filter lens to BS 679 (UK) or ANSI Z49.1 (USA). Protective clothing is required and it is normal to wear flame-resistant overalls, leather apron, gloves and industrial boots to protect the body from spatter and heat.

When welding in confined areas, good ventilation must be provided. In a small workshop this can usually be achieved by ensuring that windows and doors are kept open to allow a through flow of air. However, such a method should be supplemented by the use of at-source fume extraction in order to keep airborne contaminants below allowable limits. Exhausted fumes should pass through an appropriate filtration system before being released to the atmosphere. In an increasing number of countries and states, such fume emissions have to be within legally controlled limits.

Types of pressure regulator

Pressure regulators are mechanical devices that will reduce, to a pre-set value, the pressure of gas being delivered. They are also designed to maintain a fairly constant pressure, although the pressure at the source may vary slightly.

There are many different types of regulator, with different designs being used by different manufacturers. However, welding regulators are usually classified as either single stage or two stage, according to the number of pressure-reduction stages in their mechanisms.

Of the two types, the single stage, although cheaper, will tend to require re-adjusting from time to time, whereas the two-stage regulator, which is essentially two single-stage regulators working in series, will tend to maintain its setting.

Figure 2.22 shows a schematic view of one type of single-stage regulator. With this type, as the cylinder valve is slowly opened, gas will fill the lower tube and the cylinder pressure gauge will register. Some of the gas will then pass through the special valve and seat. This is the point where the flow of gas can be controlled. After passing through the valve, the gas then goes through the upper tube to the outlet pipe. The pressure of gas in the regulator is exerted on the diaphragm, which is connected to the control valve and seat mechanism. Another gauge, the operating pressure gauge, registers the pressure on this low-pressure side. If the pressure on this side of the regulator becomes higher than the actual pressure for which the regulator has been set, then the force on the diaphragm will tend to close the control valve against the seat, temporarily stopping the flow of gas until the pressure falls to the required level once again.

To adjust the regulator so that a required working pressure may be obtained from the outlet, the pressure is set on the spring by means of the pressure-adjusting screw; this spring exerts a force on the opposite side of the diaphragm to the side on which the gas acts. It must, therefore, be compressed until the force balances the desired gas pressure on the diaphragm. By adjusting the compression of this spring with the pressure-adjusting screw, any required working pressure at this outlet side may be obtained.

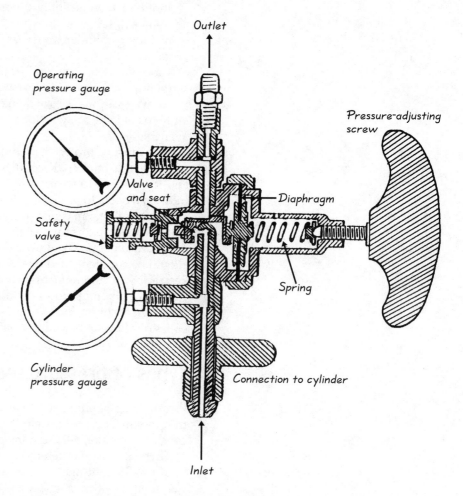

Fig. 2.22 *Schematic view of one type of single-stage regulator*

A safety valve is also fitted, in order to prevent dangerous pressures being set up in the regulator in the event of leakage.

The basic features of one type of two-stage regulator are shown in Figure 2.23. The regulator shown uses the direct principle for the first, or high-pressure stage. This employs a seat that closes against a nozzle valve and against the flow of gas (V1). For the second, adjustable stage, the inverse principle is used. The inverse system employs a seat that closes against a nozzle valve in the direction of gas flow. So in this system, the flow of gas assists in closing the seat against the nozzle (V2).

With the regulator, the gas enters at the inlet from the cylinder, going through (V1) into the high-pressure, or first-stage chamber. Pressure in this chamber is exerted on the first-stage diaphragm. This pressure tends to try and close the seat against the first nozzle valve (V1), as the two are connected by means of a pivoted lever (L1). The first-stage diaphragm is, however, pre-set under the influence of the fixed spring, which establishes the pressure in the first stage. Gas passes into the second stage through the second valve and seat (V2). This seat is, however, connected to the low-pressure, or second-stage diaphragm. Bearing against this diaphragm from the other side is the spring that is under the control of the pressure-adjusting screw. When this adjusting screw is turned clockwise, the second-stage diaphragm is depressed together with the seat of the second valve (V2), allowing gas to enter from the first stage. This gas pressure acting on the second-stage diaphragm tends to close the seat at V2.

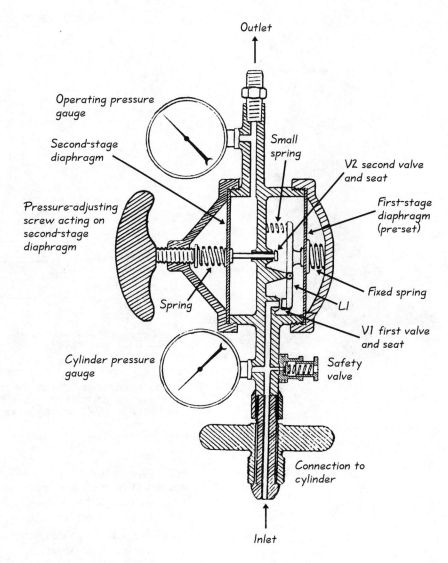

Fig. 2.23 *Schematic view of one type of two-stage regulator*

The balance in the second adjustable stage is therefore between the gas pressure acting on the diaphragm and the force exerted by the spring, which is under the control of the pressure-adjusting screw.

The gas at the regulated pressure is then available at the outlet. The two gauges show the cylinder and working pressures respectively. A safety valve is fitted and in the event of leakage building up a high pressure within the regulator, the safety valve will release, preventing this reaching a dangerous level.

Flame-traps/flashback arrestors

Flame-traps or flashback arrestors should be employed on oxy-fuel gas welding equipment. They are fitted to the outlets of both regulators (Figure 2.24) and then the welding hose is connected to the flame-trap outlet. Flame-traps are specially designed for oxygen with right-hand thread and fuel gas with left-hand thread and notched nut connector. They are clearly labelled with their

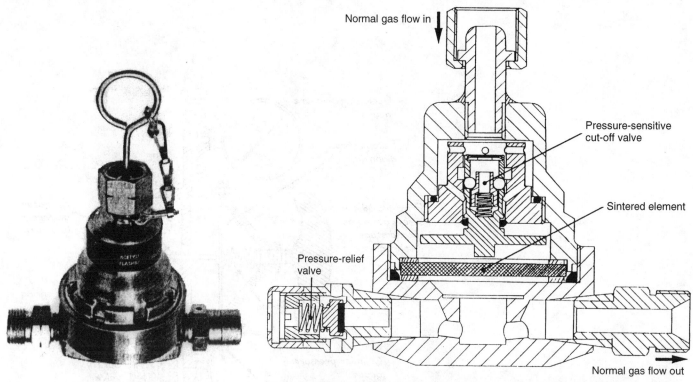

Normal gas flow in

Pressure-sensitive
cut-off valve

Sintered element

Pressure-relief
valve

Normal gas flow out

*With equipment requiring acetylene to be used at pressures between 0.62 Bar and 1.5 Bar
users must comply with Order in Council made under the Explosives Act and must obtain
approval of the Health and Safety Executive.

Fig. 2.24 *Flame-trap/flashback arrestor of the type that can be reset*
(Courtesy of Wescol Ltd.)

working pressure range, suitability of use (that is, oxygen or type of fuel gas), and safety inspectorate approval with stamp to International Standardization Organisation 9000 or country equivalent: British Standard BS 5750; USA, ANSI/ASQC Q90; CEN, European Committee for Standardization; or COPANT, Pan American Standards Commission.

Flame-traps are fitted as an added precaution, in case a flashback should occur and a flame pass from the welding blowpipe into the welding hoses, without being stopped by the hose check-valves/hose protectors (Figure 2.26).

A flashback is usually indicated by a squealing noise and popping, if the blowpipe is on fire inside. They can be caused by incorrect usage when the blowpipe nozzle is momentarily dipped into the molten pool, or when the blowpipe or nozzle is overheated, requires cleaning or is simply adjusted to the incorrect working pressures.

A flame-trap will prevent a major flashback from setting fire to the regulators and even the cylinders, by shutting off the flow of gas immediately, thus preventing an extremely dangerous situation.

One type of resettable flashback arrestor is shown in cross-section in Figure 2.24. In this type, a dense sintered stainless steel plate element is designed to arrest and quench a flame resulting from the most severe flashback. The large surface area of this element ensures flashback protection while causing the minimum interference to gas flow in normal working.

The pressure-sensitive cut-off valve is designed automatically to isolate the normal supply of gas should either a flashback or a pressure-rise occur. This cut-off valve, once activated, remains closed until it is manually reset and this should only be done once the initial cause of the flashback has been established and removed.

A pressure-relief valve allows the fumes and pressure from the flashback to be vented; it also increases the service life of the device by venting most of the carbon from the flashback, which would otherwise form a deposit on the flame arresting element and eventually restrict the gas flow. The device should be regularly serviced in accordance with the manufacturer's instructions.

In the event of the pressure-sensitive cut-off valve being activated, it will require resetting. In order to reset, the gas supply must be closed at source. The resetting pin, which is attached to the device by a small chain, should be inserted into the inlet orifice and centred until it is located against the spring-loaded valve reset mechanism, being pushed firmly until the valve resets.

As stated, with this type of resettable device, the cause of the initial flashback must be remedied, before the device is reset and welding recommenced.

Another type of arrestor device is shown in Figure 2.25. This is known as the cartridge type and incorporates a cylindrical stainless steel sintered element which is designed to quench the flame from a flashback. The large surface area of the element minimises the restriction of gas flow and, for economy in service, the element is designed to withstand a series of flashbacks.

Should a flashback occur, in most cases the sintered element will quench the flame. If, however, the flashback is so severe that a temperature of between 120°C and 150°C is reached, the solder holding the piston of the thermally activated cut-off valve will melt, releasing the piston which forces the non-return check-valve onto its seat. The non-return check-valve being held under compression will stop the gas supply.

With this type of cartridge device, once the piston has been activated, the complete unit must be replaced. Under general working conditions the non-return check-valve functions normally; this function only changes in a severe flashback situation where the solder melts and the piston-release mechanism is activated.

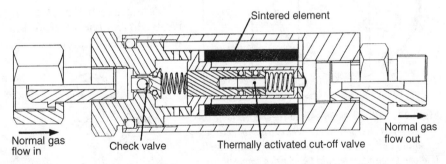

Fig. 2.25 *Section through a cartridge-type thermally activated flame-trap/flashback arrestor (Courtesy of Wescol Ltd.)*

Hose check-valve/hose protector

Hose check-valves should be fitted between the blowpipe and each hose. The connection at the blowpipe is by a left-hand threaded nut for acetylene and a right-hand threaded nut for the oxygen supply. They are fastened to the hoses by 'O'-clips.

Although a check-valve can protect the hose from damage and can reduce the risk of a flashback, it will not actually stop a flashback. For full protection against the dangers of a flashback, flame traps/arrestors must be fitted.

The hose check-valve/protector is shown in cross-section in Figure 2.26; it contains a spring-loaded valve, which will seal off the line if a backflow occurs.

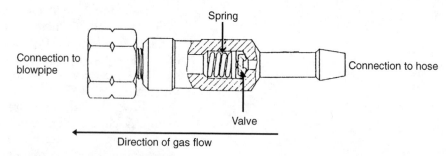

Spring

Connection to blowpipe

Connection to hose

Valve

Direction of gas flow

Fig. 2.26 *Hose check-valve/protector*

Using the equipment

Once the equipment has been assembled, all the joints should be checked for leaks using soapy water (Figures 2.27 and 2.28).

The welding blowpipe should be lit by either a spark lighter specially made for this purpose (Figure 2.29), or from a pilot light on a gas economiser as shown in *Basic Welding*, page 26.

If a welding nozzle or tip becomes blocked during use, it can be cleaned using special nozzle reamers (Figure 2.30).

Manual oxy-fuel gas welding requires training in order to reach the required skill levels. Although the process can be used on thick plate (above 6 mm), welding is slow compared with other processes, such as manual metal-arc, and a high heat input is required – see the rightward or backhand welding technique (Figure 2.33).

The process is much more economical for welds in material below 6 mm ($\frac{1}{4}$ inch) thickness and thin-walled piping of small diameter, using the leftward or forehand welding technique (Figures 2.31 and 2.32) where the speed will be adequate.

Manual oxy-fuel gas welding is very versatile, in that a skilled welder would be able to weld non-ferrous materials as well as ferrous materials and cast iron with the basic equipment and necessary filler rods and fluxes. The equipment

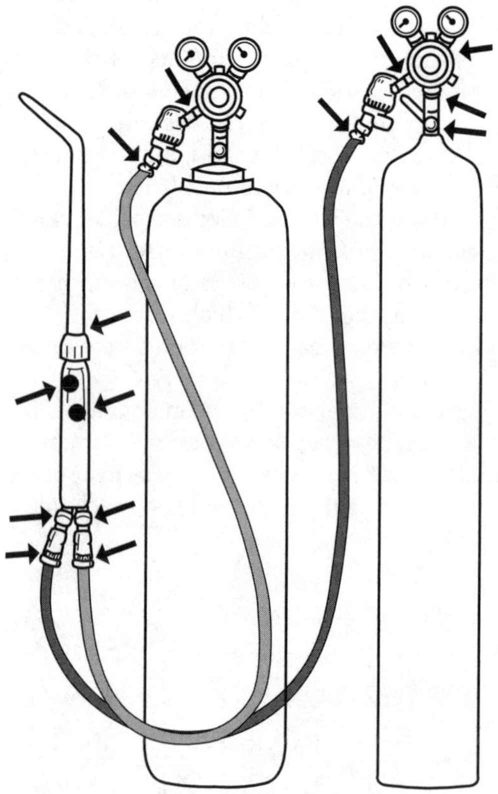

Fig. 2.27 *All connections should be checked for leaks using soapy water solution. A leak can also be detected by sound and feel, or in the case of acetylene, smell. If a leak is found, it should be rectified immediately*

Fig. 2.28 *Checking for leaks with soap and water solution*

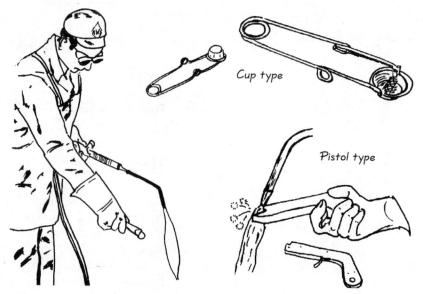

Cup type

Pistol type

Fig. 2.29(a) *A welding torch or blowpipe should always be lit by a spark lighter or pilot light. The use of matches could cause serious burns to the hands*

Fig. 2.29(b) *Two types of flint spark lighter*

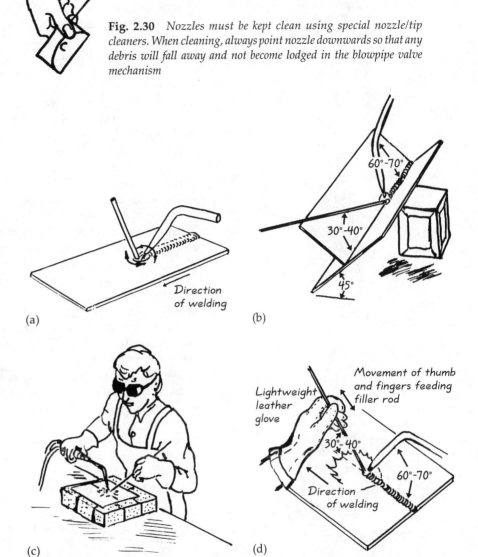

Fig. 2.30 *Nozzles must be kept clean using special nozzle/tip cleaners. When cleaning, always point nozzle downwards so that any debris will fall away and not become lodged in the blowpipe valve mechanism*

60°-70°

30°-40°

45°

(a)

Direction of welding

(b)

(c)

Lightweight leather glove

Movement of thumb and fingers feeding filler rod

30°-40°

60°-70°

Direction of welding

(d)

Fig. 2.31(a–d) *Preliminary practice in the leftward (forehand or forward) technique:* (a) *positions of blowpipe tip and welding rod when making a bead weld;* (b) *supporting a 'T' joint on a firebrick so that the joint is in the flat position;* (c) *resting work on firebricks to minimise heat loss by conduction;* (d) *details of hand movements*

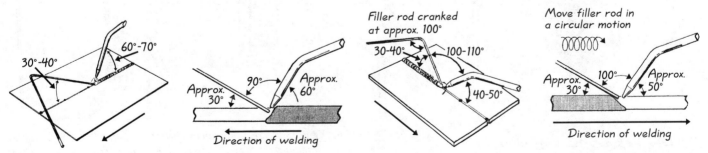

30°-40°

60°-70°

Approx. 30°

90°

Approx. 60°

Direction of welding

Fig. 2.32 *The leftward (forehand) welding technique in the flat position. This is best suited for welding material of thickness less than 6mm*

Filler rod cranked at approx. 100°

30-40°

100-110°

40-50°

Move filler rod in a circular motion

Approx. 30°

100°

Approx. 50°

Direction of welding

Fig. 2.33 *The rightward (backhand) welding technique in the flat position. This is best suited for welding material of thickness greater than 6mm*

can also be used for heating work and, with a change from a welding blowpipe to a cutting blowpipe, can be employed for oxy-fuel gas cutting. It is this versatility that makes the equipment an essential feature of most repair shops and the fact that the process is not dependent on an electrical supply can make it a useful piece of equipment when working on site in remote areas.

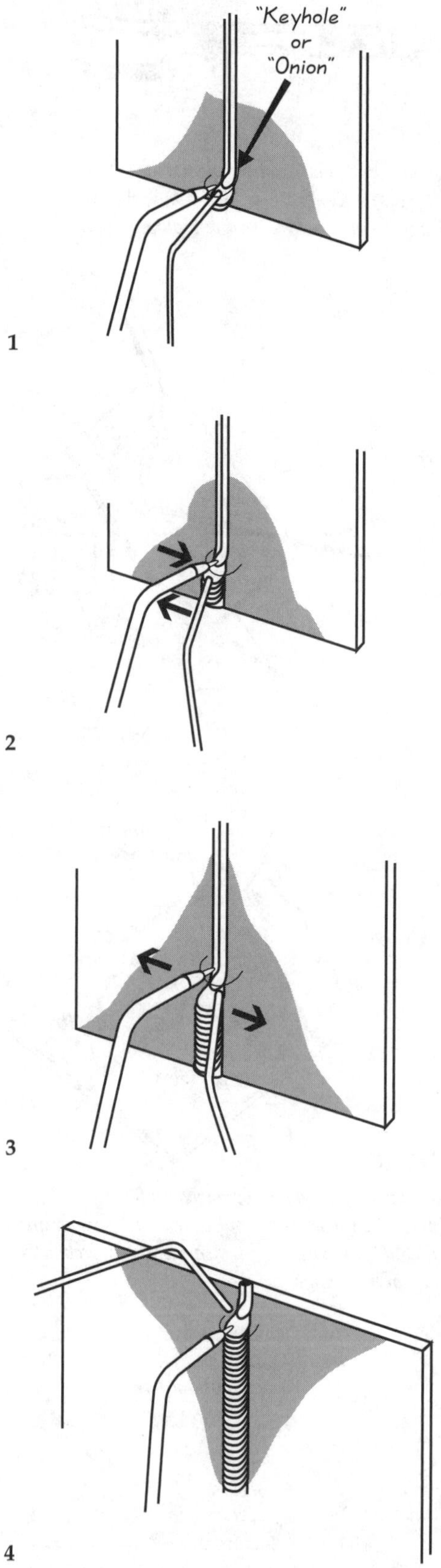

With the leftward, forehand or forward technique, the filler rod precedes the flame along the joint and if the blowpipe is held in the right hand, the weld will proceed from right to left (Figures 2.31 and 2.32).

With the rightward or backhand technique, the filler rod follows the blowpipe flame, which is pointing backwards to the deposited metal. This method is employed on material over 5 mm in thickness, 6 mm being weldable by both leftward and rightward, although speed of welding will be greater with rightward. If the blowpipe is held in the right hand, welding will proceed from left to right (Figure 2.33).

All-position rightward or backhand welding (Figure 2.34) is a modified version of the standard rightward technique, to make it suitable for welding in all positions. The angles of the flame and filler rod, together with their movement, will vary with the welding position. Sample procedures are shown in the following pages. Cranking the filler rod, as shown, not only aids manipulation, but is essential in order to keep the hand away from the heat of welding when working in the vertical or overhead position.

Figure 2.35 shows the stages in making a vertical weld in 6 mm mild steel plate using the all-position rightward or backhand vertical method.

In order to achieve a weld of maximum strength, it is necessary to ensure that a penetration bead is formed on the back or underside of the weld. This bead of penetration has to be within certain size limits specified by the particular code being worked to, but in any event, a bead that is reasonable in appearance is usually a good indication that the weld metal has fused and penetrated right through the joint, providing the added strength over a weld which does not possess full penetration.

Stage ① in Figure 2.34 indicates how the weld is started by fusing the tack weld and the plate edges to form a molten pool, and the 'keyhole' or 'onion' shape at the root. It is essential to maintain this 'keyhole' as the weld progresses, in order to ensure the correct formation of the penetration bead as extra metal is added to the molten pool from the filler rod.

Fig. 2.34 *The four stages in all-position rightward vertical welding*

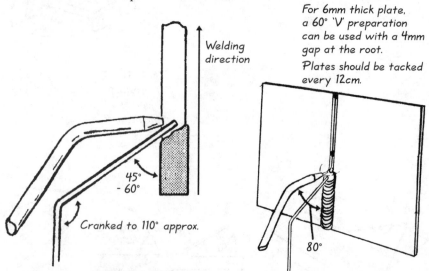

Fig. 2.35 *The welding of mild (low carbon) steel in the vertical position. The technique shown here is known as the 'all position rightward' or 'backhand vertical' method. The angle of the blowpipe nozzle should be around 80° and the angle of the filler rod between 45° and 60°. The rod should be cranked at about 110°*

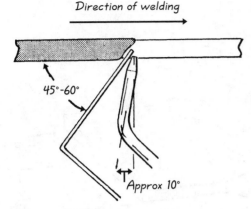

Fig. 2.36 *Side view during the welding of plate of thickness above 5 mm by the all-position rightward overhead (overhead backhand) method*

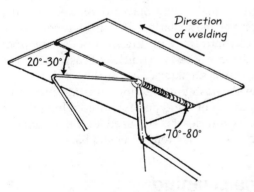

Fig. 2.37 *Welder's view of the all-position rightward overhead method*

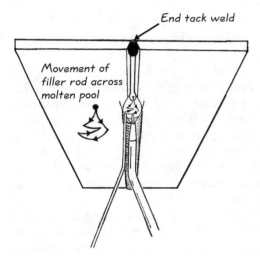

Fig. 2.38 *Welding plate of thickness up to 5mm by the overhead leftward (overhead forehand) method*

Fig. 2.39 *Large nozzles for gas pressure/pressure gas welding can be water-cooled. Circular nozzles can be in two halves to allow ease of work removal*

Stage ② shows the slight movement of the blowpipe and filler rod necessary to maintain the melting of the plate edges and the formation of the 'keyhole'. Sometimes, it is also necessary to push the blowpipe nozzle inwards slightly, towards the root, to melt the 'keyhole' to the correct size.

This movement can also ensure that the penetration metal is going through the 'keyhole'; the rod can also be pushed towards the root and even through the root, at suitable intervals.

Stage ③ shows the continued simultaneous movements of the blowpipe and filler rod as the weld progresses, giving a type of 'scissors' action, as rod and blowpipe keep crossing while travelling in opposite directions.

Stage ④ in order to build up the ends of the weld and give it the required finish, usually necessitates a change to the leftward technique. The heat input can be controlled very accurately by momentarily withdrawing the rod and flame as the end of the weld is built up, without the risk of leaving a crater or burning a large hole.

Figures 2.36 and 2.37 show the application of all-position rightward to the overhead welding of plate more than 5 mm thick and Figure 2.38 shows a typical example of overhead welding using the leftward method on plate up to 5 mm thick.

Nozzle size tables and typical edge preparations together with further details on the leftward and rightward methods can be found in *Basic Welding*, Chapter 3.

Hot pressure welding (HPW)

This term covers processes in which the ends to be joined are first heated by some means and then pressed together with sufficient force to form a weld. When metals are joined by hot pressure welding, the weld is of the solid-state type. With plastics and other materials, some melting can take place.

Pressure gas welding (PGW) or gas pressure welding – ISO 47

This process employs special oxy-fuel gas nozzles (Figures 2.39 and 2.40) to bring the ends of the components to welding heat. A system then applies a pre-determined pressure to force the heated ends together. No extra filler metal is required.

There are two variations of the process:

1. The closed-butt method, where the cleaned ends of the pieces to be joined are butted together under moderate pressure. Heat is then applied to the joint region until a pre-determined amount of joint upsetting occurs.
2. The open-butt method, in which the edges or faces to be joined are brought up to melting temperature and then forced together to produce the required amount of upsetting.

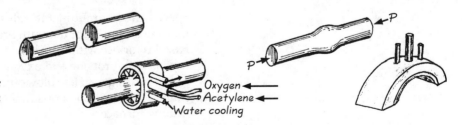

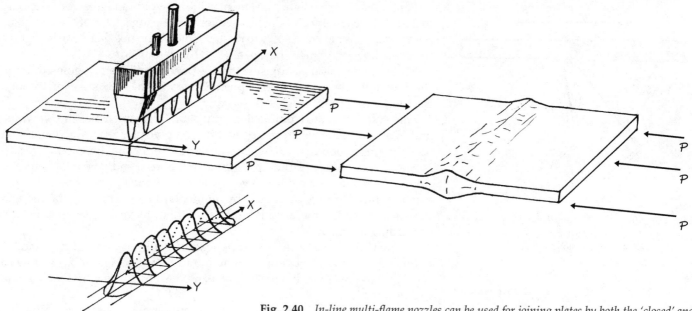

Heat flux distribution pattern

Fig. 2.40 *In-line multi-flame nozzles can be used for joining plates by both the 'closed' and 'open' joint methods*

For both methods, machines should be of rigid construction in order to maintain accurate alignment of components while supplying sufficient pressure to form a weld. Large multi-jet nozzles can be water cooled and some circular types can be hinged to allow for ease of positioning and removal. Ends for the closed-butt method have to be more accurately prepared. As-sawn surfaces are adequate for the open method, because the ends of the pieces are melted and a certain amount of metal squeezed out. Any surface contaminant such as oil or rust should, however, be removed prior to welding.

When correctly made, gas pressure welds are very strong and ductile. Because no filler metal is employed, any mechanical properties will depend on the composition of the workpieces and the rate of cooling. Machines can easily be automated and any post-weld heat treatment can usually be given by the same multi-jet heating nozzle when required.

Machine records of such process variables as pressure, gas flow rates, time and total upset distance help to give a quality control record. As well as visual inspection, magnetic particle has proved particularly useful on this type of weld. On large production runs, it is a good control measure to take samples at random and destructively test them. The nick-break test is a quick and easy method of ascertaining if the crystalline structure is satisfactory across the weld.

PGW helped to develop friction and flash-butt welding, and these processes and others have now taken over much of the type of work to which it was formerly suited. However, in some countries it is still used for the welding of small-diameter pipe and bar, such as concrete reinforcement bars.

Pre-heating and flame cleaning

Special circular multi-jet nozzles are available for pre-heating applications; they are designed for use with propane and oxygen or acetylene and oxygen. Nozzles for use with acetylene and oxygen are of a different design to those for use with propane and oxygen.

Pre-heating nozzles/blowpipes are used for applications where the size or shape of the work is not suitable for, or does not require the use of, a pre-heating furnace.

Fig. 2.41 *A multi-jet nozzle, wheel mounted, being used for the flame texturing of granite before applying road surface*
(*Photograph courtesy of BOC Gases Europe*)

Linear multi-jet nozzles, again designed for use with either propane and oxygen or acetylene and oxygen, are used for flame cleaning and surface preparation. Figure 2.41 shows a linear multi-jet nozzle, wheel mounted, being used for the flame texturing of a granite road surface.

The row of flames from the in-line nozzle can be used as a type of 'brush', for cleaning grease, paint, oxides, and so on from metal surfaces, particularly steel. Work should be carried out in the open air or with fume extraction, as any grease, oil or paint will be burnt off. Oxide tends to flake off, owing to the different expansion rates between the steel and the iron oxide when heated.

The hydrogen from the flame (acetylene C_2H_2) also tends to combine with the oxide in the iron oxide to form water vapour, the iron powder left behind being easily brushed off.

By heating metal surfaces to hand heat (about 40°C/100°F) it has been found that the viscosity of paint applied to the surfaces is lowered, and paint will tend to run into the pores of the metal, giving a firm bond with improved anti-corrosive properties, when compared with paint applied to a cold surface.

Flame heating also ensures that surfaces are dry and any water vapour is removed.

Flame straightening

Heat can be used to straighten components (see Figures 2.42 and 2.43). The heat should be applied locally, heating just one small area at a time, while the surrounding metal is kept relatively cool. The heated 'spots' should not exceed bright red heat.

Buckling in thin plate components (Figure 2.42) can be corrected by applying the flame to local spots on the convex side, starting at the centre of the buckle and working outwards symmetrically.

Angular distortion that has been caused by fillet welds (Figure 2.43) can be corrected by locally heating a narrow strip on the underside of the base plate, following the line of the joint.

Oxygen lancing and thermic lancing/drilling (OLC)

Lancing is a special technique used to pierce holes and cut thick sections of steel, concrete or brick etc., by using the heat generated by exothermic/oxidation reactions.

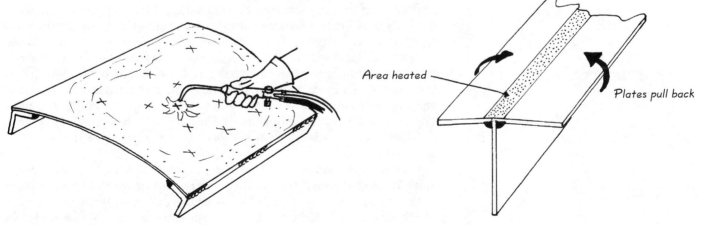

Fig. 2.42 *The use of spot heating to correct buckling in a thin sheet that has had angle iron welded to it*

Fig. 2.43 *Correcting angular distortion on a fillet weld*

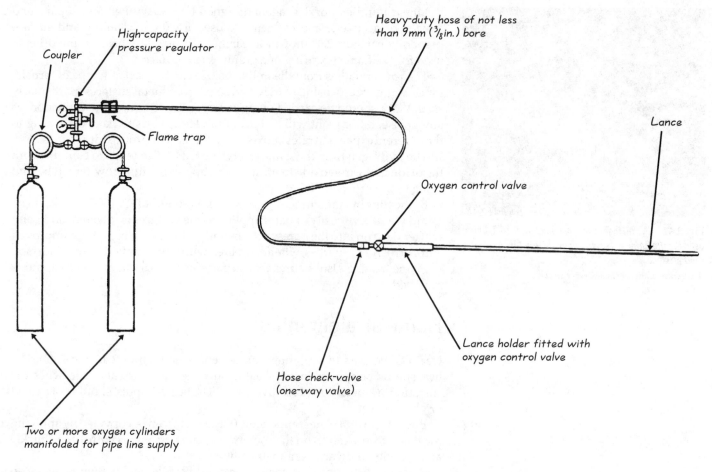

Fig. 2.44 *Typical set-up for an oxygen/thermic lance*

The oxygen lance is simply a length of small-diameter black iron or steel pipe, fitted into a holder which has a control valve for turning the oxygen on and off (Figure 2.44). Ordinary iron or steel gas pipe makes a good lance tube. It is common to use a piece around 3 metres (9 ft) long, with bores ranging from 3 mm to 12 mm ($\frac{1}{8}$th to $\frac{1}{2}$ inch), depending on the thickness and type of material being cut. The end nearer the operator should be either screw threaded or snap-fitted into the lance holder, the other end of the holder being then connected to the oxygen hose by the hose connector.

Because of the large consumption of oxygen, it is common to have two or three oxygen cylinders manifolded together. A high-capacity oxygen pressure regulator is necessary, together with heavy-duty flash-back arrestor and hose.

The lance operates in two ways. In the cutting of ferrous materials, it serves primarily to introduce oxygen into an iron oxidation reaction that has already been started. For example, when piercing a hole through a thick slab of steel, an oxy-acetylene flame would be used to bring the point on the steel up to bright red heat (ignition temperature). The oxygen lance then takes over from the blowpipe and feeds oxygen to the hot metal so that a rapid oxidation process continues.

On ferrous materials, therefore, it is primarily a method of conveniently applying pure oxygen to the point at which the cutting reaction is taking place.

When used on non-ferrous or non-metallic materials, however, the lance pipe itself is consumed, supplying the entire amount of heat necessary for the production of a fluid slag by reaction with the iron oxide.

Thermic drilling

A further development is to pack the lance tube with steel wires, which greatly increases the field of application. With this system, holes can be made in concrete or stone etc. To start this process, the end of the 'lance' is heated to a bright red heat and then the oxygen is turned on. The oxidizing end of the lance is then held in contact with the material being drilled.

Figure 2.45 shows the lancing of a large steel roller for scrap. A small steel guard has been fitted to the lance holder. Note the use of full protective clothing, including, in this instance, the breathing apparatus fitted directly to the protective helmet.

Fig. 2.45 *A large steel roller being cut up for scrap using an oxygen/thermic lance*
(*Photograph courtesy of BOC Gases Europe*)

Cutting and gouging – oxy-fuel gas methods

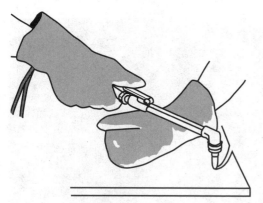

Fig. 3.1 *Oxy-fuel gas cutting, freehand. This can be made much easier by the use of guides and attachments (see Figures 3.3, 3.4, 3.5 and 3.6)*

Fig. 3.2 *The application of an additional heating nozzle when cutting very thick plate.*
(Photograph courtesy of TWI, Cambridge, UK)

An introduction to oxy-fuel gas cutting is given in *Basic Welding*, Chapter 4.

During many manufacturing processes it is often necessary to cut metal plates to size or a particular shape. In some cases, cutting processes are used to shape the edges of plates, pipes or other components with a bevel, before they can be welded. This chapter looks at the processes that can be employed for this purpose.

Gouging is a process that allows the removal of a layer of metal, forming a groove, without cutting all the way through. It is often used to remove a section of weld that contains a defect, in order that it can be rewelded.

Oxy-fuel gas cutting and gouging

This process (Figure 3.1) is suitable for cutting iron and steel, producing a clean cut with relative ease and speed. Because it is only necessary to raise the metal to a bright red heat before directing the stream of high-pressure oxygen at the hot metal, fuels that provide lower flame temperatures than oxy-acetylene can also be used. Common combinations are oxy-hydrogen, oxy-propane, oxy-natural gas and oxy-acetylene.

There are two operations involved in oxy-fuel gas cutting:

1. Heating flames are directed on to the metal to be cut, raising it to a bright red heat. This is called the ignition temperature or ignition point, of around 900°C.
2. A stream of high-pressure oxygen is directed on to the hot metal. The steel is immediately oxidized to magnetic oxide of iron (Fe_3O_4) which has a melting point much lower than the melting point of steel. Because of this, the oxide melts immediately and is blown away by the oxygen stream.

The metal cutting operation is therefore an exothermic chemical action and the iron or steel is not actually melted in this process.

Cutting machines in which one or several cutting blowpipes can be employed, are faster and more accurate than hand cutting.

Because cutting is essentially an oxidizing process, little or no steel is melted. the **kerf** (the width of cut) should therefore be quite clean and the top and bottom edges should be square. On examining melted oxides after cutting, it has been found that they contain up to 30 per cent unmelted steel, which has been scoured from the sides of the cut by the high-pressure oxygen stream. This scouring can be seen if the sides of the kerf are inspected, because drag lines will be faintly etched on the faces of the metal. For an incorrect cut, these drag lines will be more pronounced (see the figures on pages 34 and 35 of *Basic Welding*).

Figure 3.2 shows the application of an extra heating nozzle when cutting very thick plate by the oxy-acetylene cutting process.

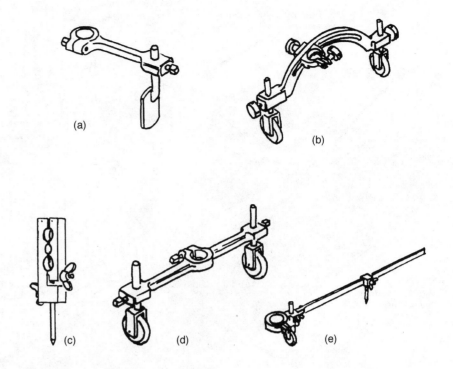

Fig. 3.3 *Cutting guides and attachments: (a) spade guide; (b) bevel attachment; (c) small circle guide; (d) roller guide; (e) radius bar for cutting large circles*

Oxy-fuel gas cutting by hand

It takes a considerable amount of skill to maintain a constant rate of travel over the work when oxy-fuel gas cutting (Figure 3.1). The general quality of cut produced with a hand-held cutting torch is therefore usually inferior to the quality of cut made with a correctly adjusted cutting machine.

Cutting guides can help to keep the torch on the correct line of cut. A roller attachment can be used to maintain the correct nozzle-to-work distance (Figure 3.3).

Stack cutting can be used to cut more than one plate at once, if the same shape is required (Figure 3.4).

See also Figures 3.5 and 3.6.

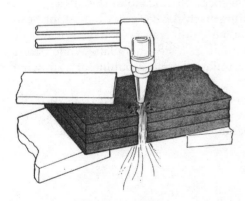

Fig. 3.4 *It is possible to cut a stack of plates if they are clamped tightly together*

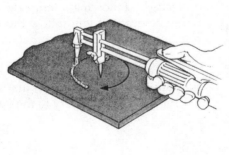

Fig. 3.5 *Cutting a small circle using the attachment shown in Figure 3.3(c) maintains the pivot point in the centre punch mark*

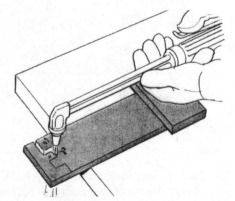

Fig. 3.6 *Showing the position of the left hand when cutting out shapes (if right-handed)*

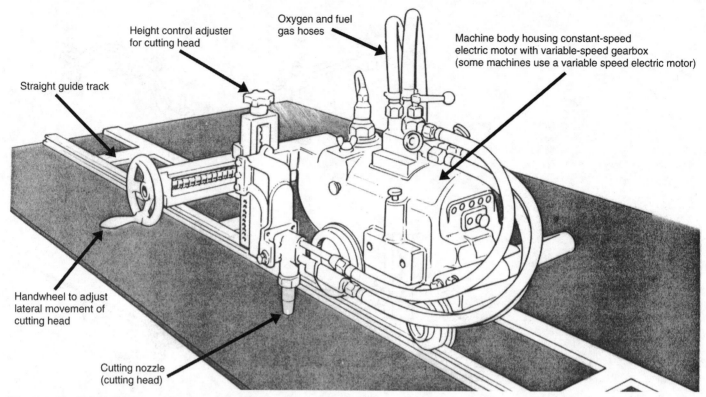

Fig. 3.7 *Typical straight line and circle-cutting machine*

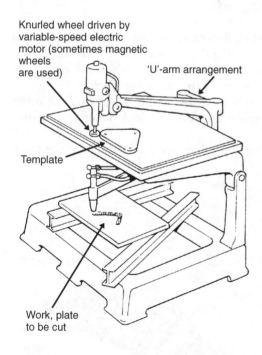

Fig. 3.8 *'U' arm type of profile-cutting machine. The wheel follows round the template on the upper table; the cutting head moves over the work on the lower table, cutting the same shape as the template. Alternatively, a photoelectric cell can be used to follow the black outline on a drawing*

Oxy-fuel gas cutting by machine

Modern cutting machines are capable of making high-quality cuts within close limits. Many machines prepare bevelled edges for welding without any additional dressing operations being required.

There are many different designs of cutting machine. Some machines have a single cutting torch, while others have many. One design moves the cutting torch or torches above the plate to be cut, while another design keeps the cutting head stationary and moves the work beneath it.

The simplest cutting machine is the *straight-line type*, which consists of a carriage, mounted on a track containing the cutting torch. The carriage is traversed over the work by a variable-speed electric motor (Figure 3.7).

Other machines often called **profiling machines** can guide the cutting head or heads by following a template. Some guiding systems have a magnetic wheel device that will follow the outline of a steel template, while others contain a photoelectric cell that will follow the black outline of a drawing (Figure 3.8).

With all these machines, you should ensure that the work is correctly supported during cutting so that it will not collapse after being cut, thus reducing the risk of injury to the operator or damage to the machine.

Gouging

Gouging is often used to remove defects from welds, to prepare a 'U' groove, or to remove the heads from rusted rivets or bolts. Flame gouging uses a

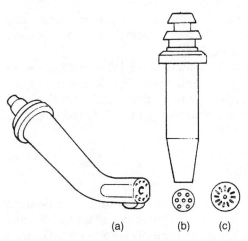

Fig. 3.9 *(a) Typical flame-gouging nozzle; (b) acetylene cutting nozzle and end face; (c) end face of a propane cutting nozzle*

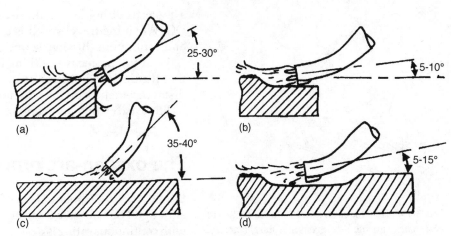

Fig. 3.10 *Nozzle angles for gouging: (a) pre-heating to start gouging at edge of work; (b) gouging in progress; (c) pre-heating when gouge does not start from edge; (d) gouging when not starting from edge*

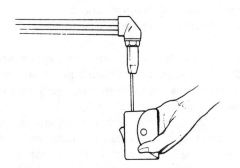

Fig. 3.11 *Always hold nozzle downwards when cleaning with nozzle reamers*

special curved nozzle (Figure 3.9) to modify the oxy-fuel cutting process into a process that will produce a groove in the surface of metal.

A cutting blowpipe is used in conjunction with this curved nozzle, designed to deliver a high volume of oxygen at relatively low velocity.

By using nozzles of different size and by varying the nozzle angle and travel speed, grooves of various width and depth can be produced. Figure 3.10, (a) and (b), shows the different angles required for starting gouging. Once the gouge commences the angle of the nozzle should be lowered (Figure 3.10, (c) and (d).

Always ensure that sparks fall in a safe area, away from other people and away from anything that may catch on fire. This is particularly important, as the sparks from cutting and gouging operations can travel a considerable distance from the actual workstation. Use nozzle reamers to keep holes clean (Figure 3.11) and hold the nozzle downwards so that any loosened debris falls away and does not go inside the nozzle or neck.

The arc-air process

A DC arc welding power source is preferred for the arc-air process, although an AC power source can be used if suitable electrodes are available. Both DC and AC power sources should have a continuous rating for the current levels to be used. An output of 450 amps is required for general-purpose arc–air cutting and gouging.

The electrode holder (Figure 3.12) is fitted with gripping jaws in a self-aligning rotating head. When the trigger is depressed, and the valve in the holder is opened, twin jets of compressed air are emitted parallel to the axis of the carbon electrode. The cable fitted to the electrode holder contains the power cable from the welding power source, and the tube carrying air from the compressor.

The electrodes are made from a mixture of carbon and graphite that has been bonded together and wrapped in a thin layer of copper. This thin copper coating reduces heat radiation and reduces tapering of the electrode. Electrodes for use with AC are designed to increase the electron emission and therefore improve the stability of the AC arc. Electrodes are made in standard 30 cm (12 inch) lengths, in sizes from 4 mm (5/32 inch) to 20 mm (3/4 inch) diameter.

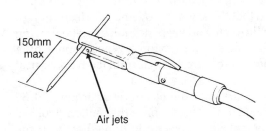

Fig. 3.12 *Arc–air electrode holder*

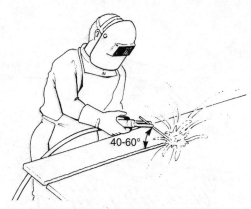

Fig. 3.13 *Hold the electrode at 40–60 degrees to the plate when cutting. When gouging, start at an angle of 30–40 degrees and then reduce to 20–30 degrees during the gouging operation*

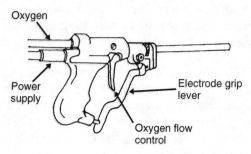

Fig. 3.14 *Gun electrode holder for the oxygen–arc process*

The electrode holder should be connected to the positive terminal when using DC. When the travel speed is correct, the process should give off a smooth and continuous 'hissing' sound (Figure 3.13).

Wear full protective clothing including safety boots. When cutting, ensure that the detached portion cannot fall, causing injury. Sparks can travel long distances; make sure they do not fall on anything that could catch fire.

Fume extraction and ventilation are mandatory if working indoors.

The oxygen-arc process

Either AC or DC manual metal-arc welding power sources can be used, although DC will give a faster cutting speed. An output of up to 300 amps with continuous rating is desirable. The oxygen supply is normally taken from a cylinder fitted with a high-pressure regulator, as used in the oxy-fuel gas cutting process.

The electrode holder is in the form of a gun (Figure 3.14) with a trigger to control the oxygen valve. The electrode is inserted through an oxygen seal washer, to ensure that there is a gas-tight seal between the oxygen supply tube and the end of the electrode.

Coated tubular steel electrodes are used. The cutting oxygen goes down the tube, with the coating helping to stabilize the arc. The usual sizes of electrode are 5 mm (3/16 inch) with 1.5 mm (1/16 inch) bore and 7 mm (5/16 inch) with 3.5 mm (1/8 inch) bore.

Portable screens should be used with both the arc–air and oxygen–arc processes. Any cutting guides used with these processes should be electrically insulated from the work. The oxygen valve is kept closed when striking the arc and then opened once the arc is established.

For cutting with oxygen–arc, point the electrode downwards and away from the body at an angle of 55–65 degrees to the surface of the plate, bringing it vertical at the finish of the cut. Keep the heel of the electrode coating in contact with the plate surface. When gouging, commence the gouge with the electrode angle at around 30–40 degrees, lowering it to around 5–15 degrees during the gouging operation. Again, keep the heel of the electrode coating in contact with the plate surface.

The manual metal-arc process

The equipment used for manual metal-arc welding can be used for cutting and gouging. There are electrodes specifically designed for this purpose, but you can use Class 1 or Class 2 mild steel electrodes that have previously been dipped in water. (Take full protection against electric shock risk.)

When cutting, hold the electrode downwards and away from the body at an angle of 60–70 degrees to the surface of the plate. Strike the arc in the normal manner. When the edge begins to melt, move the electrode down and up in a sawing movement. Make certain that the sawing movement is deep enough to cut through to the underside of the plate and keep the molten metal flowing away by the movement of the electrode and the arc force. When approaching the end of the cut, gradually increase the electrode angle until it is held vertical at the finish of the cut.

For gouging with this method, start with the electrode at around 20–30 degrees and then lower it to around 10–15 degrees once the molten pool has been established. Start the movement of the electrode immediately in the direction of the gouge, using the heat and force of the arc to push the molten metal and slag away. With this method, use a fairly rapid rate of travel, but do not attempt to gouge too deeply; the quantity of slag and molten metal can get out of control and cause difficulties.

Carbon-arc cutting

A carbon electrode connected to the negative terminal of a DC power source can also be used for cutting operations. The electrode is held at the same angles as for manual metal arc cutting (60–70 degrees during cut, bring to vertical at finish). When currents above 300 amps have to be employed with this method, a water-cooled electrode holder should be used.

Metal powder cutting – powder oxygen cutting (POC) and powder injection cutting

This is an extension of the oxy-fuel gas cutting process, allowing it to be used for cutting stainless steels / oxidation-resistant steels. A modified oxy-fuel gas blowpipe is used, which allows iron powder or aluminium and iron powder to be injected into the flame.

The powder accelerates and propagates the oxidation reaction and the melting of the refractory oxides, which are subsequently washed away by the action of the oxygen jet.

Cutting speeds for stainless steels are about the same as for oxy-fuel gas cutting carbon steel of the same thickness.

Prior to 1970, this was the main method for cutting oxidation-resistant steels. Today, however, the process competes with plasma arc cutting methods.

Other cutting and welding methods – plasma, electron beam, water jet and laser

The welding processes of plasma arc (see also Chapter 5), electron beam and laser beam, all use what is known as the 'keyhole welding technique', which can be modified into a cutting technique. For this reason, this chapter looks at the following processes for both welding and cutting applications: **plasma arc welding** (PAW), **plasma arc cutting** (PAC), **electron beam welding** (EBW), **electron beam cutting** (EBC), **water jet cutting** (WJC), **laser beam welding** (LBW) and **laser beam cutting** (LBC).

In the keyhole method, energy from the beam or plasma stream melts the parent metal, penetrating the solid parent material (cutting) or surfaces at the joint (welding), displacing molten metal to the top surface, and forming a keyhole (Figure 4.1).

As the beam or plasma arc is moved along a joint that is being welded, metal is melted at the front of the keyhole through the full thickness of the material and flows around the beam or plasma stream to solidify progressively at the rear.

In plasma welding, relatively low gas velocity from the nozzle allows the surface tension to hold the molten metal in the welded joint. When cutting with plasma, a slightly increased orifice gas velocity (Figure 4.1 (i) (ii) (iii)) will blow the molten metal away. When using beams for cutting, a gas jet concentric with the beam can be introduced in order to blow the molten metal away (see Figure 4.18 later in the chapter, for laser cutting).

These methods do not depend on a chemical action to produce a cut and they can therefore be used for cutting a range of materials that are unsuitable for cutting with the oxy-fuel gas technique. Materials such as aluminium, stainless steel, copper and nickel are readily cut by these methods.

Figure 4.2 shows a comparison between the gas tungsten arc process and the plasma arc process.

Two arc modes are employed in plasma welding and cutting: the transferred and non-transferred arc systems. These are illustrated in Figure 4.3. As the name implies, the transferred arc 'transfers' from the end of the tungsten electrode to the work. With this system, the work becomes part of the electric circuit and heat is produced by the formation of an anode spot on the workpiece, as well as the plasma jet.

In the non-transferred system, the arc is maintained between the tungsten electrode and the surface of the constricting orifice nozzle. The arc plasma is pushed through the orifice by the orifice (or plasma) gas. With this system the work does not become part of the circuit and heat for welding or cutting is only obtained from the plasma jet.

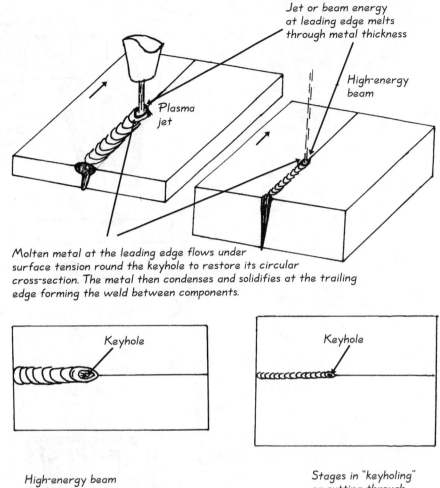

Jet or beam energy at leading edge melts through metal thickness

High-energy beam

Plasma jet

Molten metal at the leading edge flows under surface tension round the keyhole to restore its circular cross-section. The metal then condenses and solidifies at the trailing edge forming the weld between components.

Keyhole

Keyhole

High-energy beam or plasma

(i)　　(ii)　　(iii)

Beam with gas jet introduced or plasma with slightly increased gas velocity

Stages in "keyholing" or cutting through solid plate.

(i)　Energy from beam or plasma melts surface

(ii)　Crater forms as molten metal pushed around circumference

(iii)　Beam/plasma penetrates through thickness to prevent welding; increased gas velocity will blow molten metal away with plasma or a gas jet introduced when using a beam

Fig. 4.1　*The keyhole method of welding and cutting with plasma arc and high-energy beams*

The transferred arc system is the method usually used for welding as more energy is transferred to the workpiece. The non-transferred arc system is used for cutting or joining workpieces that are non-conductive. It is also useful for welding applications where a relatively low energy input to the workpiece is desirable.

The plasma welding and cutting process can be used with either a hand-held torch or a machine-mounted torch. Figure 4.4 shows a hand-held torch being used to prepare the end of a stainless steel vessel, while Figure 4.5 illustrates a machine-mounted torch.

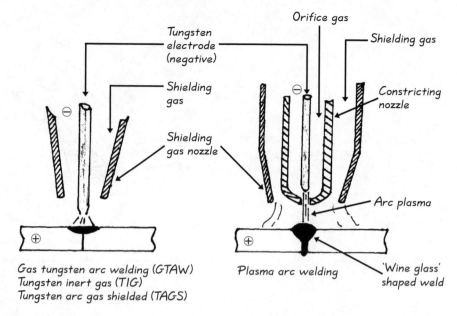

Fig. 4.2 *Illustration comparing gas tungsten arc and plasma arc processes*

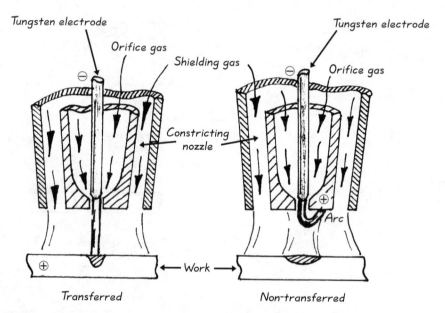

Fig. 4.3 *Transferred arc and non-transferred arc plasma systems*

Plasma arc welding (PAW) employs a non-consumable tungsten electrode as in gas tungsten arc welding (GTAW) [tungsten inert gas welding (TIG); tungsten arc gas shielded (TAGS)] and extra filler metal can be added if necessary in the form of a filler rod or wire. The PAW torch (Figure 4.2) contains a constricting nozzle, creating a gas chamber around the tungsten electrode. The arc heats up the gas that is fed into the chamber to a temperature high enough for it to become ionized and conducting electricity. The ionized gas is termed **plasma** and issues from the orifice in the nozzle at a temperature of approximately 16 700°C (30 000°F).

An early plasma system was developed by Schonherr[1] in 1909. In this system, gas was blown tangentially into a tube and then an arc was struck through the tube. The centrifugal force of the gas stabilized the arc along the

Fig. 4.4 *A hand-held plasma torch cutting stainless steel in a proprietary pressure vessel manufacturing facility.*
(Photograph courtesy of AWS, Miami, USA)

Fig. 4.5 *A Messer Griesheim cutting machine fitted with a plasma torch. With this machine, the operator controls the process via an infra-red beam at the set-up station. This eliminates the need for the operator to return to the control console, giving shorter idle times.*
(Photograph courtesy of AWS, Miami, USA)

tube axis by forming a low-pressure axial core. The system had no practical use, but proved a very useful device for studying arcs.

It took until 1953 before a similarity was observed (by Gage[2]) between an extended electric arc and a gas flame. Attempts to control the velocity and heat intensity brought about the development of the plasma torch.

The first practical torch was designed for cutting, introduced in 1955. This torch was similar to a gas tungsten arc welding torch and used a 'plasma' gas. The tungsten electrode was recessed in the torch nozzle and the arc was constricted by forcing it through a small orifice in the nozzle. The normal gas tungsten arc welding circuit was supplemented with a device to allow for a pilot arc circuit in order to give arc initiation. It was not until 1961 that plasma arc surfacing equipment became available, and plasma arc welding was introduced in 1963.

Both types of plasma system have certain advantages over gas tungsten arc welding:

1. Welding speeds can be increased in some applications, because the energy concentration is greater.
2. A lower current is required to produce a given weld size. This results in less shrinkage and large reductions in distortion levels (up to 50 per cent in some instances).

3. Arc stability can be improved, so the plasma stream can give greater directional stability. The non-transferred system is not affected by 'arc-blow'.
4. The typical 'wine-glass' shape of a plasma weld (Figure 4.2) gives a high depth-to-width ratio for a given amount of penetration, again resulting in less distortion.
5. The torch is usually held further away from the work (greater stand-off distance), allowing greater visibility and ample room for addition of filler metal, without the danger of electrode contamination.
6. Variations in stand-off distances are not as critical as with gas tungsten arc.

Some limitations of the process are as follows:

1. Because the arc is usually narrow, the joint for welding has to be aligned to close tolerances.
2. Costs of equipment and maintenance are usually higher.
3. Manual plasma torches are generally more difficult to manipulate than the same size of gas tungsten arc torch.

Health and safety considerations

Plasma arc cutting can produce high levels of fume and noise. One method of reducing both fume and noise levels is to use a special work table filled with water up to the work surface and a water shroud attachment around the torch nozzle. In another method, the actual cutting operation takes place under about 75 mm (3 inches) of water. This requires a special plasma torch which has compressed air fed through while it is under water but not cutting, in order to keep the inside of the torch water-free.

When cutting in air, ear muffs and normal cutting tables with down-draught exhaust systems are usually adequate. As with any fume extraction device, it may be necessary to fit fume removal or filtering devices before venting to the external atmosphere, in order to comply with air pollution control regulations.

Electron beam welding (EBW), ISO 76 and electron beam cutting (EBC)

In X-ray technology, the heating effect produced in a material struck by a beam of electrons is well known. In practice, the heat generated in the target of an X-ray tube is unwanted and it is necessary for the material to be water cooled.

For EBW/EBC, the electrons are raised to a high-energy state by accelerating them to velocities within the range of 30–70 per cent of the speed of light.

The welding gun is very similar to a television picture tube, with the main difference being that a television picture tube uses a low-intensity electron-beam to constantly scan the surface of a luminescent screen, producing the picture. The electron beam welding gun uses a high-intensity beam to bombard the joint to be welded or the area to be cut. Such high-intensity bombardment raises the energy to the level of heat input required to cut or make a fusion weld.

Pirani, an Italian working in Berlin, produced a 'cathode ray' using an electric current of around one-millionth of an ampere at a potential of about 15 kilovolts (kV). Pirani used this beam to fuse powdered tantalum and patented the process in 1907. The Carl Zeiss Foundation of West Germany produced a modified electron microscope in 1948 and used the beam developed with this equipment to drill minute holes in hard materials such as sapphire.

The apparatus employed at the time consisted of an electron gun with focusing coils, developing a beam current of 10 milliamps at 100 kV accelerating potential. At about the same time a device using an electron beam method of zone refining was brought out in the UK.

In the USA, an electron beam was used to weld an alloy of zirconium and, in November 1957, Dr J.A. Stohr first made public, at a technical symposium on fuel elements, details of an electron beam welding process. Dr Stohr had developed the process for welding fuel elements for the French Atomic Energy Commission and had taken out a patent in January 1956.

By 1959, Hamilton Standard had acquired the technical rights from Zeiss for North America. Hamilton Standard were soon producing their own production version of the Zeiss machine and the need to further develop machines as production tools led them to purchase World Rights to the Zeiss Electron Beam Cutting and Welding Patents in 1963. Hawker Siddeley Dynamics, who had developed long-standing links with Hamilton Standard, took out a licence to manufacture and sell EBW/EBC equipment in the UK, together with the Commonwealth, South Africa, Norway and Sweden. Figure 4.6 shows a section

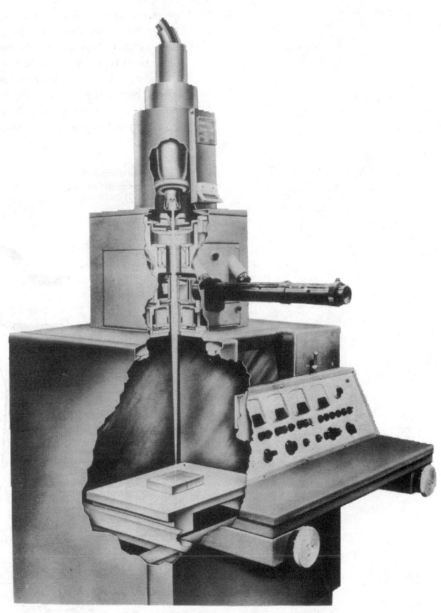

Fig. 4.6 *Section of a typical electron beam welding machine, based on the Hamilton Standard/Hawker Siddeley systems of 1959*

through one of these early Hawker Siddeley EBW machines based on the Hamilton Standard systems.

Figure 4.7 shows a simplified representation of a triode electron beam system. Electrons are generated by heating a negatively charged tungsten filament (the cathode). The electrons are emitted thermionically ('boiled off') from the hot tungsten wire cathode and are attracted to the positively charged anode. The grid or bias cup control electrode is precisely shaped, in order to provide the electrostatic field geometry that accelerates and shapes the electrons into the beam.

Accelerated by a high voltage applied between the cathode and the anode, most of the electrons pass through a hole in the anode. They are then focused by magnetic or electrostatic focusing coils. In a diode gun (cathode–anode), the emitter and beam-shaping electrode both have the same electrical potential and together are referred to as the cathode. In a triode (cathode–grid–anode) gun system, as shown in Figure 4.7, these two electrodes are at different potentials. This allows the bias cup control electrode to be at a slightly more negative value than the emitter, in order to control the beam current flow. When this is the case, just the emitter is called the cathode (or filament) and the electrode that shapes the beam is called the 'grid' or bias cup.

In both types, the anode is part of the electron gun and the generation of the beam takes place independently of the workpiece.

As they leave the gun, the electrons accelerate within the range of 30–70 per cent of the speed of light with electron gun voltages of 25–200 kV. The beam starts to spread as it leaves the gun; this is due to the radial velocity of

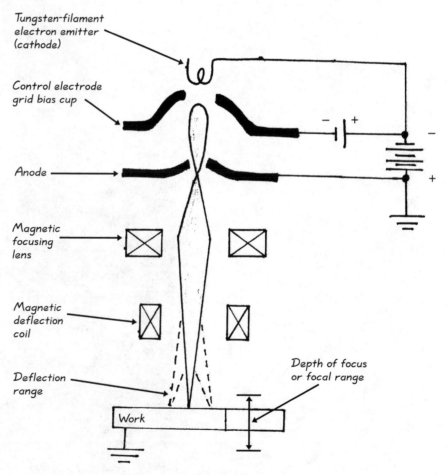

Fig. 4.7 *The essentials of a triode electron gun beam*

the electrons (because of their thermal energy) and the effects of mutual electrical repulsion. To counteract this effect, an electromagnetic focusing lens is employed to converge the beam, causing it to focus at a point on the workpiece.

The beam can be moved accurately in both the X and Y axes by adjusting the current flowing through the deflection coil. Precision alignment of the beam to the joint, or start of cut, can be achieved by viewing a tracer beam with the optical viewing system (Figure 4.8), which employs magnifying optics via a prism or mirrors. The work chamber can also be viewed through a leaded glass window (leaded to protect from X-rays). High-voltage machines are lined with lead sheet as well as having chambers manufactured from thick steel plate (up to 25 mm (1 inch) thick in some cases) in order to prevent the escape of X-rays.

The whole of the gun and chamber are maintained under vacuum. For very high-quality work, a high vacuum of up to 10^{-3} torr (mm of mercury) is used. Less critical work can be welded in a low vacuum machine, requiring less 'pumping down' time. Some applications are satisfactory without the use of a vacuum, being welded in air, while others require the use of an inert gas protective atmosphere. These systems allow higher production rates by eliminating the time required in creating the vacuum. This lowers costs on applications where the depth-to-width ratios obtainable in vacuum conditions are not necessarily required and a small amount of inclusion/weld

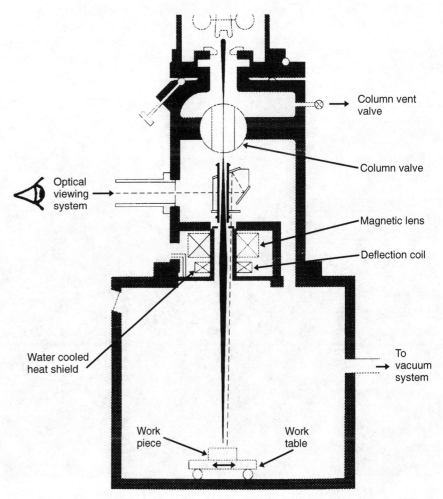

Fig. 4.8 *Cross-section through a high-vacuum machine of the Hamilton Standard/Hawker Siddeley design*

contamination can be accepted within the required code for the component(s) in question.

A vacuum system prevents dispersion of the beam, which happens when collision with air molecules takes place. It also prevents high voltage discharges between the anode and cathode, together with the elimination of oxide contamination to the weld and work surface.

The most important of the high voltage discharges, with regard to welding, are gun discharges, which can cause serious weld defects. This problem is overcome by pumping the gun to a higher vacuum than the chamber.

The beam welds by the 'keyhole' method. If it is switched off suddenly, the metal will solidify rapidly and the hole is frozen *in situ*, creating a defective weld. Beam power is therefore sloped-out gradually, aiding the formation of a smooth, defect-free weld.

In circular welds, it is normal to make a slight overlap, to make certain that fusion has been completed through the whole 360°.

As stated earlier, with a triode machine, the beam power can be controlled by regulating the current. This can be done by altering the potential difference between the cathode and the bias cup (Figure 4.7), which has the same function as a grid in a triode thermionic valve. This enables smooth control to be maintained, allowing beam power of a few watts to several kilowatts for setting the required slope and the power level for obtaining the desired weld penetration.

Because the operating variables of EBW and EBC systems are directly controllable, the processes can be adapted to computer numerical control (CNC). This allows the pre-programming of the gun or workpiece movement, beam deflection, beam current and chamber pressure.

The precision welding of large components requires the availability of a large vacuum chamber. Figure 4.9 shows a large chamber machine and Figure 4.10

Fig. 4.9 *Air-to-air continuous-strip EB welder. Process speed 12 metres/min, twin guns, each 10 kW, 150 kV. The machine gradually increases the vacuum via two chambers prior to the main chamber, and then gradually decreases it via two chambers after the main chamber.*

(*Photograph courtesy of Precision Beam Technologies Ltd, Peterborough, UK*)

Fig. 4.10 *The electron beam in action, penetrating through 100 mm (4 inches) of carbon steel while welding by the keyhole method.*
(Photograph courtesy of TWI, Cambridge, UK)

an electron beam in action, welding material of 100 mm (4 inch) thickness. This latter machine employs a beam power of 100 kV. Experimental systems have been built with beam powers up to 300 kV. The Welding Research Institute of Osaka University, Japan, has a 300 kV machine, which is being used to investigate single-pass EB welding of very thick sections.

EB welding guns in use today are variations of the original triode gun which was developed initially for generating X-rays. One version is known as the **Steigerwald** gun, named after its inventor and also known as a **telefocus** gun, because the cross-over point of the beam is below the anode. The other type of gun is known as the **Pierce** gun, again after its inventor, Dr J.R. Pierce of the Bell Telephone Laboratory. The Pierce gun is a low-voltage type of less than 30 kV, which is generally regarded as the threshold voltage, below which dangerous X-rays are not generated.

Applications

EB welding equipment is a relatively high-cost investment and in the high-vacuum mode would be considered for specialized very high-quality work, with the emphasis more on quality than production rates. In some instances, EB may be the only welding system capable of carrying out a specialized welding/cutting task. In the low or no-vacuum modes, the emphasis can swing to high production rates or components not requiring the very high standard of quality offered with the high-vacuum process.

The process has several advantages:

1. Deep penetration welds are obtainable with very low heat input, reducing distortion to a minimum.
2. With full vacuum welding, there can be no contamination except from within the metal being welded (provided pre-weld cleaning has been thoroughly carried out).

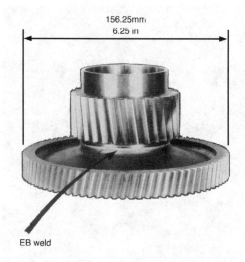

156.25mm
6.25 in

EB weld

EB weld

Fig. 4.11 *Example of electron-beam-welded cluster gear assembly*

3. Because the power can be closely controlled, along with beam oscillation and travel speed, the heat input can be regulated within very exact limits. This provides accurate control of the size of weld and the penetration.
4. Results are reproducible, making the process ideal for repetitive work and computer control, as welding parameters for a particular application can be stored for re-use.

Refactory metals (metals with high melting points – refractoriness – determined by heating a sample initially the size and shape of a Segar cone until it starts to melt; the rate of heating for the test should be an increase of 50°C (122°F) every 5 minutes, with a minimum temperature of 1500°C usually required) can be welded by this process, mainly because the high power density of the beam permits welding with minimum heat input, which is not only important for achieving minimum distortion, but helps in reducing the heating effect of welding in metals such as tungsten and molybdenum, where too much heat input can raise the ductile-to-brittle transition temperatures to above room temperature.

Niobium, tantalum and the reactive (which absorb oxygen and nitrogen rapidly) metals – titanium, zirconium, beryllium and uranium – can all be welded, with the high vacuum process usually being the best method, although inert gas shielding may be adequate for some applications using titanium.

The process can be controlled very accurately and this makes it suitable for welding very thin sections of material and very small components. The low level of heat input can allow the welding together of finished machined components, such as gears to shafts and cluster gear assemblies (Figure 4.11). It can also allow welding without damage to neighbouring components, which would be impossible with conventional arc processes.

By switching the beam on for intervals of short duration, small spot welds can be made. This approach has been employed for the welding of titanium alloys, high nickel heat resisting alloys, aluminium alloys, chromium steels and austenitic chromium–nickel steels.

As discussed earlier, the weld is made by the progression of a 'keyhole' along the joint. At higher levels of power, this hole produced by the beam does not 'self-heal' as the beam progresses. This allows the process to be used for drilling holes and cutting materials, as well as for welding.

Figure 4.12 shows several typical joint designs, some of which (vi and viii) illustrate the capability to make welds in components which cannot be accessed externally.

Weld quality

In some applications it may be necessary to add filler metal to the weld using wire-feeding equipment. This may be to eliminate underfill or concavity, or in order to achieve the desired alloying or metallurgical characteristics in the weld metal.

Hot or cold cracking can occur in EB welds in materials that are prone to these types of cracking. Hot cracks can occur because of the presence of low melting point constituents, as the weld metal solidifies. Cold cracks form after solidification, as a result of the high internal stresses produced by contraction. A build-up of hydrogen, or some other imperfection (or both) can create a stress-raiser. The crack can then propagate through the grains. Cold cracking can be overcome by designing joints to minimize stress concentration and pre-heating quench-hardenable steels to control the formation of martensite

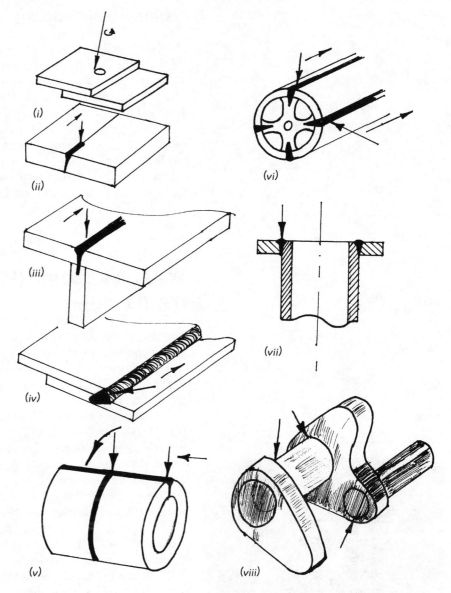

Fig. 4.12 *A selection from the wide range of joint designs that can be used with electron beam welding*

in the heat affected zone. However, because of the low level of contraction stresses with EB welding, less trouble is usually encountered with this problem than when other fusion welding processes are employed.

The problem of porosity is encountered only when it is generated by the metal being welded in one of the following ways:

1. the volatization of high vapour constituents, such as zinc in brass;
2. the releasing of gases dissolved in the material;
3. the decomposition of oxides and nitrides and other compounds.

Shrinkage voids can sometimes form if the face and root surfaces freeze before the weld centre. Increasing the volume of molten metal and slowing down the solidification rate by oscillating the beam and/or reducing travel speed can sometimes eliminate or lessen shrinkage voids, if the consequence of a wider fusion zone is acceptable.

Safety considerations

All the normal precautions of working using high-voltage equipment should be observed. Wiring and insulation should be checked by qualified personnel at regular intervals and at least every 12 months.

Lead sheeting and leaded viewing glass are incorporated into the chamber to shield operators from X-radiation.

Viewing glass and optics should also include optical filters to protect the operator from visible light radiation from the molten metal, when a weld or cut is taking place.

The workshop and general area around electron beam equipment should have adequate ventilation and lighting.

In most countries and states there are statutory regulations covering the above items.

Water jet cutting (WJC) (hydrodynamic machining)

The use of an abrasive combined with water dates back to the Ancient Egyptians, who used sand and water for mining and cleaning operations.

Sandblasters using a pressurised stream of water and sand (up to 3400 kPa (500 psi) – hydroblasting) have been in use for many years for cleaning and paint removal.

The latest variation – WJC – involves the use of very high pressures (207–414 MPa (30 000–60 000 psi)), and abrasive particles (garnet is often employed). The water stream is prevented from breaking up after leaving the nozzle by adding a long-chain liquid polymer, such as polyethylene oxide, to the water stream.

The patent for a very high-pressure water jet cutting system was taken out by Franze in 1968. Before its use in industry, high-pressure water systems had been employed for cutting in the mining and forestry industries.

With the correct transverse speed, the cut is clean, on both metals and non-metals, although the kerf width can be slightly wider at the bottom. This tapering can be minimized by adding long-chain polymers to the water stream.

There are no heat problems to cause distortion and no dust is created. Water jet cutting now has many industrial applications and these are automated or robotized in many instances.

The main item that has to be replaced in the system is the nozzle. With a pure water system a man-made sapphire nozzle could last up to around 200 hours, however, with an abrasive addition, a carbide nozzle usually only lasts from three to four hours.

Safety

Because the water jet or abrasive jet system can cut through most materials with ease, it has to be treated with the same care and respect as a moving bandsaw. Noise level is usually around 90 decibels, but can exceed 100 db.

Many installations have safety guards and enclosure systems to protect operators from the cutting operation. Although these systems also reduce sound levels, as well as giving physical protection, ear protection should also be worn.

Because of the potential danger of a high-pressure water pipe bursting, shielding should be employed and the use of pressure-sensing systems that will immediately shut down the whole system in the event of a pipe failure. Figure 4.13 shows the ease with which materials can be cut by this system.

Fig. 4.13 *Using an abrasive water jet to cut titanium plate.*
(Photograph courtesy of TWI, Cambridge, UK)

Laser beam welding (LBW) and laser welding, ISO 751

Laser beam cutting

LASER is an acronym, that is, it uses the first letter of each word of the phrase which describes the working of the device, giving: Light Amplification by Stimulated Emission of Radiation.

Laser light was first suggested in 1957 by two American scientists, Charles Townes and Arthur Schawlow, however the first laser was built by another American scientist, Theodore Maiman, three years later.

Maiman made his laser using a rod of synthetic ruby crystal and encased it in a spiral flash tube (Figure 4.14b). Each time the spiral flash tube gave an intense flash of ordinary light it created a laser beam which could be focused by a lens as it was emitted from one end of the ruby crystal (the other end of the ruby rod being fitted with a mirror).

The early lasers were not continuous, as one pulse of light produced a laser pulse. At this stage, if a continuous laser beam was required, more than one laser tube was employed, set to fire pulses in a staggered fashion, but being focused on the same point of work.

Since these early beginnings, research has shown that many materials can be made to give off laser light, and that they can be stimulated in other ways as well as by light. There are now many different kinds of laser available, each giving a different type of beam which allows lasers to be used for a variety of applications.

Three main types of lasers are now employed, these being solid-state, liquid and gas. Solid-state lasers are made from rods of solid, transparent synthetic ruby and emerald. The very first laser (Figure 4.14) was a ruby laser. The rod has a fully reflective mirror at one end and a partially reflective mirror at the other to allow the laser beam out. These lasers are excited by a brilliant flash of light, either from a flash tube coiled around the crystal, or a tube mounted next to the crystal (Figure 4.14a).

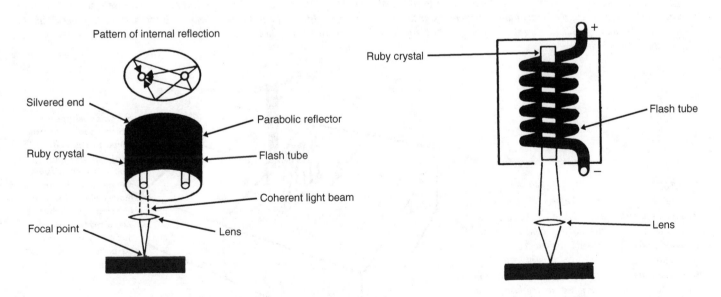

(a) Laser welding: principle of the ruby laser

(b) Laser welding: another type of ruby laser

Fig. 4.14 *Two types of ruby laser. Type (b) is the sort that Maiman designed and made*

Other solid-state lasers are made from neodymium in glass (Nd-glass) and neodymium–YAG (Nd-YAG). [Neodymium has energy levels which make it suitable for use in lasers. It is put into a host material or materials capable of transmitting the laser light produced – YAG (yttrium, aluminium and garnet) and glass being the most commonly used materials at present.]

These lasers are used in industry for welding, cutting, drilling and engraving. Liquid lasers use a liquid that has been coloured with a dye. The beam produced provides a broad range of light which can be split with a prism and then 'tuned' to give an appropriate wavelength. Liquid lasers are excited by an intense flash of ordinary light or by another laser.

The gas laser consists of a glass tube filled with gas, carbon dioxide (CO_2) being common for industrial lasers, although other gases such as helium and neon (He and Ne) are used. The glass tube which contains the gas has a fully reflective mirror at one end and a partly reflective mirror at the other, to allow the laser beam out (Figure 4.15).

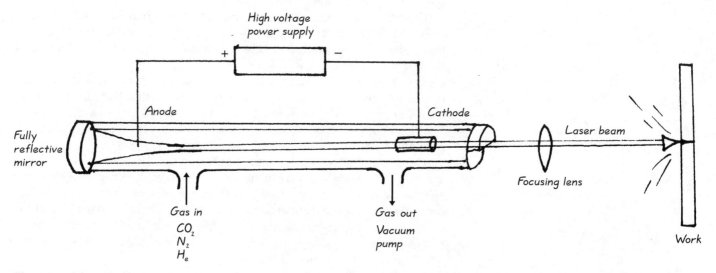

Fig. 4.15 *Schematic diagram of a slow axial flow gas laser, of output power up to 1000 watts. By incorporating a blower and heat exchangers, fast axial flow can be obtained, increasing the available output to 6000 watts. Up to 25 kW can be obtained with transverse flow lasers, where a blower is used and an electrical discharge is maintained perpendicular to both the gas flow direction and the laser beam's optical axis. With this system, the volume of the resonator chamber is large but has only a short length, enabling mirrors to be placed at each end in order to reflect the beam many times, very rapidly, before discharge*

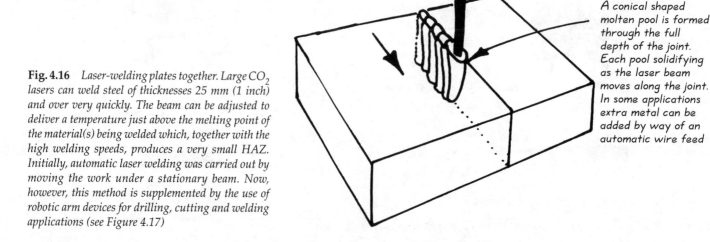

Fig. 4.16 *Laser-welding plates together. Large CO_2 lasers can weld steel of thicknesses 25 mm (1 inch) and over very quickly. The beam can be adjusted to deliver a temperature just above the melting point of the material(s) being welded which, together with the high welding speeds, produces a very small HAZ. Initially, automatic laser welding was carried out by moving the work under a stationary beam. Now, however, this method is supplemented by the use of robotic arm devices for drilling, cutting and welding applications (see Figure 4.17)*

A conical shaped molten pool is formed through the full depth of the joint. Each pool solidifying as the laser beam moves along the joint. In some applications extra metal can be added by way of an automatic wire feed

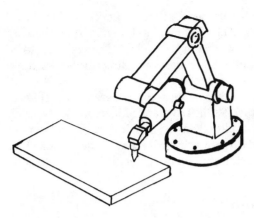

Fig. 4.17 *Laser mounted on robot arm*

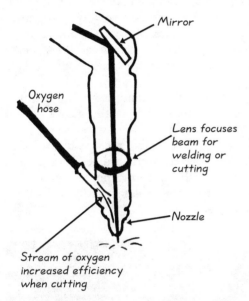

Fig. 4.18 *Laser equipment can make use of mirrors and/or fibre optics to transmit and guide the beam*

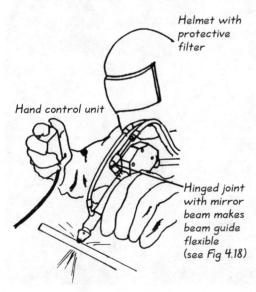

Fig. 4.19 *Cutting with an oxygen-assist hand-held laser*

The gas laser uses an electric current to excite the atoms in the gas and these, in turn, 'emit' photons (light energy). Some of the emitted photons will hit other atoms that are in the excited state, which will cause them to give off photons. This is known as 'stimulated emission of radiation'. The amplification occurs because, as a photon hits an excited atom, it produces another photon identical to itself. Both photons can then hit other excited atoms, producing more photons, which in turn make further protons, and so on.

Figure 4.16 shows how the conical shaped molten pool, which is formed through the full depth of the joint, progressively solidifies to form the weld as the beam moves along the joint.

The beam can be used for very accurate drilling or cutting, and fibre optics or a system of mirrors (Figure 4.18) can be employed to transmit the beam. Oxygen-assist increases cutting efficiency and equipment can either be hand-held (Figure 4.19), or automated – work moving under stationary beam or laser connected to the robot arm (Figure 4.17).

High-power CO_2 and YAG lasers are increasing in use for welding applications on a worldwide scale. They have always been popular for the welding of very small components, such as microchip protective cases (Figure 4.20d), pacemaker capsules and other miniature capsules (Figure 4.21). However, as a greater range of power is available, more applications are opening up and precision welding of thicker sections can be undertaken.

Evaluations are now taking place to compare other welding processes with laser for certain applications, and laser is coming out as the first choice in many instances. As examples, a US manufacturer has changed from the MIG and TIG (GMAW and GTAW) welding of domestic heater water tanks to laser welding, using 5 kW CO_2 lasers. A lap joint is employed and each tank is completely welded in 16 seconds. Quality figures indicate 99.8 per cent with any defect corrected by manual TIG (GTAW) welding.

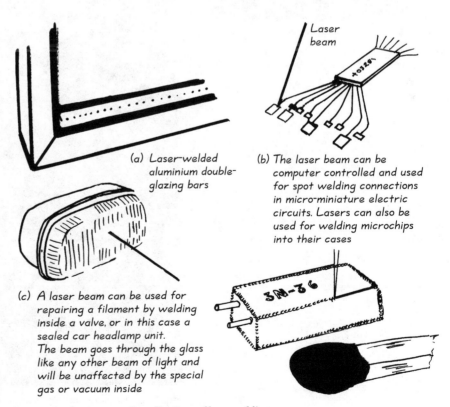

Fig. 4.20 *Some unusual applications of laser welding*

61

Fig. 4.21a *Heart pacemaker encapsulated using a 300 W solid-state laser*

Fig. 4.21b *Laser-welded pressure capsules for control equipment*

(Photographs courtesy of TWI, Cambridge, UK)

Laser beam welding is also now being used for the production of aluminium double glazing bars (Figure 4.20a). The bars have to be welded all the way along the centre seam, which is visible through the glass. The weld has to be of high strength and cosmetically acceptable. Laser welding fulfils both these criteria.

The welding of stainless steel tubing, both cirfumferencially (Figure 4.22) and longitudinally, areas in which TIG (GTAW) and high frequency welding are used extensively, is now being undertaken, in some instances, by laser.

The car industry has also started to employ laser welding in the manufacture of body and door panels and the welding of panels together in body manufacture.

Advantages of laser welding

1. Because of the nature of the conical shaped molten pool forming through the full depth of the joint (Figure 4.16), single-pass welding procedures are the norm and have been qualified in materials up to 32 mm ($1\frac{1}{4}$ inch) thick (AWS). The time required to weld thick sections can therefore be reduced, together with the need for filler wire (in most cases) and intricate edge preparation.
2. Beam heat input can be adjusted to near the minimum amount required to melt and fuse the weld metal, thus minimizing metallurgical effects in the HAZ (which is minimal) and distortion.
3. LBW is a non-contact process, so there is no tool wear and no risk of contamination. The workpiece can be located in an evacuated or controlled atmosphere and hermetically welded within this atmosphere (Figure 4.21).
4. Laser beams can be directed by mirrors or fibre optics around obstacles to allow welding in awkward places.

Fig. 4.22 *Laser welding of stainless steel tube.*
(Photograph courtesy of TWI, Cambridge, UK)

5. Control is very accurate, allowing the beam to be focused on a very small area, and the welding of very small components in close proximity (Figure 4.20b).
6. A large selection of materials and combinations of different materials can be welded by the laser process.
7. The laser lends itself to automated and robotic high-speed welding, with computer control systems.
8. The process of laser welding is not affected by magnetic fields, as is the case with electron beam welding and certain electric arc welding processes, particularly when using DC supplies.
9. No vacuum or X-ray shielding is required but an inert gas shield can be added if necessary.
10. Using fibre-optics or a system of mirrors, one beam can be shared between workstations, allowing full utilization of beam time.

Limitations

1. Joints must be presented to the beam focal point with accuracy.
2. Problems can be caused by the high reflectivity and conductivity (thermal) of metals such as aluminium and copper alloys.
3. Because laser welds solidify very rapidly, they can be prone to a certain amount of porosity and in some cases brittleness.
4. The laser is not a very efficient method of converting energy and, generally, the energy conversion rating is less than 10 per cent.

References

1. Plasma arc systems – Schonherr, 1909. In *Encyclopedia of Physics*, XXII, 300. Springer-Verlag, Berlin, 1956
2. *AWS Welding Handbook*, Vol. 2, 8th edn, Chapter 22

Further reading

Welding and Metal Fabrication, Vol. 63. No. 6, June 1995

The welding arc

The basics of electricity and magnetism

An introduction to current, voltage, resistance and Ohm's Law, together with simple direct and alternating current welding circuits, is given in the companion volume *Basic Welding* (ISBN 0–333–57853–8). It is advisable to be familiar with these basics before continuing with this chapter.

Characteristics of the arc

There are many different kinds of welding arc, designed for specific applications with a particular process, and characteristics vary depending on factors such as gaseous conduction medium, type of current (AC or DC) and type of electrode(s).

This section looks at some of the general characteristics common to most welding arcs.

For all practical purposes, the American Welding Society (AWS) states that a welding arc can be considered "a gaseous conductor which changes electrical energy into heat." An arc gives off radiation as well as heat and in certain welding processes (GTAW / TAGS / TIG) is also employed for the removal of surface oxide.

The welding arc is a high-current, low-voltage discharge, its range of operation, across processes, being generally from 10 to 2000 amperes at from 10 to 50 volts.

The arc is a system in which electrons are evaporated (emitted) from the cathode (thermionic emission), transferred through a region of hot, ionized gas to the anode, and there condensed (collected). The arc can be divided into five parts (Figure 5.1).

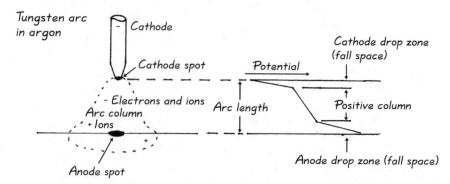

Fig. 5.1 *Example of the tungsten welding arc (TIG – tungsten inert gas (GTAW)), showing how the arc can be divided into the five regions: cathode spot, cathode drop zone, arc column, anode drop zone and anode spot*

1. The *cathode spot,* from where the electrons are emitted.
2. The *cathode drop zone* (or fall space), in which a sharp drop in potential occurs (voltage drop), approximately twice as much of a drop as that at the anode.
3. The *arc column,* which is the bright visible part of the arc. The column is highly conducting and consists of highly ionized gas called **plasma**. There is only a small drop in potential within the column.
4. The *anode drop zone* (or fall space), which is a gaseous region next to the anode where a further steep drop in potential occurs.
5. The *anode spot,* which is the zone on the positive side of the arc at which the electrodes are collected (absorbed).

Types of cathode spot

There are three types of cathode spot:

(a) Fixed, for example, a sharply pointed tungsten (approximate current density 10 000 amps/25 mm² (10 000 amps/sq.inch).
(b) Mobile cathode spot, which travels at high speed and there may be more than one spot. For example, when welding aluminium with a tungsten electrode, argon or helium gas shield and alternating current, the tracks of the cathode spots cut through the oxide, breaking up the oxide film. The pieces of oxide cut by the spots are lifted on each alternate half-cycle (Figure 5.2), leaving the molten pool free of oxide. The process is repeated as the weld progresses, providing oxide removal without the need for a flux.
(c) The ill-defined cathode spot, where the whole of the molten electrode tip appears to be the cathode.

The arc plasma

The arc plasma is the gas, usually in the centre column of the arc, that has been heated to either a partially or fully ionized state, enabling it to carry the arc current.

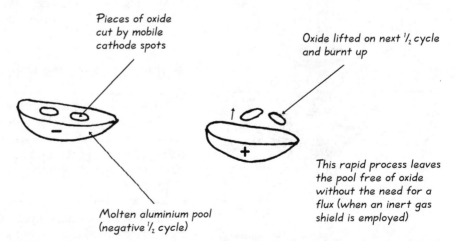

Fig. 5.2 *Showing how mobile cathode spots can remove oxide film during the welding of aluminium*

65

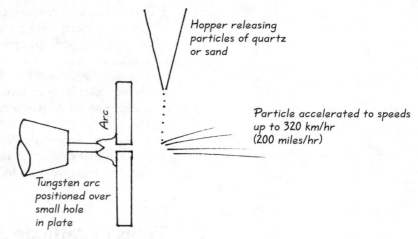

Fig. 5.3 *Experiment demonstrating the velocity of a plasma jet*

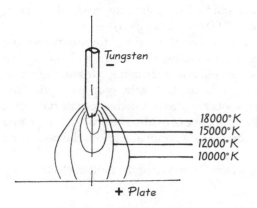

Fig. 5.4 *Typical isothermal readings based on a 200 amp DC tungsten arc shielded with argon*

The constriction of the arc causes the formation of plasma jets. Figure 5.3 shows one experiment used to demonstrate the velocity of a plasma jet. An arc from a tungsten electrode in argon is held over a small hole in a metal plate while particles of sand or quartz are released from a small hopper. As the particles pass behind the small hole they are accelerated to around 320 km/hr (200 miles/hr). Higher velocities are attained when special nozzles, designed for plasma welding and cutting torches, are used.

Arc temperatures

Arc temperatures vary according to the nature of the plasma stream and, consequently, the current conducted by it. The coatings of some covered electrodes contain sodium and potassium which are easily ionized. The maximum temperature of a shielded metal arc obtained with a flux-coated electrode is around 6000° K. A tungsten arc in an inert gas can give axial temperature readings around 30 000° K in the region close to the tungsten, when the tungsten is the cathode. Even higher readings, up to around 50 000° K have been obtained in very high power arcs. Such high temperatures are usually measured using spectrographic analysis equipment that is capable of measuring the spectral radiation emitted from different regions of the arc.

Figure 5.4 shows an isothermal map of a tungsten arc in an argon gas shield.

Further electrical characteristics

Magnetism has a major influence on certain arc characteristics. In certain instances, magnetic fields either induced or permanent can deflect the arc causing 'arc blow'. This is usually more of a problem when welding using direct current, as the magnetic effect on the arc is lessened when using alternating current, because of eddy currents being induced in the workpiece (Figure 5.5)

Metal transfer and arc deflection are examples of other characteristics influenced by magnetic fields. Controlled arc deflection in the direction of travel can be used when a uniform weld with shallow penetration is required. Such a technique might be beneficial for the high-speed welding of very thin sections.

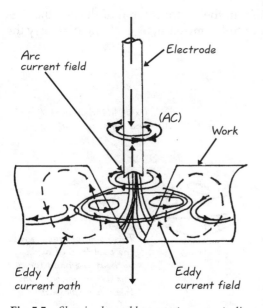

Fig. 5.5 *Showing how eddy currents can neutralize a magnetic field induced by an AC current (after AWS Welding Handbook, Vol. 1, 8th edn)*

The arc is surrounded by invisible circular magnetic fields which tend to act on the plasma stream like a constricting hose and cause an oscillating motion of the arc's column, which can often be observed when studying high-current arcs. Lochte-Holtgreven[1] gives an analogy in a mechanical device that consists of a slightly bent piece of rubber tubing attached to a water tap. The momentum of the stream of water forces the tube in a straight line, however inertia effects accentuate this action and a swinging motion results. A similar action occurs with the plasma of an electric arc. The lines of magnetic force holding the arc together take the place of the rubber tube and the running water is replaced by the streaming plasma, which moves in the magnetic field produced by the current in the electrodes. The Lorentz forces that result tend to push the plasma back to the axis, however, again, inertia accentuates this action and a swinging motion results (see also Figure 5.5). The effect of a magnetic field holding the plasma together can be clearly seen with the use of photography, and the motion of the plasma is very much like the stream of water from a tap. There are two reasons for this motion, these being the magnetic forces and the evaporation of the electrode material.

Although the exact mechanisms that take place at the arc terminals are still under research to permit greater understanding, the following is generally accepted as a basic working theory for welding arcs:

Negative electrons are emitted from the cathode, flowing along with the negative ions of the plasma to the positive anode (Figure 5.1). At the same time, positive ions are flowing in the reverse direction. (A negative ion is an atom that has gained one or more electrons above the number required to balance the positive charge on its nucleus – thus producing the negative charge. A positive ion is an atom that has lost one or more electrons – thus producing the positive charge.) The main flow of current in the arc is, however, by electron travel, the same as in a solid conductor.

The heat at the cathode area is generated mainly by the positive ions hitting the cathode surface, while the electrons generate most of the heat at the anode. This is because they have been accelerated by the arc voltage as they have travelled through the plasma and then give up their energy as heat when hitting the anode.

The hottest part of the plasma is the central column, where electrons, atoms and ions are in an accelerated state of motion and constantly colliding with each other. The outer part of the arc is relatively cooler, consisting of the recombination of gas molecules that were dissociated in the arc central column.

Changing the arc length has a major effect on the heat distribution and voltage within the regions of the arc and the arc plasma stream. A change in shielding gas can alter the heat balance between the anode and cathode. With flux-coated electrodes, the addition of salts of potassium or sodium will increase ionization and therefore reduce the arc voltage.

With some automatic welding heads, the arc length can be controlled by an electrical system that will automatically speed up or slow down the wire feed drive motor as any differences in the pre-determined arc length start to occur. This is called the **controlled arc system**.

When manual metal-arc welding with a flux-covered electrode (SMAW), irregularities in arc length are difficult to avoid and could lead to changes in heat input. In order to avoid such heat input changes and yet cater for variations in arc length, it is usual to use a welding power source that has a steeply sloping characteristic. Figure 5.6 shows such a steep sloping characteristic, where a change in arc length can represent a change in arc volts from V_1 to V_2 (or vice versa) with only a very small change in amperage. (Theoretically, a steep sloping characteristic is required, however, in practice, sometimes a slightly less steep curve is beneficial in order to give less spatter and allow

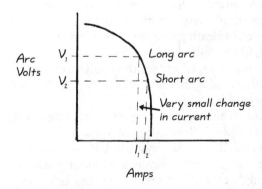

Fig. 5.6 *Volt–ampere curve of a 'sloping characteristic' welding machine*

slight cooling of the molten pool when 'drawing off' on gaps or when positional welding.

In MIG/MAGS (GMAW) the requirements of the power source are somewhat different. In most semi-automatic machines, a system known as the **self-adjusting arc** is employed. In this system, the wire electrode is fed at a pre-set constant rate through the flexible conduit to the hand-held torch or gun. If the arc length is increased, the voltage is increased with a corresponding decrease in current, which causes a decrease in electrode burn-off rate and restores the arc length. Likewise, if the arc length is reduced, the process is reversible; an increase in burn-off rate occurs, again restoring the required arc length. This system therefore easily compensates for the variations in electrode-to-work distances (arc length) that occur with a hand-held welding torch

This self-adjusting arc feature is possible because MIG/MAGS welding power sources have a **flat characteristic** (Figure 5.7). With this system, if the arc is disturbed by a surface irregularity and the arc shortens by a small amount, the arc voltage will be decreased from V_1 to V_2 (Figure 5.7), resulting in a considerable rise in current from I_1 to I_2. The burn-off rate of the electrode increases, so that the wire tip establishes the original arc length. The response rate of the flat-characteristic machine is therefore extremely rapid. Likewise, if the arc tends to lengthen during welding, this results in an immediate current drop. The electrode then burns off more slowly, again retaining the original arc gap.

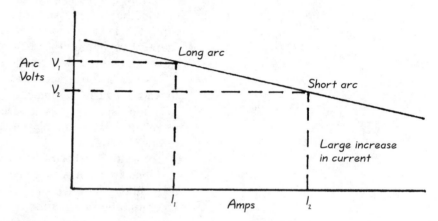

Fig. 5.7 *Volt–ampere curve of a constant-potential welding machine showing a 'flat characteristic'*

Differences in the heat generated at the anode and cathode can determine the use of a particular type of arc. As an example, a tungsten electrode in argon gas can handle around ten times more current without melting when it is connected as the cathode (negative), indicating that in this instance the anode (positive) generates more heat than the cathode.

This is not the case, however, in submerged-arc welding, where more heat is generated at the cathode than the anode. This is proven by higher melt-off rates when the electrode is connected negative.

Figure 5.8 shows the functions of the flux coating on metal transfer in metallic arc welding with a flux-coated electrode (SMAW). With an AC (alternating current) arc, an essentially similar characteristic is produced during each half-cycle. However, the arc must re-ignite after the zero current point and a voltage substantially higher than the normal arc voltage is required to re-initiate the process. For this reason, easily ionizable elements are added to the flux coating, helping initiation and stability when using AC.

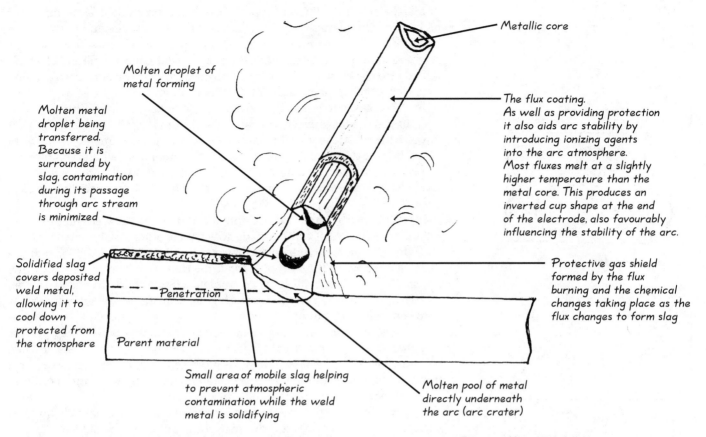

Molten droplet of metal forming

Molten metal droplet being transferred. Because it is surrounded by slag, contamination during its passage through arc stream is minimized

Metallic core

The flux coating.
As well as providing protection it also aids arc stability by introducing ionizing agents into the arc atmosphere. Most fluxes melt at a slightly higher temperature than the metal core. This produces an inverted cup shape at the end of the electrode, also favourably influencing the stability of the arc.

Solidified slag covers deposited weld metal, allowing it to cool down protected from the atmosphere

Penetration

Parent material

Protective gas shield formed by the flux burning and the chemical changes taking place as the flux changes to form slag

Small area of mobile slag helping to prevent atmospheric contamination while the weld metal is solidifying

Molten pool of metal directly underneath the arc (arc crater)

Fig. 5.8 *Shielded metal-arc welding: the functions of the flux coating on metal transfer in metallic arc welding (SMAW)*

Table 5.1 *Electron thermionic work functions and ionization potentials*

Metal	Work function (electron-volts)	Ionization potential (electron-volts)
Aluminium	4.0	5.98
Barium	4.2	5.21
Calcium	2.2	6.10
Cobalt	4.1	7.86
Copper	2.5	7.72
Iron	4.0	7.86
Nickel	4.5	7.62
Potassium	2.3	4.35
Sodium	2.4	5.13
Tungsten	4.4	8.00

With arc welding, both the ease or striking or initiating the arc and the arc stability are related to the ionization potential created by the elements in the arc. As shown in Figure 5.8, with a flux-coated electrode, the arc atmosphere will consist of flux materials and metal vapours, together with gases produced by the burning flux in order to form a shield. It is thought that arc stability in helium or argon atmospheres is obtained through the transitions from the metastable (excited state) state to the ionized state.

The **thermionic work function** is the energy required to evaporate one electron from a solid surface. The ease with which an arc can be started and maintained is related to this function. The representative thermionic work functions and ionization potentials for a number of elements are listed in Table 5.1. The values given in the table are approximate, in most cases there is a range of results.

Metal transfer and metal recovery

Arc welding processes in which the electrode is consumable are employed extensively because deposition rates and therefore efficiency are usually far greater than with systems where the electrode is non-consumable and the filler metal is fed in separately.

The filler metal needs to be transferred from the tip of the electrode in such a way as to give minimum disturbance to the arc, with uncontrolled short-circuiting being removed or minimized in order to reduce spatter losses.

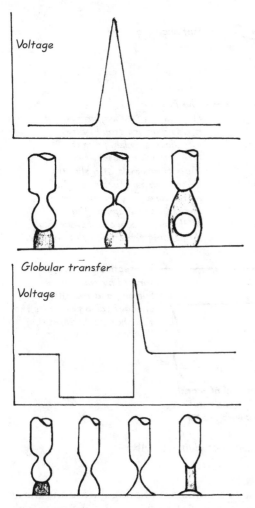

Fig. 5.9 *Metal-transfer methods proposed by Sack in 1939. The idealized voltage graph is shown above each process*

Metal transfer in different types of arc has been studied with the use of high-speed photography and oscilliograms of arc voltage and current. The latter method is most useful in arcs that are visually obscured by smoke or submerged under flux. The actual droplet detachment from the end of a flux-coated electrode is very often difficult to observe, because of restricted vision caused by the 'capping' nature of the flux coating.

In 1939, Sack suggested the systems of globular and 'bridging' or short-circuiting as metal-transfer mechanisms (Figure 5.9). Theoretically, transfer through the arc stream of a flux-covered electrode is either mainly in the form of single large drops (globular) or as a large number of small droplets (spray). In practice, transfer usually comprises a combination of both, with a bias towards one system or the other depending on electrical, chemical and physical characteristics of the particular arc system.

Metal transfer with the semi-automatic and automatic welding processes MAG/MIG (GMAW) can be made to vary considerably by changing the shielding gas and/or voltage and current ranges (Figure 5.10). With argon gas shielding, when the current is above the transition level (Figure 5.10), the transfer will be a fine axial spray of droplets with no short-circuits. However, with helium or carbon dioxide shielding, the transfer mode will be globular with some short-circuiting.

A complete short-circuiting metal-transfer mode has been developed for this process as it has many advantages when welding thin materials or when positional welding, as the weld pool is more controllable. Spatter is kept to a minimum by incorporating an electrical inductance in the circuit in order to control the rate of current rise when the electrode is in contact with the pool. This results in the peak current value at short-circuit being relatively low and the average current levels are reduced by the use of small-diameter electrodes. A schematic representation of short-circuiting metal transfer is given in Figure 5.11.

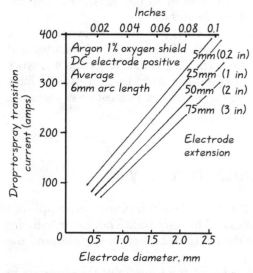

Fig. 5.10 *Influence of stick-out distance (extension) and electrode diameter on globular-to-spray transition currents for mild steel (based on drawings from AWS, Miami, USA)*

2mm electrode wire, direct current electrode positive
CO_2 consumption 12 l/min. Electrode feed rate 156m/hr.
28V, 280A   Welding rate 25m/hr

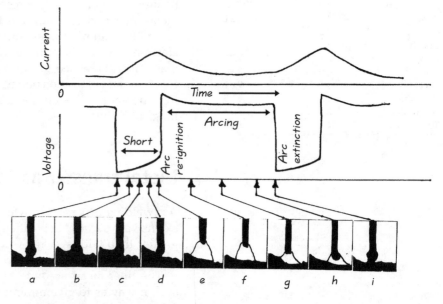

Fig. 5.11 *Short-circuiting mode of metal transfer with oscillogram, for a CO_2 shielding gas (after E.O. Paton Institute of Electric Welding, Kiev and AWS, Miami, USA)*

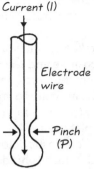

Current (I)

Electrode wire

Pinch (P)

A number of mechanisms or forces affecting metal transfer have been considered. These include:

1. surface tension, which can tend to hold the molten drop in position;
2. gravity, which will assist or hinder metal transfer, depending on welding position;
3. the pressures generated by gases at the electrode tip;
4. several electromagnetic forces (Lorentz Laws and 'pinch effect' at the tip of the electrode) (Figure 5.12);
5. Hydrodynamic force, a flowing effect – plasma jet forces dragging the electrode droplet;
6. Other forces.

Fig. 5.12 *The concept of magnetic 'pinch' is an important factor exerting control on metal transfer. The strength of the pinch is directly related to the strength of the magnetic field set up around the electrode wire by the current flowing through it. For any electrode diameter, the pinch (P) is proportional to the square of the current (I^2); for example, if the current is doubled, the pinch effect will increase four times. The spatter produced during welding with short-circuiting transfer is a consequence of the pinching action on the short-circuiting 'bridge' of metal. The amount of spatter is determined by the rate of pinch*

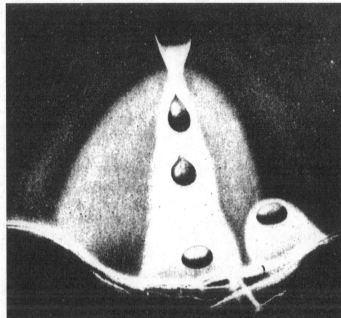

Fig. 5.14

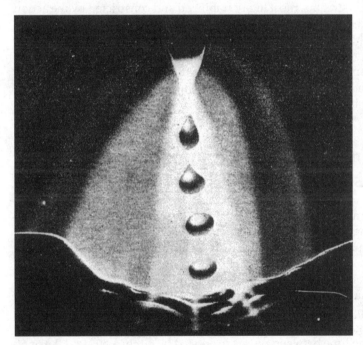

Fig. 5.13

Fig. 5.13 *Showing metal transfer (steel) in an argon atmosphere with wire electrode positive (after Gilette and Breymeir[2])*
(Courtesy of TWI, Cambridge, UK)

Fig. 5.14 *Showing the bright parts of the arc that contain metal vapour. Restrictions of the arc by field forces can be seen, and in the restricted parts the droplets are accelerated. The droplet thrown out shows the drift of ionized vapour (after Gilette and Breymeir[2])*
(Courtesy of TWI, Cambridge, UK)

Fig. 5.15 *Metal transfer in a MIG (GMAW) arc with wire electrode negative. The cathode is too cold, therefore there is spot formation that sprays off material. The second arc at the edge of the electrode tip is of greater stability, pushing the large drop sideways (after Gilette and Breymeir[2])*

(Courtesy of TWI, Cambridge, UK)

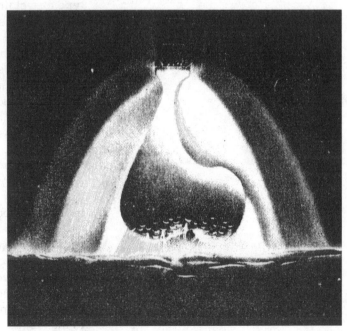

Fig. 5.15

The two forces having the major effect on transfer are the hydrodynamic and the electromagnetic. Photography has demonstrated the connection between metal transfer and plasma streams.

Figures 5.13, 5.14 and 5.15 show photographs taken by Gilette and Breymeir[2]. In Figures 5.13 and 5.14 the wire electrode is positive and the centres of the arcs are coloured by the metal vapour ejected from the electrode tip. In Figure 5.14 one of the droplets has been ejected from the arc column and the vapour cloud, which is ejected from its surface, begins to ionize rapidly, following the force of the field more rapidly than the droplet itself.

The spray of droplets in Figure 5.13 show by their elastic deformations that they have been pulled from the liquid wire end, which is hidden from view by the bright flare.

Mantel[3] states that the entire process appears similar to drops of water emerging from a tap and that the same impression was given by a film made by the DVS (German Welding Society) and shown at Essen in 1961.

The average velocity of droplets shown in Figure 5.14 and 5.15 is given as approximately 100 metres/second, however, since the velocity of the droplets will remain practically unchanged as they traverse the arc, it can be assumed that the accelerating force is active only while the droplets are being torn off the wire. Forces of such magnitude can only be understood as the effect of high-velocity gas streams, most likely stabilized and constricted by the action of the field forces on the plasma.

Figure 5.15 shows the MIG (GMAW) welding of steel with the wire electrode negative. Here, a large, heavy liquid drop is held together by its surface tension and still has a small connection to the end of the electrode.

Slow-motion film shows such a droplet rolling and deforming while being kept in suspension by plasma currents rising from the anode weld pool.

Metal transfer from a consumable electrode using pulsed current – pulsed transfer

A very precise and controllable form of transfer is achieved with a special type of welding unit designed to pulse the welding current between the globular and spray current ranges at pre-determined intervals (see Figure 5.16).

The arc is maintained with a low background current of around 50 to 100 amps. This is sufficient to maintain a weld pool and for a small amount of electrode heating to take place. Pulses of high current are then superimposed on to this background current, taking the arc characteristics briefly into the spray range. The time between pulses produced by either a 50 Hz (UK) or 60 Hz (USA) power source is short enough to prevent globular transfer at all current levels, however the pulse is long enough to allow a single droplet to be transferred before the current is returned to the background level.

The shape of the pulse and its frequency can be varied over a considerable range and, by employing microcomputer and solid-state systems within modern power sources, greater control of both the dip and pulsed-transfer systems is possible. The output from such machines can be controlled in response to instant feedback from the situation at the arc. With such machines, a relationship can be established between the voltage, current, waveform and wire feed speed, with constant automatic correction being made throughout the welding operation. Such machines are known as 'synergic-MIG'. Very easy to use 'one-knob' welding units can be used when a system is in place to maintain a proper relationship between the pulse rate and the wire feed rate.

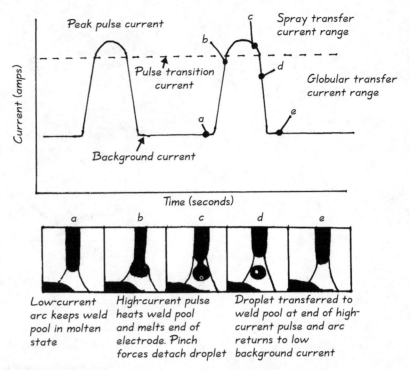

Fig. 5.16 *The metal-transfer sequence or pulsed MIG/MAG (GMAW)*

As well as aiding positional welding, a further advantage of pulsed-transfer equipment is the ability to decrease the current and deposition rate to allow the manual welding of sheet down to 1 mm (0.04 inch), and even below this thickness with mechanised systems.

In summary, the various types of metal transfer in MIG/MAG (GMAW) so far covered are:

Short-circuiting: This uses a low welding current and arc voltage. The electrode is fed at a constant speed and dips into the molten pool (hence the alternative name of 'dip transfer'), which causes a short-circuit. The current from the welding power plant increases and heats the electrode wire to a temperature where the end melts off, again creating an arc between the electrode wire end and the work (see Figure 5.11). These characteristics produce a small, quickly solidifying weld pool that is well suited for joining thinner sections in all positions, or for root runs in thicker sections.

Globular transfer: This takes place with current and arc voltage between the short-circuiting and spray transfer ranges, regardless of the shielding gas. In this form of transfer, a large droplet around two to four times the electrode diameter is produced with CO_2 shielding, the repelling forces of the arc acting towards the wire tend to hold the droplet in place until it finally transfers with gravity.

Spray transfer: With an argon-rich shielding gas, metal transfer will change from globular to spray as welding current increases. Metal is transferred across the arc in small droplets, axially directed at a high rate to the workpiece.

Pulsed spray transfer: This employs a low-level background current to maintain an arc and a superimposed pulse of high current above the spray transition level. Metal is transferred to the work only during this high-current pulse. The ideal is to transfer one droplet of molten metal during each pulse. In order to achieve high filler metal deposition rates (up to a maximum of 18 kg/hr (40 lb/hr)) a further type of transfer called:

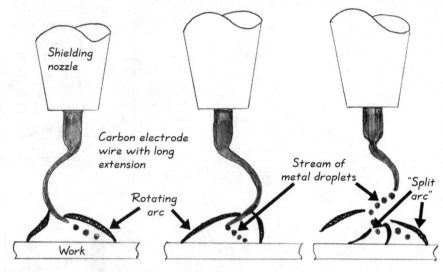

Fig. 5.17 *Rotational spray transfer (after AWS[4])*

High-current-density spray transfer can be employed. The arc characteristics of this system are further divided into rotational and non-rotational spray transfer.

High-current-density rotational spray transfer: This system employs solid, carbon steel wire electrodes fed at a high speed and with a long extension (sticking-out distance). Argon/CO_2 or argon/oxygen shielding mixtures are generally used. Because of the long stick-out distance, resistance heating of the wire electrode takes place, causing the end of the electrode to become molten. Electromechanical forces created by the current flow in the wire cause the softened wire and molten wire end to rotate in a helical path (see Figure 5.17). Different shielding gases will alter the rotational transition current by changing the surface tension at the end of the molten electrode.

High-current-density non-rotational spray transfer: When shielding gases with high content of CO_2 or helium are used, they will raise the rotational spray transition current, suppressing the tendency of the electrode to rotate. The arc is similar to that in conventional spray transfer but more elongated and diffused. The plasma stream is axial and narrower than when rotational spray transfer is used. The greater heat concentration can give greater depth of fusion than the rotational method.

Surface tension transfer (STT)

This type of metal transfer uses a new type of welding machine design, which operates neither in the constant-current nor constant-voltage mode. Instead, STT uses a high-frequency (wide bandwidth) current-controlled machine, with the power to the arc being based on the instantaneous arc requirements and not on an 'average DC voltage'.

Such machines are capable of delivering and changing current to the electrode in the order of microseconds. Designed for semi-automatic applications, STT operates in the short-circuiting welding mode.

In the circuit, the rate of change in resistance is measured by a change in voltage per unit of time. The circuit produces a signal when the rate of change of the shorted bridge voltage is equal to, or exceeds, a specific value.

Such a signal is an indication that the short is about to separate. (In conventional short-circuiting transfer it is this re-establishing of the arc that creates spatter.) In STT, this signal is used to reduce the current quickly, so that when separation occurs, it does so at a low current, typically 50 A, producing minimal spatter.

References/further reading

1. W. Lochte-Holtgreven, *The Electric Arc Between Carbon and Iron Electrodes*, University of Kiel, Germany
2. G.H. Gilette and R.T. Breymeir, Some research techniques for studying arcs in inert gases, *Weld Res. Suppl.*, S.151, March 1951
3. W. Mantel, Uberlegungen uber die Bedeutung der Pays. Vorange im Schweisslichtbogen (On the physics of welding arcs). In *Schweissen und Schneiden*, No. 8, August 1956
4. ANSI/AWS C5.10-94 *Recommended practices for shielding gases for welding and plasma arc cutting*. American Welding Society, 1994

Electric arc welding methods

This chapter looks at **carbon-arc welding** (CAW), **atomic hydrogen welding** (AHW), **bare metal-arc welding** (BMAW), and **shielded metal-arc welding** (SMAW) (manual metal-arc welding). Chapter 7 considers **gas tungsten arc welding** (GTAW) (tungsten inert gas TIG/Tungsten arc gas shielded TAGS), while Chapter 8 looks at **gas metal arc-welding** (GMAW) (metallic inert gas MIG/Metal arc gas shielded MAGS), and automatic processes including **submerged-arc welding** (SAW) and **flux cored arc welding** (FCAW).

Before proceeding, it is recommended that the reader first look at the introduction to these processes in *Basic Welding*, Chapters 5, 8 and 9. Also, it might be beneficial to look again at the previous chapter in this volume, should the reader need clarification on details of the welding arc and metal transfer.

Carbon-arc welding (CAW), ISO 181

In the carbon-arc process using a single carbon electrode, the arc is established between the non-consumable carbon (graphite) electrode and the work. There is also a variation which uses an arc between two carbon electrodes, a process known as 'twin carbon-arc welding'.

Other variations, such as **shielded carbon-arc welding**, which employed either paste or powdered flux, and **gas carbon-arc welding**, which used a small jet of reducing gas directed at the arc area or the burning of an impregnated rope in the vicinity of the arc, are no longer utilized in industry.

The single carbon-arc process, which is sometimes called the **Bernardos' process** (see Chapter 1), is still advantageous for certain applications. In the

Fig. 6.1 *Sketch of manual carbon-arc welding*

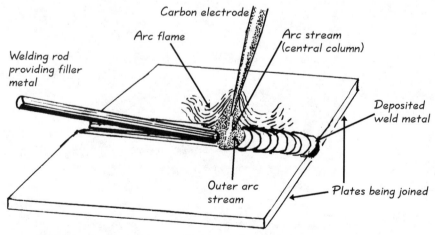

Fig. 6.2 *Welder's eye view of carbon-arc welding*

Fig. 6.3 *Motor generator DC arc welding machine*

mechanized form, the process has now been replaced by other welding methods, however manual carbon-arc welding, employing the hand-held carbon electrode holder, can have advantages for welding thin sheet (below 3 mm thickness), casting repairs and brazing applications (Figure 6.1).

From the welder's eye view (Figure 6.2) it can be seen that the carbon arc is used as the source of heat and that a welding rod can be employed, if necessary, when welding material above 3 mm (1/8th inch) in thickness. The process therefore has characteristics similar to gas tungsten arc welding GTAW, (tungsten inert gas welding TIG), where a tungsten arc is used as the source of heat.

Again, from Figure 6.2, it can be seen that there are three regions within the carbon arc. These are, on the outside, the yellow arc flame, and just inside this region is the outer arc stream which is violet in colour and has a temperature of around 2500°C (4500°F). The core of the arc contains the white central column, which, depending on the amount of current employed, can develop temperatures within the range of 3870–4980°C (7000–9000°F) (AWS).

Any standard DC arc welding machine of either the motor-generator or engine-generator type (Figures 6.3 and 6.4) or the rectifier type (Figure 6.5) can be used for carbon-arc welding, with the carbon electrode connected to the negative terminal. In use, the carbon electrode becomes white hot and

Fig. 6.4 *250 amp diesel-driven welding generator suitable for CAW, SMAW and GMAW applications, on site*
(Photograph courtesy of Lincoln Electric Company)

Fig. 6.5 *DC rectifier type arc welding machine*

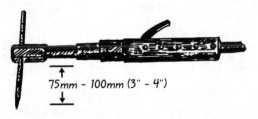

75mm – 100mm (3" – 4")

Fig. 6.6 *Typical air-cooled carbon electrode holder*

special electrode holders must be used. Figure 6.6 shows a typical air-cooled carbon electrode holder for smaller-diameter carbons. For larger-diameter electrodes and higher currents, water-cooled holders can be used. One type, which also includes a small shield to protect the hand, is shown in Figure 6.7. With higher direct currents (Figure 6.7a), induced magnetic fields can be created in the work, causing 'arc blow' or a 'wandering arc' (Figure 6.7b). One method of overcoming this, and creating a concentrated arc, is to incorporate a magnetic copper coil within the nozzle. If this coil is made from copper tube it can also be used to circulate cooling water, as in Figure 6.7c.

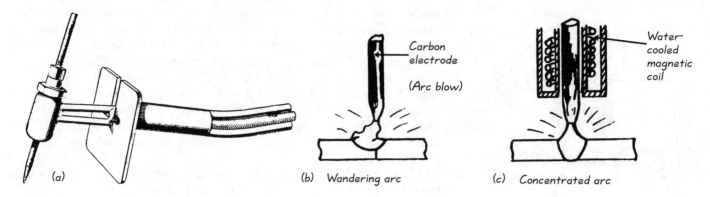

(b) Wandering arc
(c) Concentrated arc

Fig. 6.7 *One type of water-cooled carbon electrode holder*

(a)   (b)

Fig. 6.8 *Carbons: (a) as purchased; (b) prepared on emery wheel with a long, tapering point*

Table 6.1 *Recommended current values for carbon and graphite electrodes when using DC electrode negative*

Diameter of electrode		Carbon electrodes (amperes)	Graphite electrodes (amperes)
mm	inch		
3.2	1/8	15–30	15–35
4.8	3/16	25–55	25–60
6.4	1/4	50–85	50–90
7.9	5/16	75–115	80–125
9.5	3/8	100–150	110–165
11.1	7/16	125–185	140–210
12.7	1/2	150–225	170–260
15.9	5/8	200–310	230–370
19.0	3/4	250–400	290–490
22.2	7/8	300–500	400–750

Carbon electrodes are usually 300 mm long (12 inches) and are available in diameters from 3.2 mm (1/8 inch) thick to 22.2 mm (7/8 inch). Electrodes can be of either the baked carbon type or pure graphite. Graphite electrodes have the higher current-carrying capacity but are more expensive.

For welding with currents above 400 amperes, a 30 degree taper is satisfactory (Figure 6.8a). However, for currents below this, a long tapering point will maintain a more controlled arc and will remain sharp for much longer (Figure 6.8b). Such a point can be prepared on an emery wheel, fitted with adequate dust extraction.

Less current is required with the carbon arc than for a metallic arc on the same thickness of work. When adjusted to the recommended current settings, as shown in Table 6.1, the arc should be silent, with absence of the crackling sound that a metallic arc produces.

Particles of carbon can be transferred through the arc, however, because the carbon burns off very slowly, this should not have an adverse effect on non-critical applications, especially when filler metal is added to make the weld. On applications which do not require filler metal, such as the welding of thin sheet (Figure 6.9), or on casting repairs which may require machining after welding, then the hardening effect of the carbon may be prevented by using the longest arc possible, allowing any carbon vapour present to burn-up before reaching the weld.

As with other welding processes, any surface contamination must be removed from the joint area prior to welding. The arc may be struck by bringing the tip of the electrode into contact with the work and then raising it to the correct length for welding, or a carbon striking block may be used. The block can be placed on the work, near to the point where welding is to commence. The arc can then be started on the block and taken to the joint to be welded.

Generally, the carbon electrode should not extend more than 75 mm to 100 mm beyond the electrode holder (Figure 6.6). Should the arc be broken and require restarting, then the carbon should not be brought directly over the hot weld metal to restrike, as this could cause some carbon to be absorbed by the metal, causing a hard spot. To restart the arc, strike it on the cold metal adjacent to the weld then return it to the point from where welding is to be recommenced.

Welding steel

One of the main applications of the carbon-arc welding of steel is the making of upturned edge and outside corner welds in thin material, without added filler metal (Figure 6.9). Common applications are small tanks, machine guards

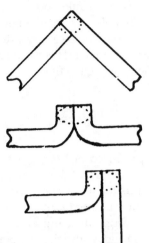

Fig. 6.9 *Some typical joint set-ups for carbon-arc welding of thin material (below 3mm or ⅛th inch thick) without using a filler rod*

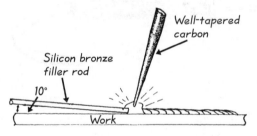

Well-tapered carbon

Silicon bronze filler rod

10°

Work

Fig. 6.10 *Method of welding galvanized sheet using a silicon-bronze filler rod and carbon arc*

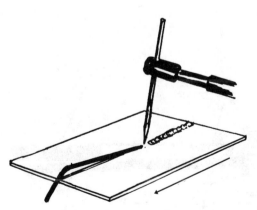

Fig. 6.11 *It is often advantageous to 'crank' the filler rod in order to obtain a low angle (around 10 degrees) to the seam, and also to help keep the hand away from the direct line of arc heat*

and sheet-metal duct work. The use of steel or copper backing bars is often beneficial, particularly when welding thin sheet steel or non-ferrous metals in order to prevent melting through.

Galvanized (zinc-coated) steel sheet can be welded by using the carbon arc and silicon bronze filler rod (Figure 6.10). With this method, lay the filler rod on the seam, holding its end at an angle of around 10° from the work. The arc should be played on the filler rod until a small pool of molten metal forms. The rod should then be fed in and melted at the leading edge of the pool as the arc is moved along the seam. This process is employed on butt, fillet and lap joints and, because the arc is always played on the molten pool of filler metal and not on the galvanized sheet, a minimal amount of the galvanizing is disturbed. Because of this fact and the minimal amount of distortion, the method is suitable for galvanized duct work. A carbon with a long taper (Figure 6.8b) should be employed and the work carried out using good fume extraction and ventilation.

When welding thicker sections of steel with a steel filler rod (Figure 6.11) the technique is very similar to gas tungsten arc welding (GTAW) (tungsten inert gas welding TIG), in that the welding rod is fed into the molten pool with one hand and the carbon arc is controlled with the other hand. Carbon-arc welding is, therefore, essentially a puddling process, with the addition of filler rod to the molten pool determining the size and shape of the weld bead. For this reason, welding is best carried out in the flat position. Joints can be achieved in the vertical and overhead positions but they will be found difficult to make.

Welding cast iron

Very satisfactory results can be obtained with the carbon-arc method using a cast iron welding rod of the type employed in oxy-acetylene welding. The technique is particularly useful for the fusion welding of castings that might be too large to repair-weld with oxy-acetylene. A cast iron welding flux can be employed and, if a machineable deposit is required, then the casting should be allowed to cool down slowly. A pre-heat of 650°C (1200°F) will be required, the same as for oxy-acetylene fusion welding. The carbon electrode should be connected to the negative terminal.

Welding copper and copper alloys

A graphite electrode is generally preferable in order to better withstand higher current levels. The electrode should again be connected to the negative terminal and a long arc length used, allowing carbon from the electrode to combine with oxygen to form carbon dioxide (CO_2), so providing a neutral protective atmosphere. A shorter arc produces a partially reducing atmosphere containing some carbon monoxide (CO).

The work should be pre-heated within the range of 150 to 650°C (302 to 1202°F), depending on the thickness of the work. If additional pre-heating equipment is not available, then the arc should be used to heat up the weld area before commencing welding, as the high thermal conductivity of the copper will take the heat away from the molten pool very rapidly. This rate of conductivity is so great with copper that it is difficult to maintain welding heat without pre-heating first.

Thin sheets may be joined without a filler rod, if the joint set-up allows. Plates over 3 mm (1/8 inch) will require the use of a filler rod. For plates 6 mm (¼ inch) thick and over, a 90 degree included angle bevel is recommended, in order to obtain root penetration. Flux is sometimes used and the filler rod, which usually contains a deoxidant such as silicon, should be held at a small angle (around 10 degrees) to the plate and fed into the molten pool in such a way as to prevent molten weld metal running ahead on to cold base metal. This can be achieved by firstly advancing the molten pool and then adding the filler.

Twin carbon-arc welding

With twin carbon-arc welding, the heat is produced by an arc between two carbon electrodes while the work does not form part of the electrical circuit. The twin carbon-arc torch (Figure 6.12) extends the amount of work possible with an ordinary AC (alternating current)-type welding machine, because the arc produced can be used for heating, brazing, surfacing and soldering operations, as well as for welding. Carbon electrodes with a thin copper coating are normally used and, although the torch is easier to use on an AC machine, it can be used with DC if the carbon connected to the positive is one size larger than the negative carbon, thus reducing the burn-off rate of the positive carbon. The gap between the carbons can be adjusted as they are slowly consumed in use, by turning the adjusting wheel on the insulated torch handle (Figure 6.12).

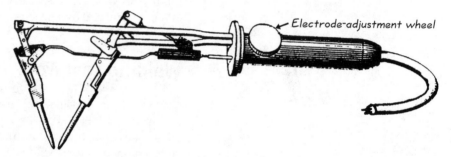

Fig. 6.12 *Sketch of a twin carbon-arc welding torch*

Atomic hydrogen welding (AHW), ISO 149

This process is now very seldom used, but it is mentioned briefly here because of its unusual characteristics and the similarity of the torch (Figure 6.13) to the twin carbon-arc torch.

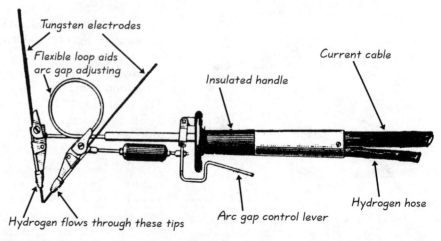

Fig. 6.13 *One design of atomic-hydrogen welding torch*

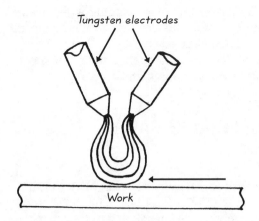

Fig. 6.14 *The shape of the atomic-hydrogen welding arc*

In this process the arc is between two tungsten electrodes and an atmosphere of hydrogen is fed through the tips of the electrode holders (Figure 6.13). The size of the arc can be controlled by the use of the arc gap control lever. The arc is fan shaped, as shown in Figure 6.14.

The process can be employed without filler on thin sections or with filler on thicker, prepared joints. At one time, it was considered the best method for welding alloy steels and was even employed on non-ferrous materials. In recent years, however, gas shielded arc processes using inert and semi-inert gas mixtures have replaced atomic hydrogen welding on most applications.

Atomic hydrogen can still be found in the workshops of some chain manufacturers and because of its high-temperature arc-flame, it is sometimes used for joining and repairing tungsten-bearing tool steels and tungsten carbide cutting inserts requiring very high strength, such as oil-well drill bits.

In its normal molecular state hydrogen is diatomic, that is, each molecule consists of two atoms. In atomic hydrogen welding, the temperature in the arc between the two tungsten electrodes reaches around 6090°C (11 000°F), which is high enough to dissociate the molecular hydrogen into its atomic form, and a large amount of heat is absorbed by the hydrogen. The atoms recombine on reaching cooler regions outside the arc and at the surface of the work. The recombination of the atoms of hydrogen back to the molecular form gives a rapid release of heat which can be controlled by the distance the arc is held from the work surface. (Increasing the distance will reduce the amount of heat input to the work, and reducing the distance will increase the amount of heat input.)

Hydrogen is a strong reducing gas, particularly in its atomic state; it therefore protects the tungsten electrodes and the weld metal from oxidation. The hydrogen also has a cooling effect on the electrodes. Any oxygen present in the outer area surrounding the arc-flame combines with the hydrogen, producing water, which is instantly turned to steam. Small amounts of hydrogen still remaining will burn off, outside the area of recombination, in the normal process of combustion.

Bare metal-arc welding (BMAW) and bare wire metal-arc welding, ISO 113

In this process an arc is maintained between a bare or lightly flux-coated electrode and the weld pool on the work. In the case of the bare wire electrode, because there is no flux coating, an electrical connection can be made at any point along its length. It can therefore be gripped in the electrode holder at any convenient position and not just at the flux-free end, as with coated electrodes (Figure 6.15).

Fig. 6.15 *Bare metal-arc welding: the bare wire electrode can be gripped at any point along its length to make electrical contact, not just at the flux-free end as with a flux-coated electrode*

Welds made in low carbon steel using bare wire electrodes will not match the quality of the parent material in strength, ductility or resistance to corrosion.

The reason for such a low-quality weld with this process is that the molten metal, during transfer from the end of the electrode through the arc, and the weld metal have no protection from the atmosphere. The molten metal is therefore subject to oxidation and nitrification. The formation of weld porosity can be detrimental to the strength and ductility. The nitrogen in the form of nitrides will tend to increase the hardness of the weld deposit at the expense of ductility.

Any water vapour present can dissociate in the unprotected arc to form hydrogen, which can cause the embrittlement of some metals and is one of the main causes of 'cold' or 'underbead' cracking in alloy steels.

Shielded metal-arc welding (SMAW) – metal-arc welding with covered electrodes – using electrodes with extruded flux coatings, has now mostly replaced the use of bare or lightly coated electrodes.

The bulk of bare wire electrodes being manufactured is as coils or spools, for use with gas shielded semi-automatic and automatic processes.

Shielded metal-arc welding (SMAW), ISO 111

Also known as **metal-arc welding with covered electrodes** or **'stick welding'**, this is an early arc welding process (1907–1908) [see Chapter 1 under developments of Kjellborg and of Strohmenger] developed from bare metal-arc welding, in order to improve the ease with which the welding operation could be carried out and also the weld quality.

To make electrode wire, the metal rod is pulled through a 'die' (Figure 6.16), with the aid of a claw, which is fastened to a chain – the other end of the chain being wrapped around a drum. As the drum rotates, the metal is pulled through the 'die', the finished size of wire being governed by the selected diameter of the hole in the die.

For bare metal-arc welding, the wire was simply cut to suitable lengths in order to fit the grip in the electrode holder.

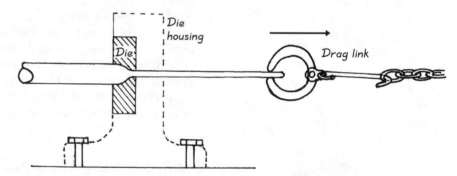

Fig. 6.16 *Wire drawing for making bare wire electrodes and core wires for flux-coated electrodes*

Dipped electrodes

It was discovered that a flux coating which was lighter than the filler metal (when molten), melted at a slightly lower temperature and tended to absorb atmospheric gases, gave vastly improved weld quality. However, the only method used for some time to apply a flux coating to the wire was by dipping it into a molten flux bath several times, allowing the flux to dry each time between the dippings. This gradually built up a coating, but made dipped electrodes very expensive because of the length of time required to produce them.

Reinforced electrodes

Fig. 6.17 *Reinforcement of the wire and string wrapped around the core wire allow the flux coating to build up faster when dipped*

By wrapping thin string and steel wire around the electrode wire in a spiral (Figure 6.17) it was found that the flux tended to build up a thicker coating in less time and therefore reduced the cost. Some special electrodes, such as those for welding cast metals, are still manufactured in this way, but most are manufactured by the extrusion process.

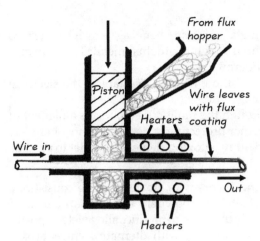

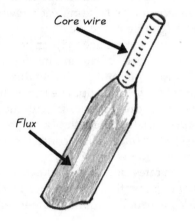

Fig. 6.18 *Method of making extruded electrodes*

Fig. 6.19 *An electric wire brush cleans a small portion of flux away so that electrodes will fit in holder and make good electrical contact*

Extruded electrodes

The wire here is automatically fed through the flux mixture, which is under pressure from the piston (Figure 6.18). The arrangement of heaters dries the flux on to the wire as it extrudes out of the machine. A guillotine cuts the electrodes to length and an electric wire brush cleans off the flux at one end (Figure 6.19) in order that the electrodes will fit in an electrode holder. Flux-coated electrodes are produced very quickly by this method.

The coating consists of a clay-like mixture of binders, usually silicate and powders such as carbonates, fluorides, oxides, cellulose and metal alloys. A typical covering contains arc stabilizers, produces gases to displace air, and forms a protective slag coating (see Figure 5.8).

Several guides to the classification of shielded metal-arc welding electrodes exist. The current British Standard is discussed on page 56 of *Basic Welding* with full details of standards being available from the British Standards Institution. The American National Standards are published by ANSI/AWS and are available from the Order Department of the American Welding Society in Miami. (There is at present, work being undertaken through the European Welding Federation, The International Institute of Welding and other bodies, such as the International Standards Organization, to rationalize information and standards appertaining to welding. Electrode Charts 1 and 2 in Appendix 3 match British and American Standard classifications for steel electrodes.

Figure 6.20 shows a typical layout for SMAW. The flux-free end of the electrode is gripped with the electrode holder, which is connected to the AC or DC (alternating current or direct current) power source by a lead (cable). The work is connected to the other power source terminal.

The arc is started or 'struck' by touching the electrode tip to the work and then lifting it away slightly (see *Basic Welding*, page 39).

Once established, the arc heat will melt the parent material (work) in the immediate area below the electrode (arc crater), forming a weld pool. The heat from the arc also melts the metal core and the surrounding flux coating (see Figure 5.8).

As the arc is moved along, the molten pool on the parent plate, together with droplets of molten metal from the core wire and metal powders in the covering, coalesce to form a weld.

The characteristics of a particular flux coating can vary depending on the type of electrode being used but, generally, a typical flux coating will provide all or most of the functions listed at the top of page 84:

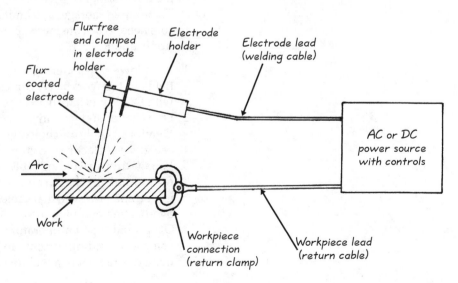

Fig. 6.20 *The layout of a typical welding circuit for shielded metal-arc welding*

1. A protective gas shield.
2. A supply of deoxidizers, fluxing agents and chemicals which can help to cleanse the weld as they rise up through the molten metal. They can also help to prevent excessive grain growth.
3. An aid for arc stability and re-ignition, helping to establish the electrical characteristics of a particular electrode.
4. A flux which, having risen through the molten metal, forms a blanket of slag over the cooling weld, further protecting it from the atmosphere.
5. Coatings that offer a means of adding alloying elements in order to provide the desired mechanical properties in a weld.

Most electrodes have a solid-metal core wire, however some consist of a thin-walled metal tube, filled with metal alloy powders. This tube is then coated with flux. These electrodes are usually designed to produce alloy weld deposits.

As mentioned in Chapter 5, when welding with alternating current (AC), the arc extinguishes and has to be re-established each time the current reverses direction. By incorporating such additions as potassium compounds into the coating, a gas is produced which remains ionized during each reversal of current. Such an ionized gas aids the re-ignition of the arc.

Increased electrode efficiency (see *Basic Welding*, page 133) can be achieved when welding some carbon and low alloy steels, by adding iron powder to the flux coating. This increases the deposition rate and the thicker flux coating forms a large cup or inverted crucible at the electrode tip. The rim of this cup can be rested on the work and the electrode dragged along, with the shape and size of the cup maintaining the appropriate arc length.

For this reason, iron powder electrodes are often called 'contact' or 'touch' electrodes and the technique is known as 'drag welding'.

The method gives very high deposition rates, but the slag takes a bit longer to solidify, making the process unsuitable for vertical and overhead welding.

Metal recovery with these electrodes is usually around 135 per cent, but can be as high as 200 per cent.

Welding power sources

Direct current power sources of the generator and rectifier type (Figures 6.29 and 6.28) can be used with this process. Also, most modern electrodes produce easily ionizable gases, making them suitable for use with an AC power source of the transformer or inverter type (Figures 6.27 and 6.31).

There are advantages and limitations for both types of current. The following factors need to be considered before choosing a particular type of current for an application:

1. Arc blow or 'wandering arc' (Figure 6.7b), caused by induced magnetic fields in the workpiece, can cause problems when welding ferritic steels with DC, but is hardly ever present when using AC.
2. DC is better and usually preferred for vertical and overhead welding. It is therefore usually the choice when pipewelding, although suitable electrodes can give satisfactory positional welds with AC equipment.
3. Voltage drop in welding cables is lower with AC. It is therefore more suitable for welding that has to be done some distance from the power source. It is important, however, that cables are not coiled, as this can cause dangerous overheating and high losses due to inductance.
4. DC provides better operating characteristics and a more stable arc when using low welding currents and small-diameter electrodes. This makes DC more desirable when welding thin sheet.

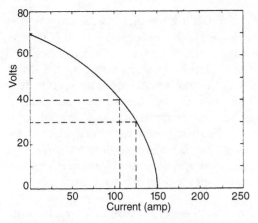

Fig. 6.21 *Typical output curve for a variable-voltage power source, adjusted for minimum current variation. This is the preferred type for vertical and overhead metal-arc welding*

(*Courtesy of Lincoln Electric Company*)

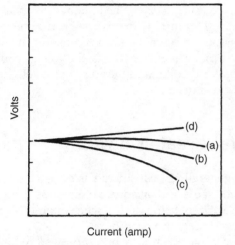

Fig. 6.22 *Typical output curves for constant-voltage power sources*

(*Courtesy of Lincoln Electric Company*)

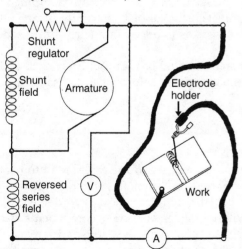

Fig. 6.23 *Connection diagram of reversed series generator (DC). By adjustment of the shunt regulator, the correct voltages for striking and welding are obtained automatically*

5. It is usually easier to strike the arc with DC. For this reason and its general ease of use on positional work etc., it is often the first choice when training welders.

6. It is easier to maintain a short arc length with DC.

Volt–ampere curves

Figure 6.21 shows a volt–ampere curve of a 'sloping characteristic' power source, or constant-current power source, which is preferred for manual welding.

This type of machine caters for variations in arc length which are likely to occur with manual welding. Quite a large change in arc length (voltage) produces only a small change in amperage.

The flat characteristic or constant-potential type machine (Figure 6.22) is more suitable for semi-automatic welding. It gives a large increase in current for a small variation in voltage, producing a 'self-adjusting arc' as it quickly increases or reduces the burn-off rate of the electrode wire, in order to maintain the arc gap at the desired pre-set distance.

The open-circuit voltage

This is the voltage set on the power source, and the voltage generated by the welding set when no welding is actually being carried out. This voltage is available to strike the arc and is generally between 50 and 100 volts. The arc voltage is determined by the arc length for any given electrode and is usually between about 20 to 40 volts.

Once the arc is struck and a welding load is applied, the open-circuit voltage will drop to the arc voltage.

Some welding machines do not have a control for adjusting the open-circuit voltage. In these machines it is usually pre-set to a level higher than the normal arc voltage maximum.

This high open-circuit striking voltage is reduced to the lower arc voltage by means of equipment designed to give a 'sloping characteristic'. This is the case with both DC generator equipment and AC welding transformers. The methods of obtaining this characteristic vary with the makes of equipment, however two typical examples are given in Figures 6.23 and 6.24.

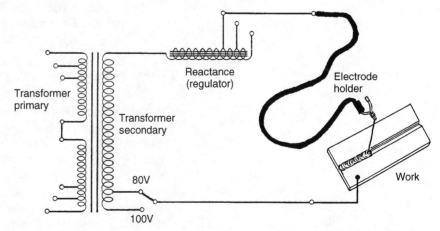

Fig. 6.24 *Diagram of connections for AC welding. The reactance coil in series with the arc produces the 'sloping characteristic'*

85

The standard theoretical method of obtaining a sloping characteristic when using a DC generator is based on reverse series windings (Figure 6.23). Here, the series coils are wound in a reverse direction to the shunt coils, therefore the field created by the series field is opposite to that created by the shunt field. In the open-circuit situation, only the shunt field is operative, giving the maximum available voltage to strike the arc. When the arc is struck, current flows through the series windings, setting up opposition to the shunt field. The resulting field strength is, therefore, less than the open-circuit field strength, resulting in a voltage drop. In this particular design, a shunt regulator is inserted in series with the shunt field and open-circuit voltage can be set to the position that will give an automatic voltage drop to the required level and the correct current value.

This method is used, with variations, by several manufacturers. In some makes the shunt field is self-excited, while in others, it is excited by an exciter fitted to the generator.

There are other designs in which the armature itself is designed to give the effect of a reversed series winding, providing a rapid sloping characteristic system.

With AC equipment the sloping characteristic can be obtained by means of a reactance coil in series with the arc (Figure 6.24). In this system, the voltage at the secondary side of the transformer remains almost constant, being available to strike the arc, however, as soon as current flows, the voltage drop across the reactance coil reduces the arc voltage to the required level. The reactance coil also controls the flow of current.

Low-voltage and power-breaker devices

The DC open-circuit voltages can generally be considered as quite safe, but with AC equipment, although the voltage is not considered dangerously high, there is always the possibility of an accident being caused when operators are required to work in confined spaces or in damp conditions.

For these reasons, low-voltage devices should be employed, which automatically reduce the voltage at the electrode within a fraction of a second of breaking the arc.

Power breakers should also be employed with all electrical equipment. These provide added protection against electrocution by instantly cutting off mains supply to the equipment if there is an overload or a wire or cable cut through.

If the breaker is triggered, all systems should be thoroughly checked and the problem rectified by a competent electrician before the device is reset.

Duty cycle

The components inside welding power sources and the cables tend to heat up during operation, as a result of the welding current flowing through them. Different methods are employed to cool the equipment down in order to keep it running at a temperature that will not damage electrical components and insulation. Some welding transformers are placed in sealed units and immersed in tanks of cooling oil, others employ an electric fan to force air over the components, while other less-expensive welding sets have side vents to allow the circulation of air by convection. Some machines employ thermal cut-outs which trip-out at a set temperature before any overheating can occur. They trip back in when the set has cooled down again.

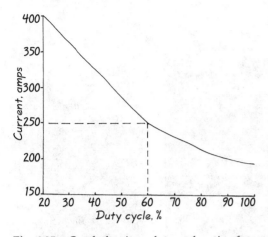

Fig. 6.25 *Graph showing a duty-cycle rating for a particular welding plant. It is shown that a 60% duty cycle means that the plant can be used at around 250 amps for 6 minutes in any 10-minute period*

The **duty cycle** expresses, as a percentage, the maximum time that the welding plant can deliver its rated output over a period of tests, without the temperature of its components exceeding a pre-determined safe limit.

In most countries, the duty cycle is based on a test interval of 10 minutes. Figure 6.25 shows a duty-cycle rating for a particular welding plant. Here, it can be seen that this unit can be used at 250 amps, giving a 60 per cent duty cycle. Over a 10-minute period, this means welding for 6 minutes with 4 minutes to cool down.

For manual welding, a 60 per cent duty cycle is usually more than adequate, as the operator has to keep stopping to change the electrode, chip off the slag, wire-brush and reposition work. For automatic welding, however, duty cycles at or approaching 100 per cent may be required.

As mentioned in Chapter 5, welding machines are classified as 'variable voltage' with a sloping characteristic or 'constant voltage' with a flatter characteristic.

A variable-voltage machine delivers a current that changes very slightly with any change in voltage. A constant-voltage machine will deliver current with only small changes in voltage, for any changes in current output. Figure 6.21 shows a typical output curve for a variable-voltage machine. This type of output is suitable for gas tungsten arc, shielded metal-arc and submerged-arc welding applications. Figure 6.22 shows typical output characteristics for constant-voltage welding machines. This shows that the voltage in the constant-voltage curve (*a*) rises slightly at the low currents and drops at the higher currents. Most constant-voltage welders are designed to have a slight downward slope, as shown in curve (*b*), with adjustments to increase the downward slope, as in curve (*c*). Some machines have a rising slope, as shown in (*d*), but this type of output is becoming less common.

AC welding machines

Transformer types

The type of transformer used for welding is of the 'step-down' design. That is to say, it changes high-voltage, low-amperage AC input current from the mains supply to a low-voltage, high-amperage AC welding supply.

The majority of transformer welding sets are designed to operate on single-phase mains supplies. The AC power produced in the UK is usually 50 hertz and in the USA it is usually 60 hertz. (1 cycle per second is known as 1 hertz (Hz), named after the German physicist who discovered electromagnetic waves.)

AC has no definite polarity, that is to say, the polarity changes at the rate of cycles per second. The current and voltage follow a curved path from zero to maximum positions. This is known as the sine curve and the waves are known as sinusoidal. Each time the polarity changes, the voltage goes through zero (Figure 6.26). This tends to create an unstable condition in the arc. Improved characteristics in the welding set and AC electrodes designed to aid re-ignition have overcome this problem.

Transformer welding machines can have different types of control systems to stabilize and adjust the welding current. One system for controlling the output current is to employ a series of taps into the secondary coil windings; another method is to use a movable/variable reactor in the output circuit. Other systems use devices that raise or lower the primary coil or a solid-iron core within the transformer windings, in order to vary the induced magnetic field and therefore the output of the machine.

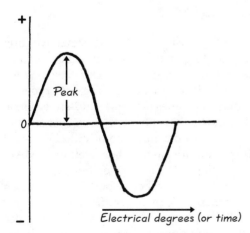

Fig. 6.26 *AC sine curve*

Fig. 6.27 *A range of AC welding transformers*
(Photograph courtesy of Murex Welding Products Ltd)

The tap system provides step control, while the reactor system and coil/core raising and lowering provide continuous stepless control (Figure 6.27).

Transformer welding machines are used widely throughout industry and small, relatively inexpensive AC welding sets are available for do-it-yourself, light industry, garage and maintenance work as well as agricultural welding applications.

Alternators

AC for welding can also be obtained from an engine-driven alternator. A petrol engine is usually used to drive the alternator in one unit and such engine–alternator sets can serve as a portable welding unit and an auxiliary power supply.

DC and AC/DC welding machines

Transformer–rectifier welding sets

Basically, the rectifier welding machine consists of a transformer welding machine with a rectifier added in order to give a DC output. Consequently, either AC or DC output can be obtained from this type of machine, by a switch which can bring in the rectifier, or take it out of circuit. The first types of machines used valve rectifiers but most machines now use solid-state silicon devices which are fitted in heat sinks and further cooled by an electric cooling fan. Output characteristics from this type of machine can be either constant or variable voltage. Machines built especially for gas metal-arc welding GMAW (MIG/MAG) contain a system for changing both the slope of the output curve and the reactance in the circuit, in order to give better performance when welding using short-circuiting transfer. Transformer–rectifiers designed for use with gas tungsten arc welding GTAW (TIG/TAGS) are fitted with a device that provides a high frequency voltage superimposed on the output voltage, in order to initiate the arc without touching the electrode to the work. When

Fig. 6.28 *Portable transformer–rectifier welding unit*

(*Photograph courtesy of Murex Welding Products Ltd*)

using AC, the high-frequency unit also helps to stabilize the arc and prevent rectification, by re-initiating the arc each time the current/voltage goes through zero.

Further units can be added or incorporated into transformer rectifier sets in order to give a pulsed arc welding facility for GMAW (MIG/MAG) and GTAW (TIG/TAGS). With these systems, the timing of the pulsed current can be adjusted very accurately. Most modern equipment is fitted with line voltage compensation, which eliminates any problems that might occur from the fluctuations in the mains voltage supply. Figure 6.28 shows a portable 630 amp transformer–rectifier unit for running off a three-phase supply. This machine has an open-circuit voltage of 65/72 volts and a current range of 8 to 630 amps. It will provide 400 amps at 100 per cent duty cycle, 500 amps at 60 per cent, and 630 amps at 35 per cent. The frequency is either 50 or 60 Hz.

DC generator welding machines

Generator welding machines can be driven by a petrol or diesel engine for site work or by an AC electric motor within a factory or workshop. Figure 6.29 shows a typical motor-generator welding machine. An armature is rotated in an electric field, and current is generated in the armature and taken off for use through the commutator (see *Basic Welding*, page 142). The arc characteristics can be controlled very precisely with a generator, which tends to give DC welding slightly more versatility than AC welding. Also, the polarity of the electrode can be changed instantly, if required. Units can provide either variable, constant-voltage, or both types of output.

The variable-voltage type of generator has a series field that causes the voltage to decrease as the current is increased. With this system, two adjustments can be made to change the welding current:

Fig. 6.29 *A typical motor-generator DC welding machine*

(*Courtesy of Lincoln Electric Company*)

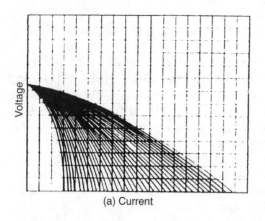

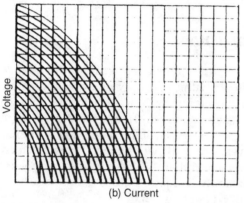

(a) Current

(b) Current

Fig. 6.30 *Output for a DC generator welder having adjustments in both the series and shunt fields. Output curves produced by adjusting the series field are shown in (a); curves produced by adjusting the shunt field are shown in (b)*

(Courtesy of Lincoln Electric Company)

Fig. 6.31 *Light compact DC inverter welding set*
(Courtesy of Lincoln Electric Company)

1. For a given voltage, the output current can be changed by adjusting the series field. This produces an output change as shown in Figure 6.30(a). It is sometimes called the 'current' control.
2. For a given current control setting, the output can be changed by adjusting the shunt field. This produces an output change as shown in Figure 6.30(b).

By combining both adjustments, output characteristics similar to Figure 6.21 or Figure 6.22 can be produced. The type of output shown in Figure 6.22 is very suitable for vertical and overhead welding, as there is a greater change in current for a given change in voltage than with the output shown in Figure 6.21. Because deposition varies with current, the operator can vary the amount of deposition, thereby exercising greater control over the molten pool with the 'flatter' output characteristic.

Inverter welding sets

In simple terms, an inverter can be classed as being the opposite of a rectifier, because it converts DC to AC. More precisely, an inverter circuit uses solid state devices such as SCRs (silicon controlled rectifiers), sometimes known as thyristors, or transistors, to convert DC into high-frequency AC, usually in the range of 1–50 kHz. Because transformer size is inversely proportional to the line or applied frequency, reductions of up to 75 per cent in the size and weight of power sources are possible using inverter technology. The solid-state devices act as very fast switches, rapidly switching on and off.

The inverter power source takes in the AC mains supply at 50 or 60 Hz. It first rectifies this, then chops it into smaller pieces using the rapid switching, transforms it and rectifies it again, using the resulting output for welding.

In simple terms, the result of chopping the 50 or 60 Hz incoming mains current into small pieces allows small amounts to be transformed at a time and therefore the transformer does not need to be anywhere near the size of a conventional welding transformer. This results in welding sets such as the one shown in Figure 6.31, which can provide 130 amp for welding, but weighs only 11 kg and therefore can be carried in one hand.

Welding techniques and procedures

Some other welding processes (GMAW and SAW, for example) can be more productive than SMAW, depending on application, but the versatility of the process and its capability of producing very high-quality welds (Figure 6.32), together with its simplicity, ensure that it is still used for a high percentage of welding work worldwide.

The process is suitable for joining both ferrous and non-ferrous metals in a range of thicknesses, although, because of the increased use of gas tungsten arc welding and gas metal-arc welding on non-ferrous materials, it is now mainly employed for welding steels and cast iron, as well as surfacing and repair applications.

It is best suited for welding thicknesses of 3 to 19 mm (1/8 to $\frac{3}{4}$ inch) (although it is capable of welding both above and below this range, other processes may give better results or be more cost effective).

From around 6 mm ($\frac{1}{4}$ inch) thick material, edge preparation is required in order to allow full access to the root of the joint and give fully fused welds (see Figure 6.33 and *Basic Welding*, page 44).

Fig. 6.32 *Construction of oil production platform using shielded metal-arc welding at Highlands Fabricators.*

(Photograph courtesy of AWS, Miami, USA)

Fig. 6.33 *Showing the preparation in the wall of a pressure vessel for a nozzle to be welded in position using multi-run SMAW.*

(Photograph courtesy of Dresser-Rand Ltd, Wythenshawe, Manchester, UK)

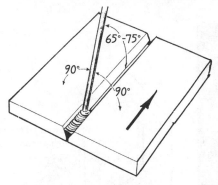

Fig. 6.34 *Making a single 'V' butt weld in the flat position. The number of runs needed is dependent on the size of the electrode and the plate thickness*

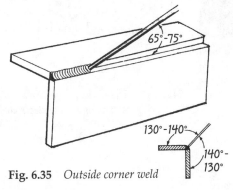

Fig. 6.35 *Outside corner weld*

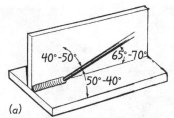

Approximate angles for making 2nd and 3rd in a fillet joint

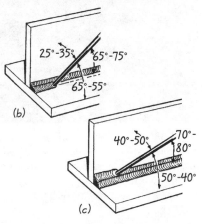

Fig. 6.36 *Technique for making a single-run 'T' fillet weld, or first run in a multi-run fillet*

Fig. 6.37 *Welding flange by SMAW. The electrode is held in position and the work is slowly rotated using a roller bed manipulator*

(Photograph courtesy of Dresser-Rand Ltd, Wythenshawe, Manchester, UK)

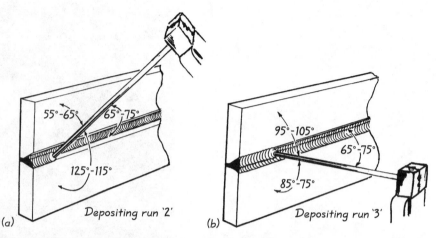

(a) Depositing run '2'

(b) Depositing run '3'

Fig. 6.39 *Method of welding a butt joint in the horizontal–vertical position using three runs. Run '1' – the root or penetration run is deposited and deslagged. Run '2' is then deposited, providing a 'shelf' or a ledge for run '3'*

Fig. 6.38 *High levels of skill are required to weld with SMAW (stick electrodes) in all positions and obtain high-quality welds. Here a student welder is practising vertical welding*

(Photograph courtesy of AWS, Miami, USA)

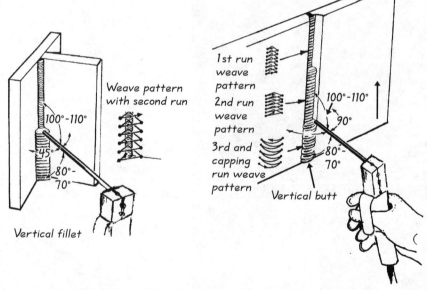

Weave pattern with second run

Vertical fillet

1st run weave pattern
2nd run weave pattern
3rd and capping run weave pattern

Vertical butt

Fig. 6.40 *Technique for vertical welding showing various weave patterns. A weave should never be greater than 3 × the electrode diameter*

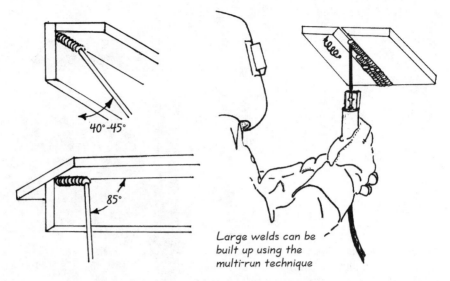

Fig. 6.41 *Technique for welding fillet and butt joints in the overhead position*

Welding is easier and faster in the flat position and Figures 6.34, 6.35 and 6.36 show recommended electrode angles for the butt, outside corner and fillet weld. Current settings can vary slightly with different makes of electrodes and advice on average settings is given on the electrode packet, with specific requirements usually being given in a manufacturer's welding procedure.

Where possible, welding is carried out in the flat position, if necessary with the aid of a table or roller-bed type manipulator. Figure 6.37 shows how the electrode can be held in one position and the work slowly rotated beneath, using a roller-bed manipulator.

If work cannot be positioned for welding in the flat, the the SMAW process is quite capable of depositing high-quality welds in the vertical and overhead positions, but the required skill is much greater and the rate of deposition slower. Figure 6.38 shows a welder perfecting the vertical technique in training school. Figures 6.39, 6.40 and 6.41 illustrate techniques for SMAW in the horizontal–vertical, vertical and overhead positions.

In all positions, welds can be made by building up single runs or 'stringerbeads', without weaving – or a combination of runs without weaving and runs using weaving can be employed. Which method to use will either be at the discretion of the welder or laid down in the welding procedure.

Because of spatter losses and stub end wastage (the small portion of the electrode left unused), only about 60 per cent of the electrode weight is deposited with SMAW, giving high material costs. Also, cleaning the slag (burnt flux) covering from the surface of the weld bead must be done after each pass, increasing the cost of labour.

Despite these facts, SMAW is a very versatile process and relatively simple to set up and operate. These facts extend its use to maintenance welding, where it is often employed for surfacing, or repair and reclamation of broken or worn parts.

Figure 6.42 shows plan and elevation views of welds made in the flat position with a general-purpose electrode under various conditions (see also *Basic Welding*, pages 61 and 62 (Difficulties and defects in welding)).

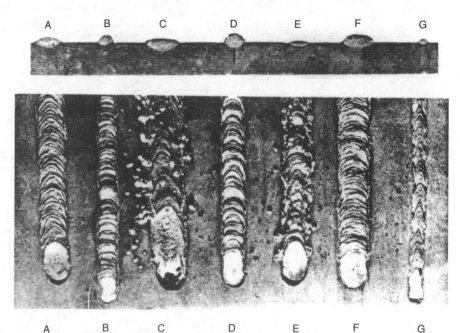

Fig. 6.42 *Plan and elevation views of welds made with a general-purpose electrode under various conditions: (A) current, voltage and speed normal; (B) current too low; (C) current too high; (D) voltage too low; (E) voltage too high; (F) speed too low; (G) speed too high*

(Courtesy of Lincoln Electric Company)

Resulting Weld Characteristics

Operating variables	Arc sound	Penetration-fusion	Melt off electrode	Appearance of bead
A. Normal Amps, Normal Volts, Normal Speed	Sputtering hiss plus irregular energetic crackling sounds	Fairly deep and well defined	Normal appearance	Excellent fusion – no overlap
B. Low Amps, Normal Volts, Normal Speed	Very irregular. Sputtering. Few crackling sounds	Not very deep nor defined	Not greatly different from above	Low penetration
C. High Amps, Normal Volts, Normal Speed	Rather regular explosive sounds	Deep long crater	Electrode covering is consumed at irregular high rate	Broad, rather thin bead – Good fusion
D. Low Volts, Normal Speed, Normal Amps	Hiss plus steady sputter	Small	Covering too close to crater. Touches molten metal and results in porosity. Electrode freezes	Sits upon plate, but not so pronounced as for low amps. Somewhat broader.
E. High Volts, Normal Speed, Normal Amps	Very soft sound plus hiss and a few crackles	Wide and crater deep	Note drops at end of electrode. Flutter and then drop into crater	Wide – spattered
F. Low Speed, Normal Amps, Normal Volts	Normal	Crater normal	Normal	Wide bead – overlap large. Base metal and bead heated to considerable area
G. High Speed, Normal Amps, Normal Volts	Normal	Small, rather well-defined crater	Normal	Small bead – undercut. The reduction in bead size and amount of undercutting depend on ratio of high speed and amps

(Courtesy of Lincoln Electric Company)

Safety

Full protection from the rays given off by the electric arc, fumes and hot metal is required. Also, all electrical equipment must be approved and regularly inspected. Ear protection must be worn in areas where noise levels exceed allowable limits. Health and safety is discussed in more detail in *Basic Welding*, pages 5–17.

Gas tungsten arc welding

In **gas tungsten arc welding** (GTAW), ISO 141 (tungsten inert gas welding, TIG; tungsten arc gas shielded welding, TAGS), an electric arc is held between a single non-consumable tungsten electrode and the workpiece, providing the heat for welding. A shielding gas surrounds the arc, the electrode and the molten weld pool. The weld is made without pressure (Figure 7.1). Depending on the thickness of the material being welded and the type of weld required, welds can be made with or without the addition of filler metal.

The process is capable of producing high-quality welds with relatively low equipment costs.

The use of helium to shield the welding arc and molten pool area was discovered in the 1920s (by H.M. Hobart and P.K. Devers), both registering US patents in 1926). However, not until World War II (1939), when the aircraft industry required materials such as aluminium and magnesium to be joined with quality welds, did the method come to the fore.

The process can either be used for making complete GTAW joints in either single or multi-runs, or for making combination welds with GTAW root runs capped with runs deposited by the SMAW process (Figure 7.2).

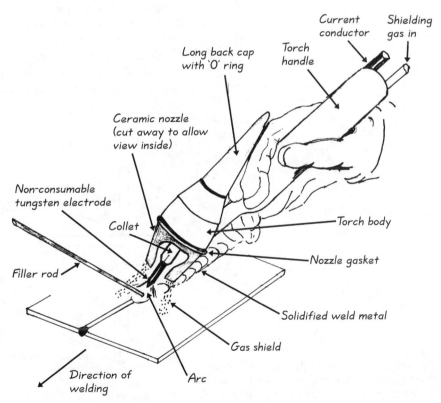

Fig. 7.1 *Showing the basics of manual gas tungsten arc welding*

Fig. 7.2 *An example of combination welds on a spherical pressure vessel (GTAW root with SMAW capping runs). The 'dual process' technique is used to produce very high-quality welds.*

(Photograph courtesy of Dresser-Rand Ltd, Wythenshawe, Manchester, UK)

Either argon or helium may be used for most applications, or mixtures of argon and helium.

For equal current settings, less heat will be put into the work with an arc in an argon atmosphere and, therefore, less penetration is obtained. This fact tends to make argon the choice as a shielding gas when manually welding thin material, in order to eliminate excessive melt-through. Argon also permits slightly improved control of the molten weld pool when welding in the vertical or overhead position.

The increased heat from an arc burning in helium can be an advantage when welding thicker plate and/or when joining metals with a highly thermal conductivity. Helium is also often the choice when high-speed mechanised welding is required.

A mixture of argon and helium can be beneficial if a compromise between the characteristics of both gases is required.

Helium is lighter than argon and, in order to provide the same level of shielding from the atmosphere, the flow rate must be two to three times that of an argon shield.

Argon and hydrogen mixtures can be used for certain mechanized applications, where the hydrogen would not be likely to cause porosity. The most used mixture is argon/15 per cent hydrogen. When used on closed butt joints in thin stainless steel sheet (up to 2 mm thick), welding speeds 50 per cent faster than those obtained using argon, and about the same speed as when using helium, can be achieved.

A typical application for the use of argon/hydrogen shielding is the mechanized welding of stainless steel beer barrels.

Determining gas flow rates

The flow rate is based on the thickness of the metal to be welded (the size of molten pool requiring protection) and the configuration of the joint (Figure 7.3). The cup or nozzle size is selected to suit the size of the weld pool and the reactivity of the metal to be welded.

The flow must be just enough to overcome the heating effects of the arc and any draughts (cross-flows of air) which may be present. Too high a flow rate can cause turbulence and allow atmospheric contamination of the weld pool.

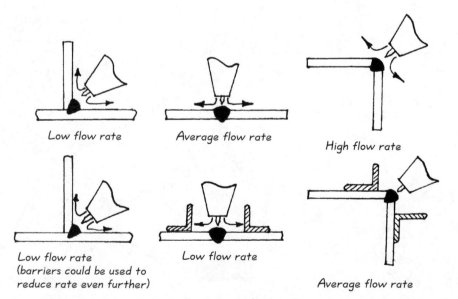

Fig. 7.3 *Use of barriers to contain the shielding gas near the junction being welded. In this case, simple angle-iron pieces laid or tacked in position would suffice*

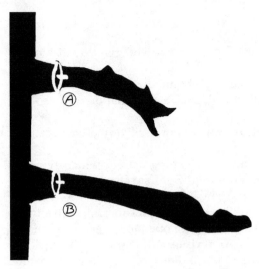

Fig. 7.4 *Representation of gas flow stream: Ⓐ from conventional nozzle; Ⓑ using a gas lens (to eliminate turbulence) – both using helium*

Protective screens or tents can be employed in order to prevent draughts or cross-winds blowing away the gas shield.

Turbulence can be virtually eliminated and the gas flow given a directional stability by using a gas lens (Figure 7.4). A gas lens has a permeable barrier of concentric fine-mesh stainless steel screens that fit into the nozzle (see *Basic Welding*, page 80).

Typical shielding gas flow rates can range from 7 to 16 litres per minute (15 to 35 cubic feet per hour) for argon, and 14 to 24 litres per minute (30 to 50 cubic feet per hour) for helium.

Pressing a control switch on the torch starts the flow of both current and gas shield (Figure 7.5). On some equipment a foot-switch is employed.

If the equipment is not fitted with automatic 'slope-in' and 'slope-out', then the foot-control can gradually increase the current at the start of the weld and gradually decrease it at the finish, making it possible to build up the end crater in order not to leave a cavity.

Pre- and post-weld gas purges can also be set on most machines. These allow the shielding gas to flow for a few seconds to clear the system of air before welding commences and then to remain on for a few seconds after the weld is completed to allow the weld to cool down slightly, before being exposed to the atmosphere.

The shielding gas from the nozzle protects the surface of the weld, but a root pass can be contaminated on the underside if it is in contact with the air. The air can be purged from this region by various methods such as back-purges and baffles. Some typical systems are illustrated in Figure 7.6.

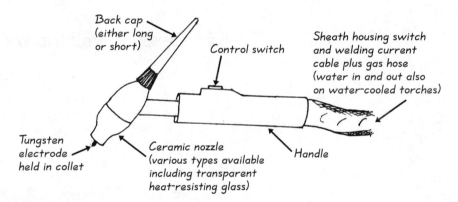

Fig. 7.5 *Torch for GTAW (TIG)*

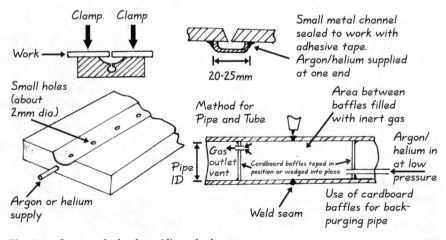

Fig. 7.6 *Some methods of providing a back purge*

Type of arc current

Direct or alternating current (DC or AC) can be used; which is selected depends mostly on the metal to be welded.

Fast welding speeds on steels can be obtained by using DC electrode negative, especially with a helium shield. Such a combination will also give deep penetration welds.

DC with the tungsten electrode positive is not used very often because it causes the electrode to overheat.

Alternating current gives a cathodic cleaning action, which removes refractory oxides from the joint and weld pool surfaces when welding aluminium and magnesium alloys (Figure 7.7; see also Chapter 5). There is, therefore, no need to use a corrosive flux with this process and, with either a machine or manually (using a skilled welding operator), very high-quality welds can be made with this process. Argon is usually chosen as the shielding gas, to allow greater control of penetration and fusion (as it allows more time to make adjustments, because of the lower heat input), when manual welding.

The arc voltage

This is measured between the tip of the tungsten electrode and the work surface (Chapter 5).

Variations to this distance, the arc current, the shape of the electrode tip, and the arc atmosphere (shielding gas) will all have an effect on the arc voltage.

Types of welding torch

There are two main types of welding torch:

1. Air-cooled or, more correctly, gas-cooled, which use the relatively cool shielding gas flowing through the torch to provide the cooling. Figure 7.8 shows the main components of a typical air-cooled torch, designed for use up to 159 amps DC at a 60 per cent duty cycle. 200 amps is generally

Fig. 7.7 *Trainee carrying out GTAW (TIG) on an aluminium test piece.*
(Photograph courtesy of AWS, Miami, USA)

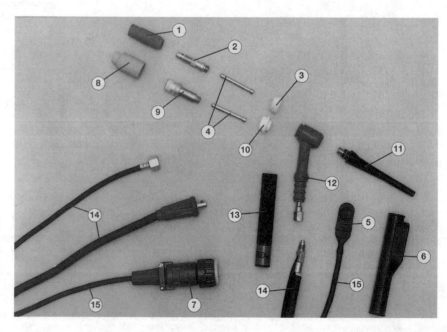

Fig. 7.8 *Main components of an air-cooled (gas-cooled) GTAW (TIG) torch for use up to 150 amps at 60% duty cycle DC electrode negative, or 90 amps at 60% duty cycle AC:* ① *ceramic nozzle,* ② *collet body,* ③ *nozzle gasket,* ④ *collet,* ⑤ *momentary switch,* ⑥ *switch boot,* ⑦ *plug,* ⑧ *gas lens holder,* ⑨ *gas lens collet body,* ⑩ *gas lens insulator,* ⑪ *long back cap with 'O' ring,* ⑫ *torch body,* ⑬ *torch handle,* ⑭ *power/gas leads,* ⑮ *torch switch leads*
(Photograph courtesy of Murex Welding Products Ltd)

considered to be the maximum working amperage for gas-cooled torches.

2. Water-cooled torches. These are designed for use up to around 500 amps, using water flowing through the torch to keep it cooled. Closed systems comprising a reservoir, pump and radiator can be employed to conserve water. Most automatic applications would use water-cooled systems. Figure 7.9 shows the main components of a typical water-cooled torch capable of use up to 450 amps.

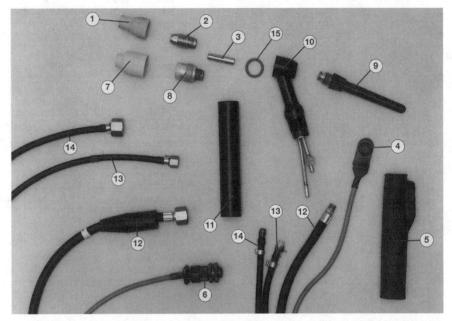

Fig. 7.9 *Main components of a water-cooled GTAW (TIG) torch for use up to 450 amps DC electrode negative/350 amps max. AC:* ① *ceramic nozzle,* ② *collet body,* ③ *collet,* ④ *momentary switch,* ⑤ *switch boot,* ⑥ *plug,* ⑦ *gas lens nozzle,* ⑧ *gas lens collet body,* ⑨ *long back cap,* ⑩ *torch body,* ⑪ *torch handle,* ⑫ *power/water cable,* ⑬ *gas hose,* ⑭ *water hose,* ⑮ *'O' ring*
(Photograph courtesy of Murex Welding Products Ltd)

Electrodes

Tungsten is chosen as the electrode material because of its very high melting point of around 3400°C (6152°F) and its high thermionic emission. Electrodes classed as 'pure' tungsten contain a minimum of 99.5% tungsten and no intentional alloying elements.

The current-carrying capacity and thermionic emission can be increased with the addition of certain metal oxides during manufacture. The most common additions are: thoria (ThO_2), Zirconia (ZrO_2) and Cerium oxide (CeO_2).

Selection of electrode

The current should be high enough to allow the arc to cover the whole area of the electrode tip. If the current is too low (electrode too large), the arc can be unstable and erratic, ejecting small particles of tungsten. If the current is too high (electrode too small), the electrode can overheat and even melt at the tip, dropping molten globules of tungsten into the weld. The arc would again become erratic and unstable.

Table 7.1 shows the recommended amperage for different sizes of electrodes for use on DC electrode negative, DC electrode positive, and AC.

Table 7.1

| Electrode size | | Nozzle size | | Direct current (amps) | | Alternating current (amps) |
mm	in.	mm	in.	electrode –ve	electrode +ve	
0.5	0.020	6.4	$\frac{1}{4}$	5–20	—	5–15
1.0	0.040	9.6	3/8	10–75	—	15–70
1.6	1/16	9.6	3/8	70–150	10–20	50–100
2.4	3/32	12.8	$\frac{1}{2}$	150–250	15–30	100–160
3.2	1/8	12.8	$\frac{1}{2}$	250–400	25–40	150–210
4.0	5/32	12.8	$\frac{1}{2}$	400–500	40–55	200–275
4.8	3/16	16.0	5/8	500–750	55–80	250–350
6.4	$\frac{1}{4}$	19.2	$\frac{3}{4}$	750–1000	80–125	325–450

Tungsten electrodes are colour coded either with bands or dots to help identify them. Table 7.2 gives the composition and identification colours of tungsten electrodes under BS EN 26848: 1991 (equivalent to ISO 6848: 1984).

Table 7.2

Composition	Identification colour	General use
99.8% tungsten	Green	AC
0.80–1.20% ThO_2	Yellow	DC
1.70–2.20% ThO_2	Red	DC
0.70–0.90% ZrO_2	White	AC
1.80–2.20% CeO_2	Grey	AC + DC

In the USA, the colour code shown in Table 7.3 is used.

Table 7.3

AWS classification	Colour	Alloying oxide*		Nominal weight of alloying oxide (%)
EWP (Electrode W Pure)	Green	—	—	—
EWCe-2	Orange	CeO_2	(Cerium)	2
EWLa-1	Black	La_2O_3	(Lanthanum)	1
EWTh-1	Yellow	ThO_2	(Thorium)	1
EWTh-2	Red	ThO_2	(Thorium)	2
EWZr-1	Brown	ZrO_2	(Zirconium)	0.25
EWG	Grey	Not specified	—	—

*(Manufacturer must identify the type and nominal content of the rare earth oxide addition).

Although non-consumable, during use, the tip of the tungsten degrades and this can make arc control and striking difficult. It is therefore necessary to regrind the electrode from time to time. With AC welding (usually carried out with a pure or zirconated tungsten), a hemispherical tip is preferred. A ball can be formed by striking an arc on a water-cooled copper block and increasing the current until the end of the tungsten just begins to melt and form a ball shape. Recommended grinding angles for electrodes being used on DC are as shown in Table 7.4.

Table 7.4

Approx. current (amps)	Recommended electrode point angle	
up to 20	30°	(not rounded)
20–100	60–90°	(can be slightly rounded)
100–200	90–120°	(can be slightly rounded)

Fig. 7.10 *Compact power source for use with SMAW and GTAW*

(Photograph courtesy of Lincoln Electric Company)

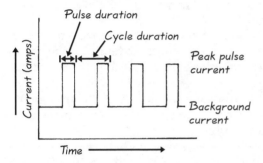

Fig. 7.11 *Typical pulsed current waveform (also known as a 'square wave' – see also Figure 7.13)*

Fig. 7.12 *Trainee on Canadian welding training scheme using pulsed GTAW (TIG) to control penetration on a pipewelding test piece*

(Photograph courtesy of AWS, Miami, USA)

Safety is very important when grinding; goggles and a face shield should be worn. A grinding machine fitted with extraction is also desirable.

The grinding of thoriated tungstens requires particular care, as the thoria is slightly radioactive. The grinding procedure should comply with safety regulations and the dust must be disposed of carefully.

Cerium and lanthanum electrodes were developed in the 1980s as possible replacements for thorium, since ceria and lanthana are not radioactive.

In order to avoid any hazard with grinding dust, chemical sharpening can be employed. This involves dipping the red-hot end of the tungsten into a container filled with sodium nitrate. There is a chemical reaction between the hot tungsten and the sodium nitrate which causes the tungsten to erode at an equal rate all round the circumference and end of the electrode. Therefore, repeated heating and dipping of the tungsten into the sodium nitrate forms a tapered tip suitable for welding.

Power supplies

Welding power sources of the constant-current type are employed with this process. Transformer–rectifier welding sets can provide for either AC or DC GTAW (TIG) welding at the flick of a switch.

AC and DC rotary generators of the kind used for SMAW (metal-arc welding with flux-covered electrodes) are also suitable for GTAW.

Advances in semiconductor technology have reduced the size of equipment and made transformer–rectifiers a popular choice for all arc processes (Figure 7.10), although generators are still used extensively on site work.

With DC, electrode negative is the system most used, as only around 30 per cent of the arc heat is generated at the cathode, with a high thermionic electrode material such as tungsten; 70 per cent of the heat is therefore generated at the anode, which is the work in this instance. The tungsten will therefore remain relatively cool while allowing deeper penetration welds to be carried out (deeper penetration than with DC electrode positive).

The positive tungsten can soon start to overheat. For this reason, DC electrode positive is not used much and, if it is, it is limited to low currents and the welding of very thin section.

Pulsed DC

Pulsed DC involves the addition of an electrical device within the welding plant that can provide adjustable pulses of current (Figure 7.11) within a range of one pulse every second to 20 pulses per second. DC electrode negative is usually used and the peak pulse current level can be set from around twice to ten times the background current level.

Machines with pulsed capability have many advantages over standard equipment, for example, greater penetration can be obtained with less heat input and dissimilar thicknesses can be more easily welded.

Very thin section can be joined and one set of welding parameters can be employed on a joint in all positions, such as a circumferential pipe weld (Figure 7.12).

The pulsed system is mostly used for automatic or machine welding but pulsing helps to give very high-quality manual welds, as the welder can count the pulses and time the movement of the torch and filler wire additions more exactly.

Another type of DC pulsed welding is known as **high frequency pulsing.** This involves rapid switching from low to high current levels. The frequency used can be around 20 kH. An arc produced with this system has increased pressure and directional stability. It is sometimes called a 'stiff' arc and is less susceptible to draughts (although the shield is still vulnerable), or arc blow. Power sources that provide high-frequency pulsing are expensive and, if the switching frequency is within the audible range, the noise can become annoying.

Alternating current

With AC, the arc undergoes reversals as the electrode and work change polarity. The frequency of reversals is fixed at the standard Hz frequency of the supply power, which is typically 50 Hz in the UK and 60 Hz in the USA.

This reversal of polarity means that the AC arc combines the oxide removal of electrode positive (Figure 5.2) with the deep penetration characteristics of electrode negative. It also prevents the tungsten from overheating, which could happen if it were positive all the time.

Some method of stabilizing the arc is required, otherwise it can cut out. With some metal combinations, a rectification effect can take place, trying to turn the AC arc into DC, again extinguishing the arc, as the process can cut out a half-cycle.

Some machines are designed to discharge capacitors at the appropriate time. Superimposed high-frequency sparks in parallel with the arc help re-ignition on each half-cycle in another type.

The latest power supply systems use electronic wave balancing, where the parameters are monitored and adjusted every half-cycle (that is 50 or 60 times per second). Such variable square wave equipment adjusts the time and current of the wave form to match a particular application. Figure 7.13 shows the characteristics of a variable AC square wave.

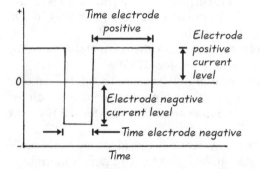

Fig. 7.13 *Illustrating the characteristics of variable square wave AC*

Methods of initiating the arc

Less-expensive or older equipment often uses the touch or scratch start, where the electrode is touched on the work and then raised a short distance to form the arc. A carbon block is sometimes used to try to minimize electrode contamination and prevent particles of tungsten being included in the initial weld deposit.

The use of a spark-gap-oscillator, which superimposes a high-voltage output at radio frequencies in series with the welding current, is another method of initiating the arc without touching the electrode on the work and risking contamination.

Adequate suppression and in some cases screening are required to prevent radio, television and computer interference with this method.

A method that does not produce high frequency uses a surge injector unit, which provides a momentary high-voltage pulse which is sufficient to ionize the shielding gas and initiate the arc. This system is often used when welding by machine.

These systems are covered in more detail in *Basic Welding*, pages 82 to 84.

One further method is to use a small pilot arc, powered by a separate power source. This pilot creates the ionized gas, enabling the main arc to be established.

Suitable applications for the process

The fact that GTAW is so flexible and capable of producing high-quality welds makes it of value to many industries, ranging from power, aerospace and food, through to repair shops. Figure 7.14 shows the process being used to repair a bronze statue. Figures 7.15 to 7.22 illustrate the recommended

Fig. 7.14 *GTAW (TIG) welding of a bronze statue at Tallix Morris Singer Ltd, Basingstoke, UK, using helium (with a small percentage of hydrogen) as the shielding gas*
(*Photograph courtesy of BOC Gases Europe*)

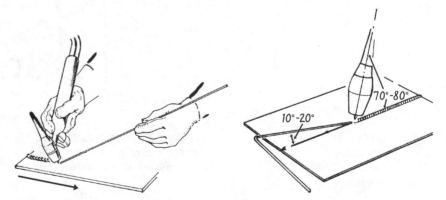

Fig. 7.15 *General technique for welding in the flat position*

Fig. 7.16 *Recommended angles for welding a butt weld in the flat position. It is often beneficial to 'crank' the filler rod as shown to keep the hand out of the line of heat*

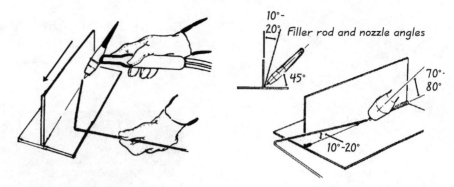

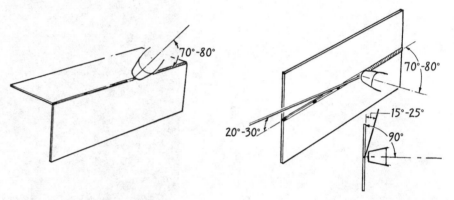

Fig. 7.17 *General technique for carrying out a fillet weld in the flat position using a 'cranked' filler rod*

Fig. 7.18 *Recommended torch and filler rod angles for a fillet weld in the flat position*

Fig. 7.19 *Outside corner welds can be completed with or without filler, depending on the plate thickness and weld requirements*

Fig. 7.20 *Recommended torch and filler rod angles for making a butt weld in the horizontal–vertical position*

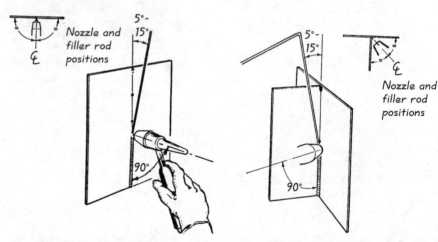

Fig. 7.21 *Vertical butt weld – again, rod can be 'cranked' to take left hand out of path of rising heat (not shown on this sketch)*

Fig. 7.22 *Recommended technique for making a vertical fillet weld*

technique with torch and filler rod angles for the manual welding of various joints. In cases where a filler rod or wire is needed, it is useful to be able to master the technique of using a 'cranked' rod (Figures 7.16, 7.17, 7.18 and 7.22), as well as a straight rod (Figures 7.15 and 7.20). By bending or 'cranking' the filler rod as shown, the hand can be kept out of the line of direct heat, making it much more comfortable to weld and therefore easier to carry out welding to a high standard. In some situations, both with this process and with oxy-fuel gas welding, it would be impossible to weld with a straight filler rod, because of either the type of joint being welded, or the amount of heat rising from the welded area.

Figure 7.23 shows the manual GTAW (TIG) welding of a stainless steel pipe butt using a square wave machine.

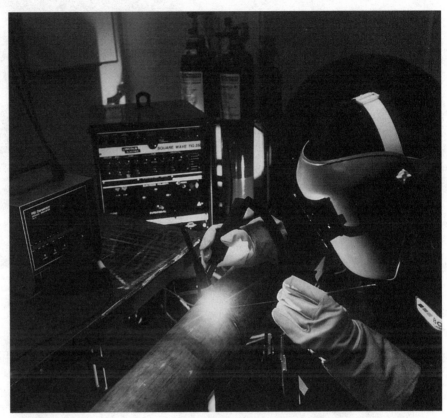

Fig. 7.23 *GTAW (TIG) welding of a stainless steel pipe butt using a square wave machine*
(Photograph courtesy of the BOC Cutting and Welding Research Centre, Guildford, UK)

Fig. 7.24 *'Cold wire' tungsten arc welding (for 'hot wire' welding, see* Basic Welding, *page 79)*
(Photograph courtesy of TWI, Cambridge, UK)

The process lends itself to automation and, by mounting a torch on a variable speed trolley and adding a wire-feed, a simple welding machine can be made.

Wire feeders consist of a spool of wire, the end of which is fed by variable-speed rollers through a tube to the molten pool.

There are two variations:

1. cold wire (Figure 7.24), where a wire at room temperature is fed into the leading edge of the weld; and
2. hot wire, where the wire is pre-heated and fed into the trailing edge of the molten pool (see *Basic Welding*, page 79).

Figure 7.25 shows the automatic orbital welding of small-diameter stainless steel tubing. The machine head clamps on to the pipe like a large pair of pliers and then automatically rotates around the joint to perform the weld. The weld parameters are pre-programmed into the machine computer.

Figure 7.26 shows a close-up of the completed weld. Note the regularity of the weld ripple obtained with the automatic pulsed technique.

Fig. 7.25 *Automatic orbital GTAW (TIG) welding of a small-diameter stainless steel pipe using Helishield H4*
(Photograph courtesy of BOC Gases Europe)

Fig. 7.26 *Close-up of the completed weld being undertaken in Figure 7.25. Note the uniformity of the weld ripple with automatic pulsed welding*
(Photograph courtesy of BOC Gases Europe)

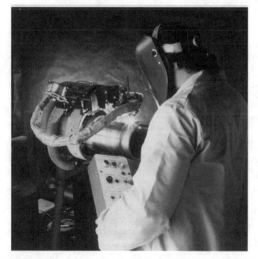

Fig. 7.27 *Larger type of orbital machine. Note spool of filler wire for automatic wire feed to the left of the picture.*
(Photograph courtesy of AWS, Miami, USA)

A larger orbital GTAW machine is shown in Figure 7.27. This device has its own track on which it can orbit the pipe. The track is so positioned to give the correct electrode-to-work distance. All parameters can be pre-programmed into the control unit of these machines and several weld procedures can be stored in their memory. Note the spool of filler wire for the automatic wire feed, to the left of the picture. Some specialist automatic applications are shown in Figures 7.28 and 7.29.

Advantages and limitations

The gas tungsten arc welding process has many advantages, some of the main ones being:

1. aluminium and its alloys can be welded without the use of corrosive flux;
2. it is capable of producing very high-quality welds;
3. with most materials, when welding there is no spatter;
4. depending on the application, it can be used with or without filler metal;
5. it allows excellent control, particularly of the root run (even greater control is given with the addition of a pulsed unit);
6. it is capable of welding almost all metals and many dissimilar metal combinations;
7. the power source and equipment are relatively inexpensive;
8. the heat input and filler-metal addition are independent;

Fig. 7.28 *Automatic GTAW (TIG) machine using pulsed DC and fitted with two water-cooled weld heads with arc-length control equipment and two 150 amp power sources. Operated under computer control with multiple stored programs. Arc-length control allows for welding round a corner radius in office-equipment manufacture*

(Photograph courtesy of Precision Beam Technologies Ltd, Peterborough, UK)

Fig. 7.29 *Automatic GTAW (TIG) machine using high-frequency pulsed DC for sealed battery manufacture and fitted with two weld heads with multi-electrode carousels and two 75 amp power sources. The rotary table indexes through four positions: (i) load/unload; (ii) surge – lid to can; (iii) weld – lid to can; (iv) test component*

(Photograph courtesy of Precision Beam Technologies Ltd, Peterborough, UK)

9. autogenous fusion welds (without filler) can be produced relatively inexpensively and very quickly.

Some limitations are as follows:

1. With the exception of (9) above, deposition rates and welding rates are generally slower than with consumable electrode processes.
2. The process would require shielding from winds or draughts.
3. High levels of dexterity and co-ordination are required, therefore more training is usually necessary in order to master manual welding. (Welders trained on oxy-fuel gas technique usually find the skills are transferable; GTAW, however, demands a greater level of skill than needed for manual GMAW or SMAW.)
4. It is less economical than consumable electrode arc welding processes as the plate / material section gets heavier. It is usual to change to other processes if this is evident. Around 10 mm (3 / 8 inch) is generally considered as being the upper limit of the process capability before it comes too uneconomical.

A variation of the process – GTAW (TIG) spot welding

This process provides an alternative method to resistance spot welding, if access can be obtained from one side of the component only and if it is not possible to fit the component between the spot welding machine arms.

With this process, the electrode is held at a pre-determined fixed distance above the surface of a lap joint (Figure 7.30). This is achieved by the use of a special spot welding nozzle which rests on the surface of the work.

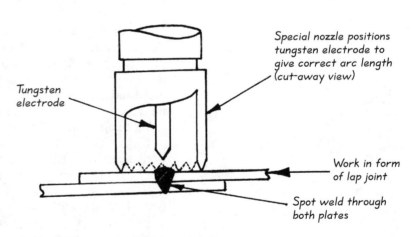

Fig. 7.30 *Basics of GTAW (TIG) spot welding*

Once positioned, an arc is established which melts a circular weld pool. This weld pool penetrates through the top sheet, across the interface into the lower sheet (Figure 7.30), forming a weld. The arcing time is pre-determined for various thicknesses and types of material but is usually quite rapid, say from 0.5 to 1 second. After this arcing time, the current is gradually reduced (sloped out), to allow the weld to solidify without forming a crater. The shielding gas supply is also continued for a few seconds after the weld is completed, to give protection as it cools down.

Gas metal-arc welding and submerged-arc welding

Gas metal-arc welding (GMAW), metallic inert gas welding (MIG), ISO 131, and metal active gas welding (MAG), ISO 135, (metal-arc welding with a non-inert gas shield)

Other common terms used for these processes are **metal-arc gas shielded welding, semi-automatic welding** and **submerged-arc welding** (SAW).

Although the concept of GMAW (MIG/MAG) was known much earlier, it was not until 1948 (see Chapter 1) that it began to be commercially developed. In the early days, the main application was for the welding of aluminium and in the UK, because an argon shield was invariably employed, the process gained the name of 'argonaut welding' (from argon automatic) as opposed to 'argon-arc', the name given to GTAW (TIG) welding. This also led to the term MIG welding – **metallic inert gas**, with the process being used with either argon or helium.

Developments introduced equipment capable of welding at low current densities and the range of weldable materials was vastly increased by the use of reactive gases. The principal gas employed was carbon dioxide (CO_2), allowing the process to be used on steels. From this point, the growth of the process was rapid and in many areas became known as CO_2 welding. This combination of inert and reactive gases and gas mixtures led to the term **gas metal-arc welding** being accepted in many countries, as it covers both inert and reactive shields.

GMAW is particularly accepted in the USA, while other countries, particularly the UK and those in Continental Europe, still tend to use the terms MIG and MAG for metallic inert-gas and metal active gas welding.

Being a semi-automatic process, it can be easily completely automated and is used extensively in both modes.

By selecting the appropriate shielding gas, consumable electrode wire and electrical parameters, the process is capable of welding all the major industrial metals such as carbon steel, low alloy steel, stainless steel, copper, aluminium, titanium and nickel alloys. Welding out of position is made easier with the use of pulsed equipment.

The basic principles of the process

The process is really a natural extension of the GTAW (TIG) process, with the non-consumable tungsten electrode being replaced by an automatically fed, consumable electrode wire, again the arc being shielded by an inert or reactive gas supplied from a cylinder, through a tube to the nozzle (Figure 8.1).

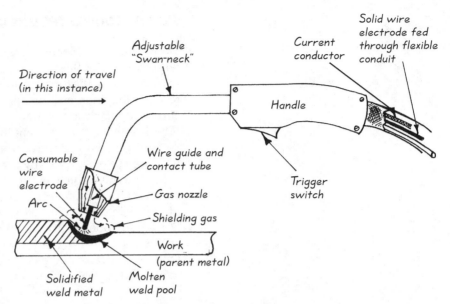

Fig. 8.1 *Basic principle of the gas metal-arc (GMAW) process (MIG/MAG)*

The spool electrode and wire feed can either be horizontal (Figure 8.2), with the wire being fed down a conduit to the gun, or vertical (Figure 8.3), again with drive rollers feeding the wire electrode down a conduit. Very thin or very soft wires, which may not feed successfully through a long conduit, can be fed over a relatively small distance using the spool-on-gun arrangement as shown in Figure 8.4 (see also *Basic Welding*, pages 92 and 93).

Once the initial settings have been selected by the operator, the machine control unit provides automatic self-regulation of the arc (see also details of the self-adjusting arc, in Chapter 5, 'Further electrical characteristics'). This means that the only manual variables with semi-automatic welding are the travel speed, direction of welding and positioning of torch. With the equipment correctly set, the arc length and current (wire feed speed) will be automatically maintained.

Fig. 8.2 *Small GMAW (MIG) unit with horizontally mounted spool and wire-driven unit*

Fig. 8.3 *Vertically mounted spool and wire-drive unit*

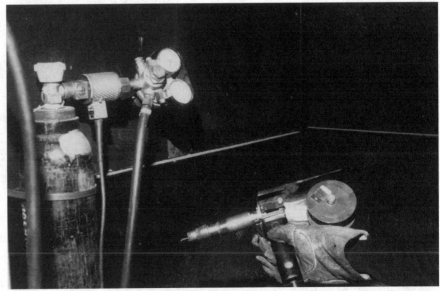

Fig. 8.4 *'Spool on gun' type arrangement. The small spool of electric wire and gun-mounted wire-drive unit allows very thin or very soft wires to be used; it would be impossible to feed such wires from a machine-mounted spool through a conduit to the gun*

109

Power source requirements

Transformer–rectifier units are generally used to provide constant-voltage DC with an essential flat volt–ampere characteristic, giving a 'self-adjusting arc'. (Any change in the torch position with the constant-potential/constant-wire-feed combination and the 'flat' characteristic, causes a change in welding current to match any change in electrode stick-out from the end of the contact tube, keeping a fixed arc length – see also Chapter 5).

An alternative method of arc adjustment, sometimes employed on machine welding systems, uses any arc voltage fluctuations to re-adjust the feeder control circuits which, in turn, change the wire feed speed accordingly.

Generator equipment is employed as an alternative to transformer–rectifier machines in some areas, particularly on site work.

DC electrode positive is used for most applications, as this combination aids in rapid melting of the relatively small-diameter electrode (compared with the diameter of the electrodes used in SMAW – MMA 'stick welding'), in order to supply sufficient weld metal.

Metal transfer

Metal is transferred from the wire electrode by one of the following mechanisms, depending on the parameters:

1. *Short-circuiting or dip transfer*
This type of transfer is found at the lower range of welding currents. The electrode wire contacts the molten pool in a range of between 20 and 200 times per second. At each short-circuit, metal is melted as the resistance current increases. The arc is re-established and the cycle repeats. Metal is not transferred across the arc gap.

In order to minimize spatter, the rate of current increase at re-ignition can be controlled by adjusting the inductance in the power source.

2. *Globular transfer*
With a DC positive electrode, this can take place at low currents with different shielding gases. When using CO_2 or helium, however, globular transfer will take place at all usable welding currents. It is recognized as a result of the drop size generally being of greater diameter than the electrode wire. Because such a large size of transfer droplet can be easily affected by gravity, this type of transfer is usually confined to welding in the flat position.

3. *Spray transfer*
Again, with DC electrode positive and using a shielding gas of argon, or one rich in argon, it is possible to produce a highly directional stream of small molten droplets, in a stable spatter-free 'axial spray' from the end of the electrode wire. Such a form of metal transfer can be used to weld most metals and alloys. However, it is not suitable for thin sheet or vertical and overhead welding.

4. *Pulsed transfer*
These limitations of thickness and welding position with spray transfer have been largely overcome with the introduction of equipment that can supply adjustable pulsed waveforms. With this system, a background current maintains the arc and a peak pulse current is superimposed. During each pulse, one or more molten droplets are transferred across the arc.

The frequency and amplitude of the pulse will control the total energy level produced by the arc and therefore the rate at which the electrode wire melts.

With this system, transfer is controllable and the average arc energy reduced, allowing the welding of thin sections and positional welding of thicker sections.

Further information on these mechanisms of metal transfer is given in Chapter 5. See also *Basic Welding*, pages 95 and 96.

Arc length and arc voltage

Arc length is very important. Too long an arc can cause arc wander and disruption of the gas shield; while too short an arc can cause it to extinguish when using short-circuiting (dip) transfer, or cause short-circuits to occur when spray transfer mode is intended and required.

Arc voltage depends on arc length, as well as other variables such as size (diameter) of electrode, composition of electrode, type of shielding gas, welding technique and, if measured at the welding plant, the length and diameter of the welding cable.

Speed of travel

This is the rate at which the arc is moved along the joint being welded. When all other factors are held constant, penetration of the weld into the parent material will be greatest at an intermediate speed.

If the speed of travel is decreased, the amount of filler metal deposited per unit length will increase. However, there comes a stage, when at a very slow rate of travel, the arc will be directed only on the molten pool, with little or no melting of the parent metal, thus creating an over-wide weld with reduced penetration.

With an increased travel speed, thermal energy will at first be increased, as the arc will be concentrated more directly on the parent metal. With greater increases in travel speed, less thermal energy will be imparted to the parent metal.

The melting rate therefore increases up to a certain limit of travel speed, after which any further increase reduces the parent metal melting rate.

High travel speeds can also cause undercutting, as a result of the deposition of insufficient filler metal.

Electrode extension

This is the amount that the wire electrode protrudes from the contact tube tip, or the distance between the end of the contact tube and the end of the electrode. An increase in this electrode extension would result in a corresponding increase in electrical resistance, causing the electrode temperature to rise and resulting in an increase in the melting rate of the electrode.

The recommended electrode extension for short-circuiting (dip) transfer is from 6 to 12.5 mm ($\frac{1}{4}$ to $\frac{1}{2}$ inch) and from 12.5 to 25 mm ($\frac{1}{2}$ to 1 inch) for the other metal-transfer systems.

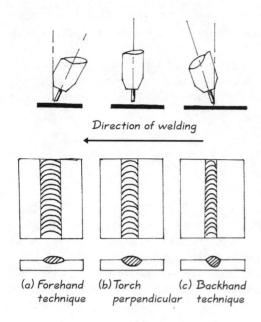

Direction of welding

(a) Forehand technique   (b) Torch perpendicular   (c) Backhand technique

Fig. 8.5 *Effect of electrode position and welding technique (after* AWS Welding Handbook, *Vol. 2, 8th edn)*

The orientation of the electrode

The position of the electrode with respect to the joint being welded affects the weld bead shape and penetration to a greater extent than either the arc voltage or travel speed.

If the electrode is pointed away from the direction of travel, the technique is known as 'backhand welding', and has a drag angle (Figures 8.5c and 8.6b).

If the electrode is pointed in the direction of travel, the technique is known as 'forehand welding', and has a lead angle (Figures 8.5a and 8.6a).

If the electrode angle is changed from the perpendicular to the forehand technique, with all other parameters remaining constant, the penetration will decrease and the weld bead will become increasingly wider and flatter.

Maximum depth of penetration can be obtained from backhand or drag welding in the flat position with the torch/electrode held at around 25 degrees from the perpendicular (Figures 8.6b and 8.5c). The weld bead will also be more convex and not as wide as the beads produced by the perpendicular or forehand methods.

For certain non-ferrous applications, such as welding aluminium, the forehand method is beneficial, as it provides a 'cathodic cleaning action' slightly in advance of the molten weld pool.

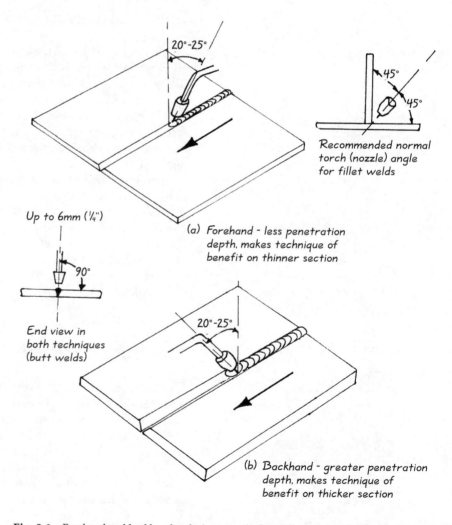

Fig. 8.6 *Forehand and backhand techniques applied to different plate thicknesses*

Position of weld

The highest deposition rates are usually achieved by using GMAW (MIG/MAG) in the flat or horizontal position, allowing spray transfer to be used. For this reason, where possible, even large fabrications are positioned to allow spray transfer to be employed. Figures 8.7 to 8.10 show the positioning of a large fabricated base during welding operations with GMAW.

Pulsed and short-circuiting (dip) transfer can be used for welding in all positions, even overhead (Figure 8.11), but welding times will be slower than for spray transfer in the flat position.

Fig. 8.7 *Bed in flat position showing how wire-drive unit can be moved to near welding operation*
(Photograph courtesy of Dresser-Rand Ltd, Wythenshawe, Manchester, UK)

Fig. 8.8 *Bed in flat position undergoing GMAW (MIG/MAG) welding*
(Photograph courtesy of Dresser-Rand Ltd, Wythenshawe, Manchester, UK)

Fig. 8.9 *Large bed turned on side*
(*Photograph courtesy of Dresser-Rand Ltd, Wythenshawe, Manchester, UK*)

Fig. 8.10 *Bed turned over to allow completion of welding in flat position*
(*Photograph courtesy of Dresser-Rand Ltd, Wythenshawe, Manchester, UK*)

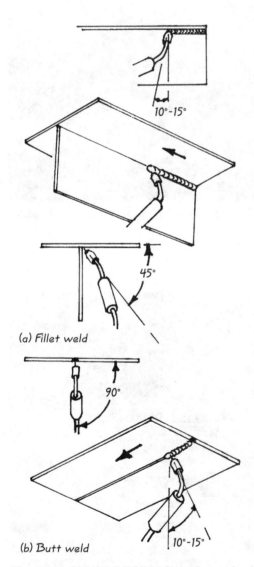

(a) Fillet weld

(b) Butt weld

Fig. 8.11 *GMAW (MIG/MAG) in the overhead position using either DIP (short-circuiting) or pulsed transfer*

To help overcome the effect of gravity on the molten weld metal when welding in the vertical and overhead positions, small-diameter wires (1.1 mm (0.045 inch) and below) are used, as the low heat input allows the molten pool to freeze rapidly, keeping it in position.

Welding vertically downwards, or 'downhill' welding, is a fast method of welding thin sections. It can also be employed for depositing the penetration or 'root' run in plate or pipe welds (see also Chapter 20, 'The welding of pipes').

Welding torches/guns

The names 'torch' and 'gun' are generally interchangeable, although it is common in many countries to term the pistol type as 'gun' and the swan-neck type as 'torch' (see Figure 8.12).

There are many designs with curved or straight necks and water or air cooling for heavy- and light-duty applications.

Figure 8.13 shows the main components of a lightweight 150 amp air-cooled torch. Note that there is usually a spot welding facility available, which involves the setting of an arc timer mechanism and the use of the spot welding nozzle.

The timer is set for the desired arc duration and the weld is carried out with the plates in a lap configuration. The spot weld melts through the top plate into the lower plate, forming a circular weld in a similar fashion to GTAW (TIG/TAG) spot welding.

Figure 8.14 shows the components of a heavy-duty 600 amp water-cooled torch with at-source fume extraction facility. The shielding gas is fed through the inner nozzle and fumes are sucked up through the outer fume nozzle (24).

The contact tip/tube is made from copper or an alloy of copper. As well as directing the wire electrode to the work, it supplies the electricity to the electrode. It should, therefore, always be kept in good condition and replaced when worn or damaged.

As different makes of torch/gun have different characteristics, it is recommended that the manufacturer's instructions are referred to in order to

Table 8.1 *Choice of gases for GMAW (MIG/MAG)*

Metal	Shielding gas	Proven advantages
Using spray transfer		
Aluminium	Argon	0–25 mm (0–1 inch) thick. Gives the best metal transfer and arc stability with least amount of spatter
	75% Helium/25% Argon	25–76 mm (1–3 inch) thick. Gives a higher heat input than argon alone
	90% Helium/10% Argon	Over 76 mm (3 inch) thick. Provides the highest heat input with minimum risk of porosity
Magnesium	Argon	Gives best cleaning action
Carbon steel	Argon/3–5% Oxygen	Gives a fluid and controllable weld pool combined with good arc stability. Good weld contour with minimum risk of undercutting. Allows higher welding speeds than argon alone
	Carbon dioxide (CO_2)	Gives high speeds with mechanized welding and low-cost manual welding
Low alloy steels	Argon/2% Oxygen	Generally produces welds with good mechanical properties. Reduces the problem of undercutting
Stainless steels	Argon/1% Oxygen	Produces a fluid and controllable weld pool with good arc stability. Gives a good shape of weld contour and minimizes the chance of undercutting on heavier sections
	Argon/2% Oxygen	Gives improved arc stability and increased welding speed. Usually better than 1% oxygen when welding thinner section
Copper, nickel and alloys	Argon	Gives a good wetting action with good control of weld pool up to 3.2 mm (1/8 inch) thick
	Helium/Argon	The higher heat inputs gained with using 50% and 75% helium mixtures allow the welding of heavier section by offsetting high-conductivity heat losses
The reactive metals (Ti, Zr, Ta)	Argon	Provides good arc stability with minimum weld contamination. Welding in a chamber filled with argon or argon backing is required to stop atmospheric contamination of the underside of the weld area
Using short-circuiting (dip) transfer		
Carbon steel	Argon/20–25% CO_2	For welding material up to 3.2 mm (1/8 inch) thick. Gives high welding speeds with no melting through (burn-through). Distortion and spatter are also reduced. Gives good penetration
	Argon/50% CO_2	For welding above 3.2 mm (1/8 inch) thick. Gives good weld pool control in vertical and overhead positions. Reduced spatter and a clean weld appearance
	CO_2	Fast welding speeds with deep penetration and low cost
Stainless steel	90% Helium/ 7.5% Argon/ 2.5% CO_2	Gives good arc stability with a small HAZ and no undercutting. Has no effect on corrosion resistance and aids minimum distortion
Low alloy steel	60–70% Helium/ 25–35% Argon/ 4–5% CO_2	Excellent arc stability, toughness and wetting characteristics combined with minimum reactivity. Good weld contour with little spatter
	Argon/20–25% CO_2	Excellent arc stability, wetting and weld shape. Little spatter. Toughness slightly less than Helium/Argon/CO_2 mix above
Aluminium, copper, magnesium, nickel and alloys	Argon and Argon/Helium	Argon is perfectly satisfactory on thin section, with argon/helium giving improved heat input on thicker section

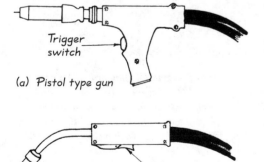

(a) Pistol type gun

(b) Swan-neck type torch

Fig. 8.12 *Although the names 'gun' and 'torch' are interchangeable, it is usual to term the pistol type as a 'gun' and the swan-neck type as a 'torch'*

select the correct size of contact tip/tube for a specific material and electrode size.

Gas nozzles should be cleaned frequently and kept free of spatter build-up which could impede the flow of shielding gas, or even fall into the molten weld.

Care should be taken not to bend the wire feed conduit excessively, as this can cause the wire to snag. The wire feeds better if the conduit is curved smoothly.

Different gas shields are used according to the application. Table 8.1 lists the shielding gases suitable for spray and short-circuiting (dip) transfer on various materials.

Automatic and robotic welding

Figure 8.15 shows the GMAW (MIG) welding of a stainless steel assembly using a hand-held torch. This is classed as semi-automatic welding and is

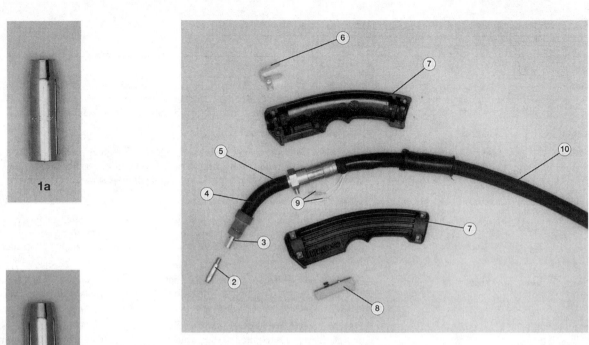

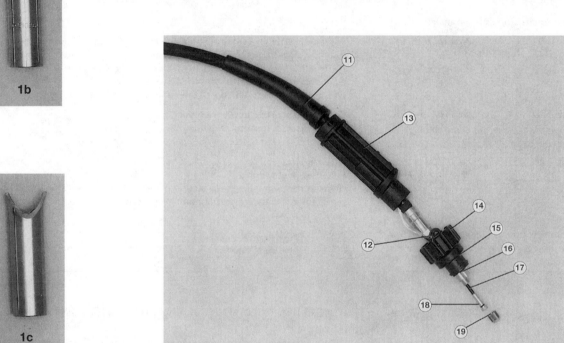

Fig. 8.13 *Small 150 amp light-duty torch, suitable for short-circuiting (DIP) transfer welding:* ① *short DIP nozzle,* ① *standard DIP nozzle,* ① *spot welding nozzle (these nozzles are not shown to exact size),* ② *contact tip,* ③ *nozzle holder/insulator,* ④ *swan-neck assembly,* ⑤ *neck insulating sleeve,* ⑥ *torch clip,* ⑦ *pair of handle mouldings,* ⑧ *switch assembly with spring,* ⑨ *switch contacts,* ⑩ *hose cable assembly,* ⑪ *support sleeve,* ⑫ *switch leads,* ⑬ *connector support,* ⑭ *connector end nut,* ⑮ *connection block,* ⑯ *contact pin,* ⑰ *liner (steel liners are recommended for solid and tubular wires; polymer liners are recommended for soft/aluminium wires),* ⑱ *liner 'O' ring,* ⑲ *liner nut*
(*Photographs courtesy of Murex Welding Products Ltd*)

116

1a

1b

1c

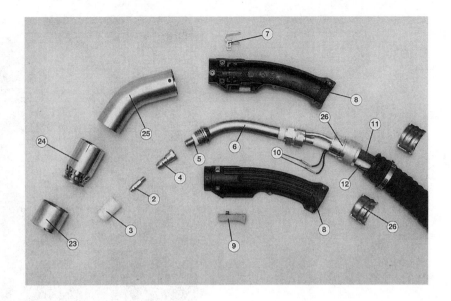

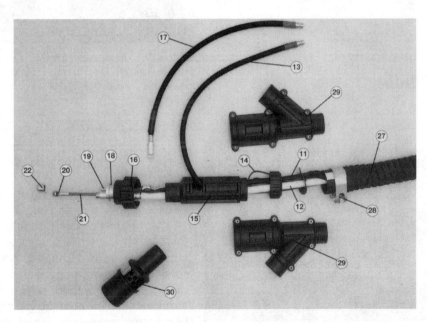

Fig. 8.14 *Heavy-duty 600 amp (at 80% duty cycle) water-cooled torch with fume-extraction facility: ⓐ heavy-duty tapered spray nozzle, ⓑ small-bore spray nozzle, ⓒ spot welding nozzle (these nozzles are not shown to exact size), ② contact tip, ③ nozzle insulator, ④ gas diffuser, ⑤ neck washer, ⑥ swan-neck assembly, ⑦ torch clip, ⑧ pair of handle mouldings, ⑨ switch assembly with spring, ⑩ switch contacts, ⑪ power cable, ⑫ conduit, ⑬ water hose, ⑭ switch cable, ⑮ connector support, ⑯ connector end nut, ⑰ water hose, ⑱ connection block, ⑲ contact pin, ⑳ liner 'O' ring, ㉑ liner (steel liners are recommended for solid and tubular wires; polymer liners are recommended for soft/aluminum wires), ㉒ liner nut, ㉓ fume nozzle funnel, ㉔ fume nozzle, ㉕ fume neck tube, ㉖ fume hose holder, ㉗ fume extraction hose, ㉘ fume hose clamp, ㉙ fume outlet assembly, ㉚ hose adaptor (with air regulator)*

(Photographs courtesy of Murex Welding Products Ltd)

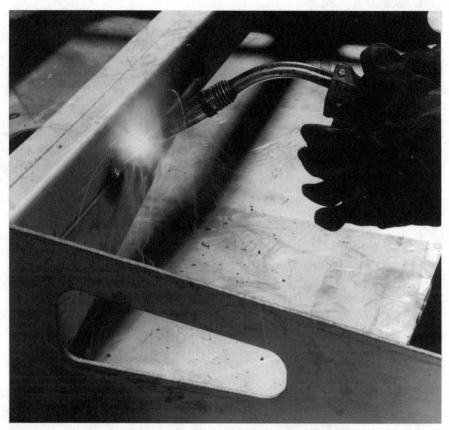

Fig. 8.15 *The gas metal-arc (MIG) welding of a stainless steel assembly for a municipal refuse cart at Johnston Engineering Ltd, Dorking, UK, using a helium–argon shielding gas*
(Photograph courtesy of BOC Gases Europe)

Fig. 8.16 *Building up an internal surface using a fixed welding head and roller bed manipulator*
(Photograph courtesy of Farrell Engineering Ltd, Rochdale, UK)

very suitable for welding different shapes of component as it has the versatility of being able to change rapidly from one welding situation to another.

Where there is repetition of the weld / component shape, then automation or even robotic welding may be considered.

The simplest type of automation can be achieved by having a fixed welding head, with the work moved at a controlled rate beneath it. This is often employed when welding the circumferential seams of vessels, if they can be rotated on a manipulator beneath the welding head. Internal welds can also be made by this method, as shown in Figure 8.16, where an internal cylindrical surface has been built up by welding using wear-resisting material. Flux-cored wire has been used and also a special elongated welding head.

Figure 8.17 shows two different types of nozzle assemblies for fully automatic welding and Figure 8.18 is a simplified diagram of a fully automatic welding facility capable of welding with solid-wire electrodes and a gas shield, or flux-cored electrodes with or without a gas shield. Using flux-cored wire and a gas shield will, of course, in most cases, provide double protection and is usually employed where very high-quality welds are needed. Extra alloying elements can also be added to the flux core with some electrodes if required.

Automatic welding heads can be mounted on a unit equipped with four wheels and driven along by a variable-speed electric motor or a fixed-speed motor through a variable-speed drive. Such a unit is known as a 'tractor' and the wheels can be made to run along a girder or some other form of track.

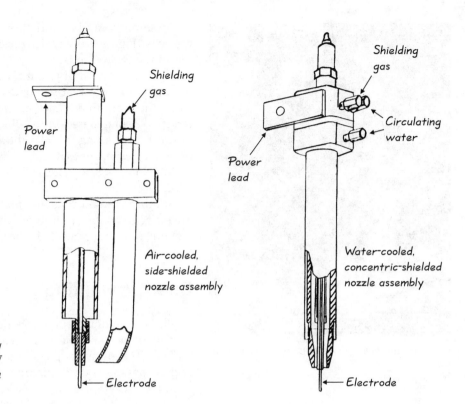

Fig. 8.17 *Typical nozzle assemblies for fully automatic welding with FCAW, GMAW and GMAW with additional gas shielding (based on drawings from AWS Miami, USA and Lincoln Electric Company*

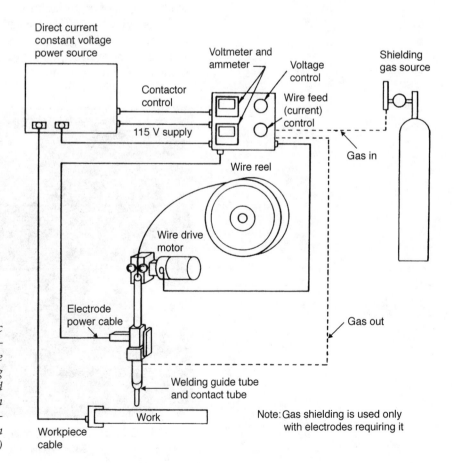

Fig. 8.18 *Simplified diagram for a fully automatic welding facility for use with: (i) flux-cored wire – flux-cored arc welding (FCAW), automatic; (ii) bare wire electrode with a gas shield – gas metal-arc welding (GMAW) (MIG/MAG), automatic; (iii) gas shielded with a flux-cored electrode, automatic. The broken lines indicate the additions necessary when the gas-shielded version is used (based on drawings from AWS, Miami, USA and Lincoln Electric Company)*

Fig. 8.19 *Automatic joint tracking using a 'seam tracker'*
(Photograph courtesy of AWS, Miami, USA)

These units can be guided by automatic joint tracking devices or 'seam trackers' (Figure 8.19), which enable the torch accurately to follow the line of the joint to be welded.

There are several different types of device available. One uses an electromechanical sensor that responds to the movement of a probe, which moves along the seam in advance of the welding head. Signals sent by the probe return to a solid-state control which, in turn, activates vertical and horizontal linear slides, having axis at right-angles to each other. These slides are driven by small electric motors and their movements, in response to the probe, will maintain the welding electrode in the required position as it moves along the joint.

Another method uses an optical system incorporating fibre optics and a microcircuit control unit, which again operates linear slides to position the welding nozzle constantly in the correct situation vertically and horizontally.

Robotic welding

The most common form of robotic welding uses the welding torch mounted on a robot arm and there are many variations of this basic concept. Figures 8.20 to 8.22 show a sample of typical set-ups and applications. In some cases, barriers or safety guards have been removed in order to take the photographs.

In these examples, the GMAW (MIG/MAG) process is being employed; GTAW (TIG/TAG) and resistance processes can also be used in conjunction with robot-arm devices.

Robotic welding was first utilized on a large scale in the automobile industry (see also Chapters 9 and 10 on Resistance welding) and then quickly spread to other types of manufacturing. It is particularly suited to mass-production methods, where repetition, speed and accuracy are needed.

Complete welding sequences can be programmed into robotic welding stations and several sequences can be stored on disk or in memory, depending on the type of equipment.

Fig. 8.20 *GMAW (MIG/MAG) torch mounted on robot arm*
(Photograph courtesy of AWS, Miami, USA)

Fig. 8.21 *Robotic welding using computer-controlled arm and positioning equipment. The safety guards have been removed in order to take the photograph*
(*Photograph courtesy of AWS, Miami, USA*)

By using the computer screen, the program can be checked visually, before welding commences and during the actual welding operation. Figure 8.23 shows a visual display as the computer maps out the paths of the component and the robotic welding arm for welding a branch-pipe to a main.

Fig. 8.22 *Welding automobile axles using torch mounted on robot arm*
(*Photograph courtesy of TWI, Cambridge, UK*)

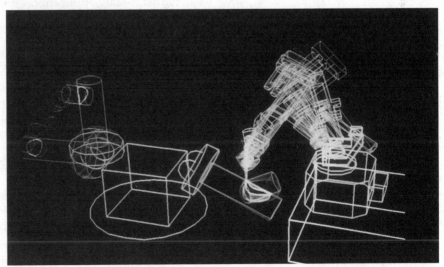

Fig. 8.23 *Three-dimensional computer graphic kinematic simulation. The computer maps out the paths of the component and the robotic welding arm for a welding operation around a pipe-branch*
(*Photograph courtesy of AWS, Miami, USA*)

Good joint fit-ups are usually required for automatic and robotic welding, although some machines fitted with seam-tracking devices can cope with discrepancies of up to 3 mm (1/8 inch) either way. Coping with such discrepancies will, however, slow down the rate of welding.

Automatic welding is defined by AWS as "welding with equipment that performs the welding operation without adjustment of controls by a welding operator." Robot welding is a specialised type of automatic welding, because, as opposed to being dedicated, robots can be flexible and capable of welding different shapes of components. This is due to the fact that the mechanical movement and manipulation of the robot arm are controlled by a computer, which can be programmed to produce a new sequence of movements.

Robots are multi-functional, in that they can perform a variety of tasks simply by changing the tool or tools attached to the manipulator 'arm'. They have the ability to manipulate and transport objects or tools in a similar manner to the human hand and arm. It is this high degree of manipulation that provides a comparison with a human worker.

Types of robot

There are many mechanical configurations for manipulators, being either hydraulic, electric, or pneumatic in operation. The type employed will influence the surrounding work area (or work envelope), and the size and shape of the components that can be welded.

The configurations can best be described in terms of their co-ordinate systems, of which there are three main types.

The cartesian system

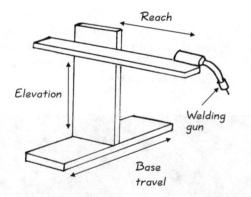

Fig. 8.24 *The cartesian (also known as the rectangular or rectilinear) co-ordinate system*

This is the simplest arrangement (Figure 8.24), in which movement is translated along any one of three perpendicular axes. The system is employed on large workpieces for carrying out long continuous welds. Such a system can also impart accurate weaving motions and can therefore be employed for precision hardsurfacing applications. It is also known as the **rectangular** or **rectilinear system.**

The polar, or cylindrical and spherical co-ordinate system

This system (Figure 8.25) is similar to a gun-turret mounting, in that the vertical column can rotate and the arm can move in and out. The arm can also move up or down.

The jointed arm type

This configuration (Figure 8.26) gives movement closest of all three systems to the human arm. Such systems have a rotating base, representing the human waist, elevation and reach, giving shoulder and elbow type movement and wrist action.

A wrist assembly attached to any type of robot arm can add an extra two or three degrees of freedom. Generally, most industrial robot welding applications will require five or six degrees of freedom.

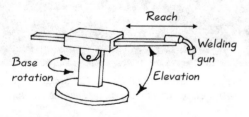

Fig. 8.25 *Polar co-ordinate robot*

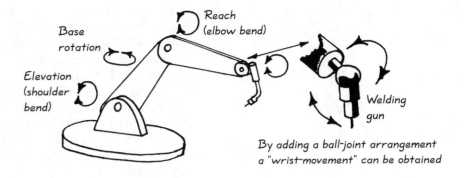

Base
rotation

Reach
(elbow bend)

Elevation
(shoulder
bend)

Welding
gun

By adding a ball-joint arrangement
a "wrist-movement" can be obtained

Fig. 8.26 *Jointed-arm robot configuration*

Submerged-arc welding

This process gets its name from the fact that the welding arc, between the end of a bare wire electrode and the work, is actually 'submerged' under a layer of granular fusible flux.

There are six different methods of striking the arc, and which method is employed will be determined by the characteristics of the work and any specific requirements. The methods are as follows:

1. *Use of a small ball of steel wool*, about 10 mm diameter (3/8 inch), which is placed in the joint directly beneath the electrode. The electrode wire is then lowered, compressing the ball to about half its previous size. Flux is added to cover the area and the current switched on. The wire wool establishes a current path and then immediately disintegrates, creating an arc under the flux.
2. *Sharp wire or chisel point start.* The electrode is tapered to a fine chisel-like point and lowered so that it touches the work. Again, when the current is switched on, the point melts to establish an arc underneath the flux.
3. *Scratch start.* With this method, the electrode wire is set to touch the work and the area covered with flux. The movement of the tractor/carriage is then started simultaneously with the welding current being switched on. The motion of the welding head prevents the end of the electrode from fusing and an arc is initiated.
4. *Molten flux start.* Once an arc is started, a molten flux pool is formed and further arcs, when using multi-electrode welding, can be started by feeding the new electrode wire(s) into the molten flux pool.
5. *Touch and retract start.* Some welding heads are designed to lower the electrode with the current switched on and underneath the layer of flux, then, on receiving a low-voltage signal, the wire feeder instantly withdraws the tip of the electrode a short distance from the workpiece, forming an arc.
6. *High-frequency start.* Here, special equipment enables a high-frequency spark to ionize a path from the end of the electrode to the work, enabling the main arc to be established under the flux.

When the arc has been established by one of the above methods, the wire feed mechanism starts to drive the electrode wire at a controlled rate and the feeder is moved either manually or automatically along the joint being welded. In the case of machine or automatic welding, depending on the application, the work can be either moved horizontally or rotated under a stationary wire feed unit. A typical example of this type of set-up is shown in Figure 8.27, where the work is rotated under a boom-mounted unit.

Flux is constantly fed from a hopper in order to keep the arc and molten pool area submerged and protected from the atmosphere. The flux that is melted solidifies to form a slag on the surface of the completed weld. Some machines incorporate a vacuum nozzle which follows behind the welding operation and returns any unused flux back to the hopper.

Submerged-arc welding can be employed in either a semi-automatic, automatic or machine mode, with each mode requiring that the flux and molten weld pool remain in position until solidification is complete. This is achieved by the use of fixtures and positioners, which are available as standard items or specifically designed and manufactured for a particular application.

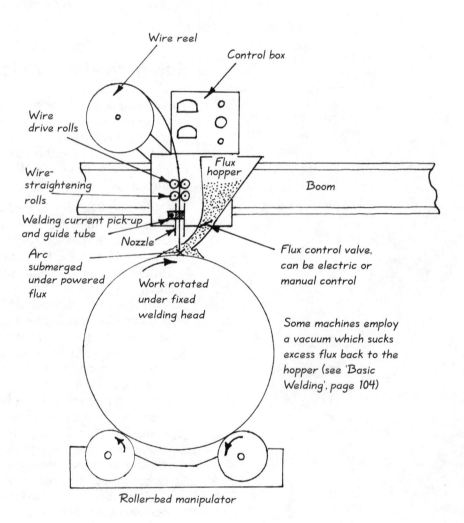

Fig. 8.27 *Schematic diagram of a boom-mounted submerged-arc welding machine*

Semi-automatic submerged-arc welding

This is carried out using a hand-held welding gun (Figure 8.28). Such a gun will deliver both the flux and electrode wire to the joint. The flux may be delivered by a gravity hopper, as shown in Figure 8.28, or it can be pressure-fed through a flexible hose from a larger hopper. Hand-held equipment uses relatively small-diameter electrode wires and cannot achieve travel speeds obtainable by machine welding. Some guns are fitted with a small drive motor, which drives through a wheel resting on the work.

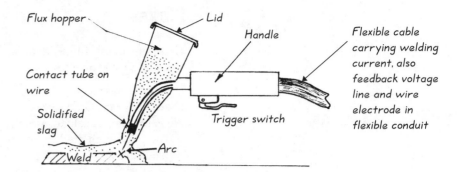

Fig. 8.28 *Schematic diagram of one type of hand-held submerged-arc welding gun*

The welding unit comprises power source, control unit, wire feed reel and wire drive motor with an extended flexible tube carrying the welding current cable, wire electrode in conduit and arc voltage feed-back wire. The arc length is automatically controlled, based on arc voltage feed-back signals. A scratch start is used and immediately an arc is initiated, the wire commences to feed. At the end of the weld, the gun is lifted rapidly, extinguishing the arc and stopping the wire feeding, but leaving the weld covered in flux until it cools down.

Automatic submerged-arc welding

This is achieved with expensive equipment that incorporates self-regulating devices. The welding operation can be undertaken with this equipment without the need for an operator to check and adjust the controls continually.

Machine welding

Figure 8.29 shows a tractor-mounted submerged-arc welding machine and power source. Such equipment will carry out the entire welding operation but will require a welding operator to be present to monitor the controls, position the machine or work, adjust the controls, start and stop the machine, and adjust the welding speed.

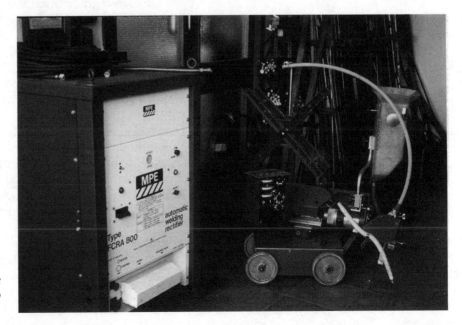

Fig. 8.29 *A tractor-mounted submerged-arc welding machine and a transformer–rectifier unit to supply constant-voltage DC power*

Fluxes

There are many fluxes available, but generally they are based around the following two varieties:

1. *Fused fluxes*

These are granulated glasses of a special type made by melting a mixture of silica, magnesia, calcite, alumina and fluorspar.

The glass is melted in an arc or gas-fired rotary furnace at temperatures between 1200° and 1400°C (2192–2552°F) and, while molten, the glass is poured into water. The glass is granulated, sieved and graded. No ferro-alloys are added, as they would oxidize.

2. *Agglomerated* (also known as bonded)

These fluxes are made by mixing the powdered ingredients, including alloys, with a binder such as sodium silicate or potassium silicate, or a mixture of both, and baking at around 300–500°C (572–932°F).

The baked flux is then crushed, sieved and graded. The advantages of agglomerated fluxes are that correction to the weld metal composite can be made via the flux.

Both types of flux must be completely dry, otherwise poor results will be obtained and the welds will suffer from porosity.

A drying temperature of 300°C (572°F) is usually adequate. Generally, submerged-arc fluxes have an acid-type slag (that is, they are rich in silica), however, it is possible to make basic fluxes of the agglomerated type and these are particularly effective for the welding of alloy steels.

The basicity or acidity is often referred to by the ratio of CaO or MgO to SiO_2. Fluxes having ratios greater than one are called chemically basic, while ratios approaching unity are classed as chemically neutral and those less than unity are known as chemically acidic

For pressure-vessel work, the weld preparation must be machined and the welding area cleaned strictly in accordance with the welding procedure prior to welding.

Some fluxes are fairly tolerant of rust and other surface contaminants. These fluxes usually have a high MnO content. Some other fluxes are suitable only for second- or third-pass welding. Coarseness of powdered flux is important and the correct grade should be used. For example, at high welding currents above 750 amps a fine dusty flux is used, while coarser-graded fluxes are used at lower currents.

Types of weld

The process is used extensively for making butt, fillet and plug welds (in a plug weld the weld is deposited through a hole or slot in one component that fastens it to another component). The AWS definition is "a weld made in a circular hole in one member of a joint fusing that member to another member." The process is also used for cladding or overlaying.

Figure 8.30 shows some example plate preparations for butt welds up to 12.5 mm ($\frac{1}{2}$ inch) thick. Above this thickness the two-pass welding technique (Figure 8.31) can be employed, although it is uncommon to use the two-pass method above 37.5 mm ($1\frac{1}{2}$ inch) thickness, as there are few applications on which the poor impact properties with these welds could be tolerated. It is more usual to use a multi-pass technique with a single-'U' or double-'U' preparation (Figure 8.32), and currents in the range of

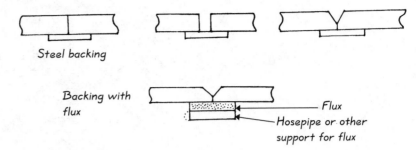

Steel backing

Backing with flux — Flux

Hosepipe or other support for flux

Fig. 8.30 *Butt welds up to 12.5 mm ($\frac{1}{2}$ inch)*

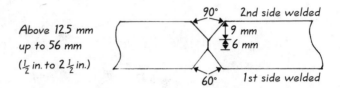

Above 12.5 mm
up to 56 mm
($\frac{1}{2}$ in. to 2$\frac{1}{2}$ in.)

90°   2nd side welded
9 mm
6 mm
60°   1st side welded

Example procedure:

1st side   900 amps   35 volts   Speed 350 mm min. (14ipm)   5 mm wire ($\frac{7}{32}$ in.) dia.

2nd side   1050 amps   35 volts   Speed 300 mm min. (12ipm)   5 mm wire

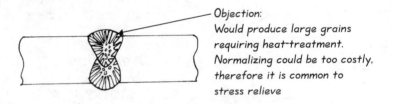

Objection:
Would produce large grains
requiring heat-treatment.
Normalizing could be too costly,
therefore it is common to
stress relieve

Such a weld would only be accepted for low-quality conditions (e.g. it would only pass low inspection standards).
Improved quality can be achieved by using a multi-run technique

Fig. 8.31 *The two-pass welding technique*

15°

R = 9.4 mm ($\frac{3}{8}$ in.)   R

6 mm to make sure of no burn through

90°

And for 75 mm (3 in.)
and above a double
"U" multi-run
method can be
employed

R

R = 9.4 mm   6 mm

Fig. 8.32 *One type of preparation for 35–75 mm (1$\frac{1}{2}$–3 inch) thickness for multi-run weld*

400–800 amps, thus ensuring refinement and better impact properties (grain refinement taking place as a result of reheating by subsequent runs in the multi-run technique).

A large amount of butt welding is carried out on the circumferential seams of pressure vessels. Figure 8.27 shows a typical general arrangement for carrying out circumferential welds from the outside using a boom-mounted submerged-arc welding machine and a roller-bed manipulator. Figure 8.33 gives information relating to welding from the outside and inside.

Fillet welds

Because submerged-arc welding gives greater penetration, weld size can be reduced slightly in comparison with shielded metal-arc welding (SMAW) using 'stick' electrodes (Figure 8.34a). Single-pass fillet welds up to 9.4 mm

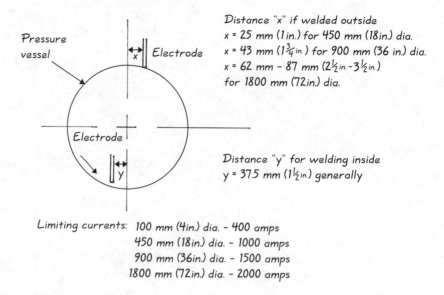

Distance "x" if welded outside
x = 25 mm (1 in.) for 450 mm (18 in.) dia.
x = 43 mm (1¾ in.) for 900 mm (36 in.) dia.
x = 62 mm – 87 mm (2½ in. – 3½ in.)
for 1800 mm (72 in.) dia.

Distance "y" for welding inside
y = 37.5 mm (1½ in.) generally

Limiting currents: 100 mm (4 in.) dia. – 400 amps
450 mm (18 in.) dia. – 1000 amps
900 mm (36 in.) dia. – 1500 amps
1800 mm (72 in.) dia. – 2000 amps

Fig. 8.33 *Information relating to welding on the circumferential seams of pressure vessels*

(3/8 inch) leg length can be made in the standing (horizontal–vertical) fillet position (Figure 8.34b), using a flux dam if necessary.

Above this size, multi-pass welds are necessary in the standing (horizontal–vertical) fillet position, but if the work can be tilted into the gravity (flat) position, single pass welds can be made with leg lengths up to 25 mm (1 inch).

Overlaying

The process is used extensively for overlaying or cladding of components. This can be depositing a hard-wearing material on to a softer component, or a corrosion-resisting material such as stainless steel on to a low carbon steel component, or many other similar applications.

With a normal welding head, the penetration can be sufficiently deep for a large amount of dilution with the base metal to occur (Figure 8.35). For example, a deposit of 18/8 stainless steel could be diluted to such an extent that it is no longer austenitic but martensitic, very hard and brittle. It is therefore quite common to use a high Cr/Ni alloy (such as 27Cr 14Ni), to allow for a certain amount of dilution.

In order to try to reduce the amount of penetration, and therefore dilution, when carrying out overlaying operations, different types of welding head have been devised. The first two of these are shown in Figure 8.36 and give

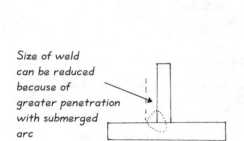

Size of weld can be reduced because of greater penetration with submerged arc

(a)

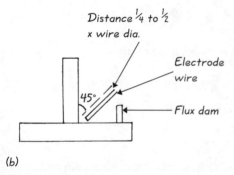

Distance ¼ to ½ x wire dia.

Electrode wire

45°

Flux dam

(b)

Fig. 8.34 *Welding fillets*

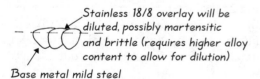

Stainless 18/8 overlay will be diluted, possibly martensitic and brittle (requires higher alloy content to allow for dilution)

Base metal mild steel

Fig. 8.35 *Overlaying with normal head*

Fig. 8.36 *Overlaying with oscillating head. Weld surface with both (a) and (b) is smooth and uniform. The main improvement over the normal head is better weld-shape*

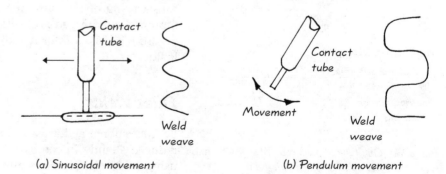

(a) Sinusoidal movement

Contact tube

Weld weave

(b) Pendulum movement

Contact tube

Movement

Weld weave

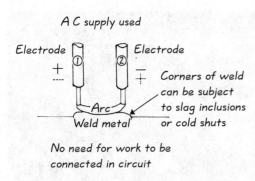

A C supply used

Electrode ① + Electrode ② −

Arc

Weld metal

Corners of weld can be subject to slag inclusions or cold shuts

No need for work to be connected in circuit

Fig. 8.37 *Series arc method of overlaying*

either a sinusoidal or pendulum movement. Both these methods give a reduction in penetration with an improved weld shape.

Another system employs two electrode wires (Figure 8.37), and an alternating current. With this system there is no need for the work to be connected in circuit. A wide, flat weld is produced but the corners can be subject to slag inclusions or cold shuts if care is not taken.

A process developed in Sweden uses strip electrodes (Figure 8.38a). This method gives the required low dilution coupled with good weld shape and is used for cladding plates before they are rolled. Two strips can be used (Figure 8.38b) but care must be taken again, to avoid slag inclusions and cold shuts.

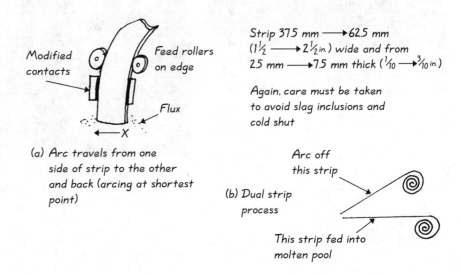

Modified contacts

Feed rollers on edge

Flux

← X

Strip 37.5 mm ⟶ 62.5 mm (1½ ⟶ 2½ in.) wide and from 2.5 mm ⟶ 7.5 mm thick (¹/₁₀ ⟶ ³/₁₀ in.)

Again, care must be taken to avoid slag inclusions and cold shut

(a) Arc travels from one side of strip to the other and back (arcing at shortest point)

Arc off this strip

(b) Dual strip process

This strip fed into molten pool

Fig. 8.38 *Strip process and dual strip process*

Submerged-arc welding with powder additions

The maximum speed of welding with submerged arc is limited by the onset of undercutting, however, multi-wire systems can be used to increase the deposition rate. When welding panels, the maximum is generally considered to be four-wire systems. The maximum arc energy/heat input is also limited by the need to achieve satisfactory weld-metal and HAZ properties.

Deposition efficiency can be increased, however, while maintaining the same energy input, by the addition of metal powder to the molten weld pool. With this system, the excess heat that is normally used up in melting the parent plate (and giving dilution as high as 70 per cent) is used up in melting the extra added powder.

Increases in weld deposition rates of up to 50 per cent can be obtained with this modification.

There are two techniques employed for adding the powdered metal. In the first, the 'bulk-welding system', the powder is fed at a controlled rate, through a tube and into the joint directly ahead of the flux and welding arc.

In the second system, powder is fed so that it falls directly on to the wire, where it is held by the strong magnetic field generated by the current flowing through the wire electrode. It is then carried through the flux layer into the arc and weld pool.

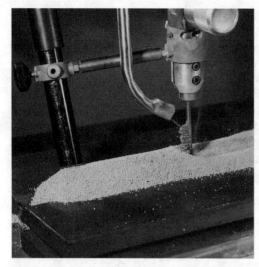

Fig. 8.39 *Use of powder additions when submerged-arc welding a butt weld*
(Photograph courtesy of TWI, Cambridge, UK)

Fig. 8.40 *Use of iron-powder additions on fillet welds*
(Photograph courtesy of AWS, Miami, USA)

Figure 8.39 shows this system being employed on a butt weld and Figure 8.40 shows a fillet welding application. In both cases the powder can be seen to be held to the wire electrode by magnetic forces.

Some benefits that can be achieved are as follows:

1. a reduction in weld-metal penetration/dilution;
2. increased tolerance to root gap variations;
3. an increase in weld size.

130

Resistance welding processes: Part I – Spot, seam and projection

An introduction to resistance welding is given in *Basic Welding*, Chapter 10, page 106. See also Chapter 10 of this book.

In Chapter 10 of *Basic Welding* it was discussed how the amount of heat generated when an electric current passes through an electrical resistance depends on the amount of current, the amount of resistance and how long the current flows. The basic resistance welding principle is expressed by the formula:

$$H = I^2RT$$

in which H = the heat generated in joules
I^2 = the current flow in amperes squared
R = the resistance in ohms
T = the time the current flows in seconds.

During welding, heat is lost to the adjacent base metal material and the electrodes, with heat travelling both horizontally and vertically by conduction.

As mentioned in *Basic Welding*, a factor can be added to the formula to represent these heat loses (see Figure 9.1 and *Basic Welding*, page 106).

There are many methods of joining sheet-metal components but the lap joint is favoured for the majority of joints as it adds strength and rigidity as well as often being the easiest to accomplish.

Among the methods employed, pop rivetting, soldering, self-tapping screws and adhesives are all used to some extent but where the joint is subjected to

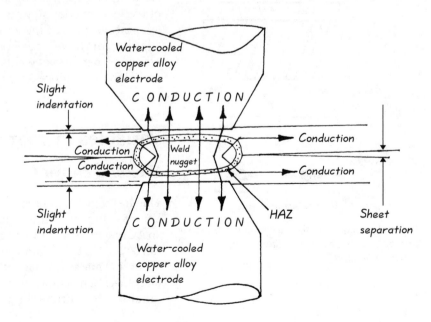

Fig. 9.1 *Showing heat dissipation by conduction into base metal (sheets being welded) and water-cooled electrodes*

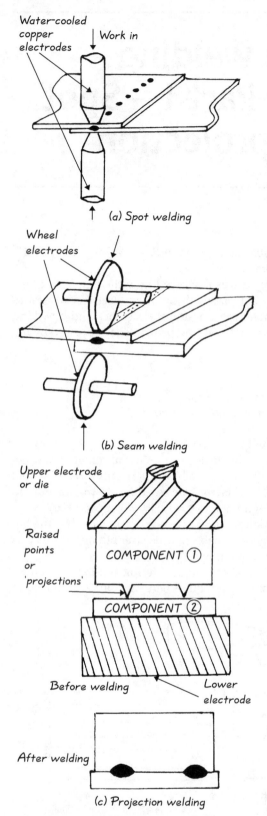

Water-cooled copper electrodes — Work in

(a) Spot welding

Wheel electrodes

(b) Seam welding

Upper electrode or die

Raised points or 'projections'

COMPONENT ①

COMPONENT ②

Before welding — Lower electrode

After welding

(c) Projection welding

Fig. 9.2 *Simplified sketches showing the basic principles involved in resistance spot, seam and projection welding*

loading, then resistance welding of one type or another is often chosen, because of the speed of the processes and the strength possible with these types of weld.

Sometimes combination joints are used, for example, on some vehicles, bonnet and boot-lid stiffeners are attached by both adhesives and spot welds. Figure 9.2 shows simplified sketches of the three resistance welding processes: **resistance spot welding** (RSW), ISO 21, **resistance seam welding** (RSEW), ISO 22 and **resistance projection welding** (RPW), ISO 23.

These processes are employed to a major extent in various manufacturing industries, for example, automobile manufacture, the aerospace industry, the production of domestic equipment such as refrigerators, freezers and washing machines, the food and drinks industries, together with general catering equipment, and many more. These processes are popular because they are versatile, fast in operation, and can reproduce quality welds time after time when maintained and supervised appropriately. It is for these reasons that resistance welding methods are often incorporated within mass-production/assembly-line techniques.

In spot welding, a weld nugget is produced directly between the electrodes, however two or more welds can be made simultaneously by employing multiple sets of electrodes.

Figure 9.3 shows a set-up on a car production line, where a single resistance spot welding unit is positioned on each side of the line, attached to a robot arm. The computer-controlled units weld the car roof assembly to the rest of the body.

Figure 9.4 illustrates a more complex welding station, where elaborate fixtures and jigging combine with three sets of spot welding electrodes, mounted on three robot arms, to produce a major section of an automobile body.

Projection welding takes place in a similar way to spot welding, except that the weld nugget location will be determined by the position of the projection(s) or embossment(s) on the faying surface (Figure 9.2c), or at the point of intersection of components in the case of welding wires or rods to form a mesh.

Because a weld takes place at the point of each projection, two or more projection welds can be made at the same time using just one set of electrodes.

Projections are either pressed into one of the components as a separate operation prior to welding, or forged in manufacture as in the case of captive nuts (see *Basic Welding*, page 108).

Seam welding (Figure 9.2b) is a variation of spot welding, using wheel electrodes. Seam welding is usually employed to produce welds consisting of a series of overlapping nuggets, giving a continuous leak-tight seam. Typical applications include the welding of fuel tanks and silencer boxes for cars. By adjusting the travel speed of the electrodes and the timing between welds, a series of separate spot welds can be made with a seam welding machine. This variation to the process is known as **roll spot welding**.

Variables in spot, seam and projection welding

Welding current

This has a greater effect on heat generation than either resistance or time. Variation in power line voltage or in the impedance of a secondary circuit in an AC machine can cause variation in welding current. Changes in impedance can be caused by variations in the geometry of a circuit or by the introduction of varying masses of magnetic metal within the secondary loop of the welding machine. DC machines are not usually affected in these ways. Current density can also vary if current shunts through a contact point other than at the position

Fig. 9.3 *A single resistance spot welding unit on a robot-arm assembly joins one side of a car roof to the rest of the body*
(Photograph courtesy of AWS, Miami, USA)

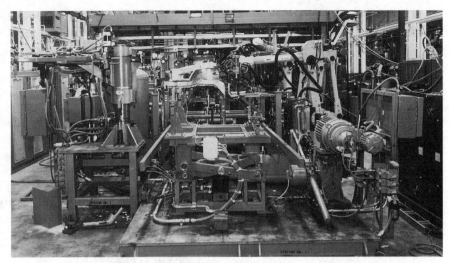

Fig. 9.4 *Three spot welding heads mounted on three robot arms weld a major automobile body section (Photograph taken over the safety barrier)*
(Photograph courtesy of AWS, Miami, USA)

of the weld being made. Large decreases in current density, and therefore welding heat, will be obtained if the contact face of the electrode area is increased, in the case of spot and seam welding, or if the size of the projection is increased when projection welding. Such alterations could cause a significant decrease in weld strength.

Sufficient current density is required to overcome heat losses and to be capable of producing fusion at the welding point on the interface.

Excessive current density can cause a molten area that is unstable, with the result that molten metal can be expelled from the weld area, resulting in internal voids and possible weld cracks, together with reduced mechanical strength. Excessive weld current can cause overheating of the base metal, giving deep surface indentations which will damage the components being welded and also cause rapid electrode wear, as a result of electrode overheating.

Weld time

The weld time is taken as the length of time during which the welding current is switched on; the total heat developed is therefore proportional to the weld time.

Heat losses by conduction into the surrounding base metal and electrodes, with a small heat loss due to radiation, increase with an increase in weld time and metal temperature. These heat losses have to be accepted as an uncontrollable side-effect in producing a satisfactory weld nugget. However, an optimum, minimum time is required in order that a melting temperature is obtained at a suitable current density.

Too long a weld time can greatly exceed the required melting point, causing internal pressure and, again, molten metal can be expelled from the joint.

Excessive weld times can have the same effect as excessive amperage on the base metal and electrodes. They can also cause weak, over-large heat affected zones.

The total weld heat may be altered by adjusting either amperage or weld time. The transfer of heat is, however, a function of time and the production of a required nugget size will require a minimum weld time, irrespective of the amperage.

For the resistance welding of thicker material, the welding current can be applied in short pulses without removing the electrode pressure. This allows a gradual build-up of heat at the interface, between the material being joined.

Welding pressure

The amount of pressure applied will affect the contact resistance, since on a microscopic level, the surfaces being joined together will appear like a lunar landscape, covered with peaks and valleys. If the pressure is light, the surfaces may only be in contact at the peaks. These having a relatively small surface area, the contact resistance will be high. If pressure is increased, these peaks will be depressed and the actual metal-to-metal contact area will increase, decreasing the level of contact resistance.

Electrodes

The electrodes should always be well maintained as they play a major part in the heat generation, being the means of conducting the welding current to the work. They must have good electrical conductivity but also sufficient hardness and strength to minimize the deformation caused by repeated applications of pressure.

Condition of surfaces being joined

The surfaces of the components being joined should be clean, as any contaminant such as oil, scale, oxide or paint can cause a variation in contact resistance and other weld-defect problems such as porosity and inclusions.

Metals being joined

The electrical resistance of the metals being welded will, of course, directly affect the resistance heating during the formation of the weld. With very high-conductivity metals such as silver or copper, very little heat will be developed even under high current densities, and any that is generated will be instantly conducted away into the electrodes and the surrounding work.

The composition of a material will determine its melting point, specific and latent heat, and thermal conductivity, all of which will determine the amount of heat required to melt the material in order to produce a weld.

The amount of heat necessary in joules per gram (Btu per pound) to raise the majority of commercial metals to fusion temperature is almost the same. A typical example is aluminium and stainless steel, both of which require the same input to reach fusion temperature, although they have completely different welding characteristics.

Aluminium has a conductivity about ten times larger than that of stainless steel and therefore the welding current for aluminium must be correspondingly greater, to compensate for the heat loss.

The welding cycle

As shown in Figure 9.5, the welding cycle consists of four basic phases:

1. *Squeeze time*. This is the time allowed for the electrodes to establish full, predetermined pressure on the work before the welding current is switched on.

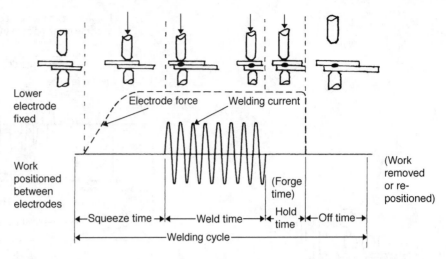

Fig. 9.5 *Single-impulse resistance spot welding cycle (also applicable to seam and projection welding)*

2. *Weld time.* This is the duration that the welding current is applied to the work with single-impulse welding.
3. *Hold time* or *forge time.* This is the length of time the electrodes maintain pressure on the work after the welding current has been switched off. This gives the weld nugget time to cool and solidify. The electrodes remain in position until the weld nugget is calculated to have adequate strength.
4. *Off time.* This term is generally used when the welding cycle is repetitive. It refers to the length of time the electrodes are off the work while the work is being repositioned or the electrodes are being moved to the next weld location.

Other features can be added to this basic welding cycle. Some examples are as follows:

(a) an initial pre-compression force in order to seat electrode and workpieces prior to squeeze time;
(b) a pre-heat may be necessary to give a gradual thermal gradient with some metals – this is positioned just before weld time;
(c) a forging force during the hold time;
(d) quench and temper cycles can be added to give desired weld properties with hardenable alloy steels;
(e) a post-heating cycle in order to refine the size of the weld grain in the steels;
(g) gradual reduction of current (current decay) to slow down the cooling rate on aluminium welds.

For some applications, the current is pulsed intermittently during the weld time. The sequence of operations for a more complex resistance welding schedule is shown in Figure 9.6. This is known as 'multiple impulse welding' as opposed to 'single impulse welding' (Figure 9.5).

Resistance welding machines

Machines are available that will provide either AC or DC for spot, seam and projection welding. The welding machines transform the supply to low-voltage, high-amperage welding power. Direct current is employed on

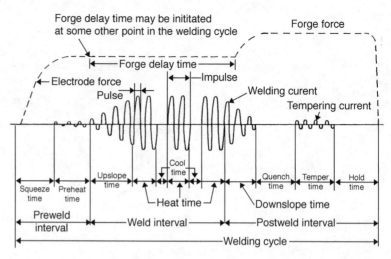

Fig. 9.6 *Multiple-impulse resistance spot welding schedule*
(*Courtesy of AWS, Miami, USA*)

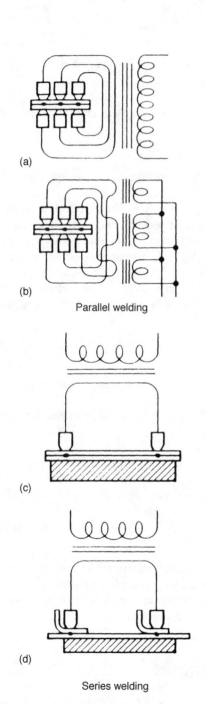

Parallel welding

Series welding

Fig. 9.7 *Typical arrangements for multiple spot welding*
(*Courtesy of AWS, Miami, USA*)

applications that require high amperage as the load can be balanced on a three-phase power supply.

The control systems of resistance welding machines are all based around three functions:

1. switching the welding current on and off at the required times;
2. controlling the amount of current;
3. activating the electrode pressure mechanisms for the various cycles, and releasing them at the correct times.

Depending on the type of machine, the electrode force can be applied by hydraulic, pneumatic, magnetic or mechanical means.

Multiple spot welding can be achieved by parallel or series welding, examples of which are given in Figure 9.7. In parallel welding, the secondary current is divided and conducted through the electrodes and workpieces in parallel paths, forming spot welds simultaneously. In Figure 9.7a, the welding current originates from the single transformer fitted with multiple electrodes in parallel in the secondary circuit. Figure 9.7b shows a parallel system with a three-phase primary. This type of system would be limited to three workstations.

Series welding is represented by (c) and (d) in Figure 9.7. Here, the secondary circuit welding current is conducted through the electrodes and work in a series electrical path, forming multiple spot welds simultaneously at each electrode location. In series welding, equal resistance values are required at each welding point in order to give equal heating. With two electrodes in series, some of the welding current will travel through the work from one electrode to the other as a 'shunted' current that does not contribute to the welding operation. This current loss must be taken into account when a welding procedure is being developed.

Direct and indirect welding methods

In direct welding, the welding current and electrode force are applied to the work by directly opposed electrodes (Figure 9.8a, b and c).

Indirect welding is a variation (Figure 9.8d, e, f and g) in which paths of the welding current are through the workpieces. Backing plates are present

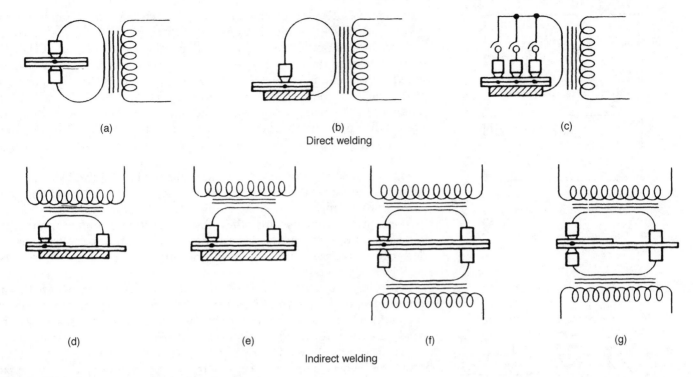

Fig. 9.8 *Typical arrangements for single spot welds*
(Courtesy of AWS, Miami, USA)

in (d) and (e) to provide both pressure and a current path when the plate is made from conducting material. If the plate is non-conducting, then it will only provide pressure and the current will travel along the lower workpiece to the return connection further down the joint.

The arrangements in Figure 9.8f and g are for spot welding high-resistance materials that require higher voltages. The two secondary circuits are in series and connected to two transformers. The primary circuits can be connected either in series or parallel. The two secondary circuits will provide the sum of their respective voltages at the point of the spot welds.

Electrodes

The electrodes in resistance welding perform four functions:

1. They conduct the welding current to the work, and in spot and seam welding they also fix the current density at the weld. The current density in projection welding is determined by the number of projections and their size and shape.
2. They have to transmit a force to the workpieces.
3. They dissipate some of the heat from the weld zone by conduction.
4. They help to maintain the relative alignment of the components and the positioning of the workpieces in projection welding.

Because the electrodes have to apply pressure without deformation, as well as transmit electricity, they are usually made from copper alloys. Spot welding electrodes are usually hollow, to allow for water cooling. Figure 9.9 shows some different shapes of electrodes; the sizes and shapes are usually determined by the thickness of the sheet and the type of material being welded. Figure

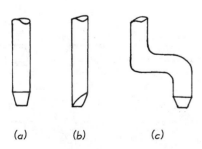

Fig. 9.9 *(a) Standard, (b) straight offset and (c) irregular offset electrodes*

137

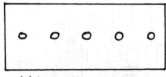

(a) Less effective spacing

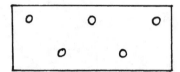

(b) More effective spacing of
spot welds or projection welds

Fig. 9.10 *Spacing of spot welds*

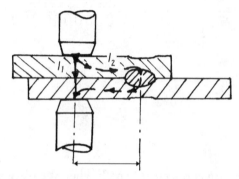

Ratio of $\frac{I_1}{I_2}$ will depend on distance
between welds

Fig. 9.11 *Shunting of current through a previous weld*

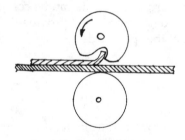

Fig. 9.12 *Specially shaped wheel electrodes used for welding workpieces with an obstruction*

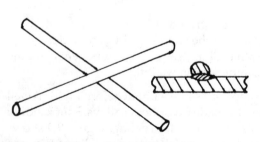

Fig. 9.13 *Cross-wire projection welding*

9.10 illustrates how the spacing of spot or projection welds is important, the type of pattern in (b) being far more effective than in (a). The distance between welds is also important to avoid 'shunting' taking place (Figure 9.11). This occurs when the previous spot weld provides an alternative low-resistance path for the welding current and some of the current is shunted along this route. The total weld current is therefore shared between the new weld being made and the existing spot weld. The relative proportions will depend principally on the distance between the two points measured at the interface. In these circumstances, the second weld can be smaller for the same nominal current.

Figure 9.12 shows an example of specially shaped wheel electrodes for seam welding workpieces that have some form of obstruction in the welding path.

Very often, projection welding electrode patterns can be specially shaped to fit around components during the welding operation. Common applications of projection welding include the securing of captive nuts or bolts. These are designed with either a pressed or forged projection in each corner (see also *Basic Welding*, page 108).

Figure 9.13 illustrates the projection welding of cross-wires. This makes an almost ideal projection weld, because the two contacting radii form a point of contact which allows the weld to be easily formed. Fabrications made by cross-wire welding can be seen almost everywhere. They include wire baskets, shelves, protective screens, barriers, gratings and reinforcing mesh for concrete structures. When square wire is used, the sharp edges are placed in contact with each other for welding.

Upset welding (UW) – resistance butt welding, ISO 25

In **upset butt welding** (Figure 9.14), the metal pieces being welded are brought into contact, a welding force applied, and electric current (the welding current) passed through them. Resistance heating builds up from the contact surfaces. An upset force is then applied and the current switched off. The upset force is then released and the weldment unclamped. The process is different from **flash welding** (Figure 9.15) in that constant pressure is applied during the heating process, keeping the end faces together and eliminating arcing.

The heat to form the weld, generated at the point of contact, therefore results entirely from resistance.

The upset speeds up recrystallization at the interface, while at the same time, forces some metal outwards from this area. This tends to rid the joint of oxidized metal.

The process has two variations:

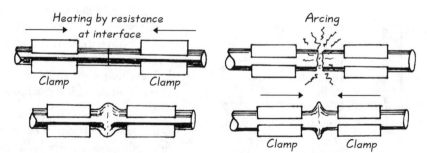

Fig. 9.14 *Upset welding (UW) – resistance butt welding*

Fig. 9.15 *Flash welding (FW) – flash butt welding*

1. it is used for the joining of components of the same cross-section with an end-to-end butt joint;
2. in the manufacture of pipe and tubing, it is used for the continuous welding of butt-joint seams.

Variation 1 can also be carried out by flash welding and friction welding, while variation 2 is also done using high-frequency welding.

Flash welding (FW) – flash butt welding, ISO 24

Flash welding is a resistance welding process that employs both resistance heating and arc heating to form the weld. The two parts to be joined are gripped in the electrode clamps (Figure 9.15), which are connected to the secondary coil of a resistance welding transformer. Voltage is then applied as either both components are advanced towards each other, or one component is held stationary and the other is advanced, depending on the type of machine.

When contact first takes place, at the high points of surface irregularities, resistance heating occurs at these locations. The high amperage causes rapid melting and vaporization of the metal at these points of contact. Once the metal is removed at the points of contact, small arcs begin to form and this action is termed 'flashing'. As the parts continue to be moved together, the faying surfaces become covered with a layer of molten metal and the components reach forging temperature for some short distance from each surface. The weld is made by bringing the molten surfaces into contact and applying a forging pressure.

The metal on each side of the joint begins to bulge out, forming an upset, and any solidified metal expelled from the interface is called 'flash'. The welding current is switched off as soon as the upset starts to form. **Flash butt welding** is a term sometimes used to distinguish the welding of heavier section, where the cycle of bringing the components together and separating them can be repeated, in order to build up the required welding heat.

Figure 9.16 shows a sketch of a machine flash welding a large steel ring.

The three most common types of flash weld application, for bar and pipe, mitred weld and ring weld, are illustrated in Figure 9.17.

Fig. 9.16 *Sketch of a machine flash-welding a large steel ring*

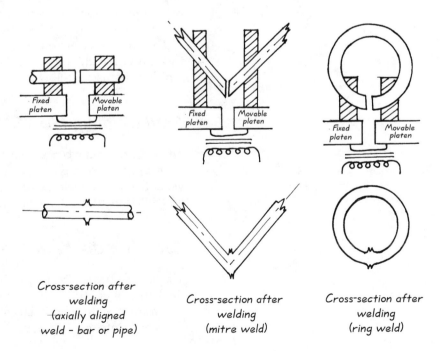

Fig. 9.17 *Common forms of flash welds*

Cross-section after welding (axially aligned weld – bar or pipe)

Cross-section after welding (mitre weld)

Cross-section after welding (ring weld)

Fig. 9.18 *Flash welding machine being used to butt-weld $1\frac{1}{2}$ m (56 inch)-diameter high-strength steel pipe at the E.O. Paton Electric Welding Institute, Kiev* (Photograph Courtesy of AWS, Miami, USA)

Figure 9.18 illustrates the flash welding of large-diameter pipes at the E.O. Paton Electric Welding Institute in Kiev. The process has been used for many years for the joining of pipes. See also Figure 1.16 (Chapter 1), which shows the welding of the PLUTO with a flash welding machine on board the tug *Britannic*.

Percussion welding – capacitor discharge welding (PEW) and stud welding, ISO 78

Percussion welding produces heating with an arc resulting from a rapid electrical discharge. The process gets its name from the fact that pressure is applied in a percussive manner during or immediately after the electrical discharge. The process is used in the electronics industry for joining wires and contacts to flat surfaces.

The process can also be used for joining a metal stud to a surface and when this is the case, the process is known as **capacitor discharge stud welding**, which is discussed below.

Stud welding, ISO 78

There are two main types of stud welding employing electrical energy. These are capacitor discharge stud welding and arc stud welding. In the last few years, the friction welding process can also be applied to the welding of studs, and this is mentioned in Chapter 21 on 'underwater welding'.

Capacitor discharge stud welding

This technique derives the welding heat from an arc resulting from a rapid discharge of electrical energy that has been stored in a bank of capacitors. During or straight after the electrical discharge, pressure is applied to the stud, pressing it into the molten pool on the workpiece.

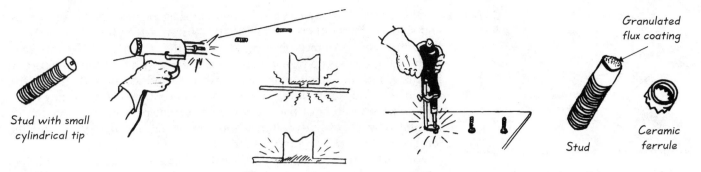

Fig. 9.19 *Capacitor discharge stud welding*

Fig. 9.20 *Arc stud welding*

The arc can be formed by rapid resistance heating and vaporization of a small cylindrical tip on the end of the stud (Figure 9.19), or by rapidly forming an arc as the stud is momentarily lifted away from the work. In the first method, the arcing time is about 3–6 milliseconds and in the second method it will range from 6 to 15 milliseconds. Because of this short arcing time and the small amount of molten metal expelled from the joint, this method does not need a protective ceramic ferrule. The system is particularly suited for component manufacture requiring small to medium-sized studs.

Arc stud welding

In this process, the heat for welding is developed from a DC arc between the end of the stud and the work. The power can be supplied by either a DC motor-generator or a transformer–rectifier unit. The welding operation is pre-set and then timed automatically. It is common for the end of the stud to be coated with granular flux (Figure 9.20), which acts as an arc stabilizer and deoxidising agent. The studs are available in a vast variety of shapes and sizes for different applications. A ceramic ferrule is also employed, which helps to concentrate the arc heat, acts, together with the flux, to restrict air from the molten weld, confines the molten metal to the weld area, acts as a shield against the arc rays, and prevents charring of the material (if any) through which the stud is being welded.

To carry out a weld, a stud is loaded into the chuck of the gun and a ferrule positioned over the stud. When the trigger is depressed, the current activates a solenoid which lifts the stud away from the plate, causing an arc to form, which melts both the end of the stud and a small area of plate (the work) beneath the stud. The pre-set timing device shuts off the current at the calculated time. The solenoid releases the stud and a spring action plunges the stud into the molten pool, making the weld. The whole operation is usually completed in less than one second. The ceramic ferrule is not used with some variations of the technique or when welding some non-ferrous metals.

High-frequency welding – high-frequency resistance welding (HFRW), ISO 29 and high-frequency induction welding (HFIW), ISO 74

The two main processes that use high-frequency current to produce the heat for welding are **high-frequency resistance welding** (HFRW) and **high-frequency induction welding** (HFIW). (Induction heating can also be used in the diffusion welding process as the heat source – see the next section

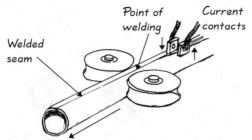

(a) High-frequency resistance welding of tube seam

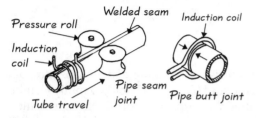

(b) High-frequency induction welding of tube seam and pipe butt

Fig. 9.21 *Showing essential differences between HF resistance welding and HF induction welding methods*

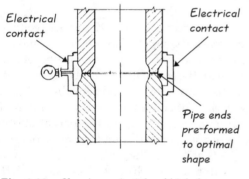

Fig. 9.22 *Showing principle of high-frequency resistance (HFR) heating when applied in the SAG forge process. This process involves heating and hydrogen flushing, followed by forging*

('Diffusion welding'). It can also be employed as the heat source for brazing and soldering operations.) In HFRW, the current is conducted to the work through electrical contacts physically touching the work surface (Figure 9.21a). With HFIW, there is no physical electrical contact with the work; the current is induced into the work by employing an external induction coil (Figure 9.21b).

In high-frequency welding, the current is concentrated at the surface of the part and the location of the concentrated current path in the work can be controlled by the relative position of the parts being welded, together with the location of the current contacts or induction coil.

A much lower current is required in order to reach welding temperature than when either low-frequency or direct-current resistance welding is employed.

The high-frequency current is very concentrated and because it heats up only a very small volume of metal at the point where the weld is to take place, it is extremely energy efficient. The process is capable of very high welding speeds, these usually only being limited by the materials handling operations before and after the welding operation. The minimum welding speed will be determined by the type of material being welded and the quality of weld required.

Fit-up and surface preparation are important for high-quality welding. Flux, although not normally used, can be brought to the weld area by using an inert gas stream. Inert gas as a shield is usually only needed when reactive metals such as titanium are being welded.

One variation of the process is **shielded active gas forge welding (SAG–forge welding)**. This process uses a hydrogen shield or 'flush', and much development work has been undertaken in Norway, for applying the method to pipe welding.

The SAG forge process takes into account the production of isostatic pressure (a function of geometry and temperature gradient) during the forging operation, by pre-shaping the joint to the desired shape. The development of optimal shape for different welds is given by the correlation of practical experiments and non-linear finite element computer calculation.

A high temperature gradient generally gives a better stress pattern, a shorter forging distance and therefore a better finished surface. Buckling during forging is the main problem to be avoided.

In order to obtain a good temperature gradient, the contact method of **high-frequency resistance heating** is used (Figure 9.22). Here, the current follows the route of least reactance, not the least ohmian path. Therefore, current flows at the mating surfaces of the joint, giving excellent heating conditions.

Diffusion welding (DFW), ISO 45 and diffusion brazing

Other terms which have the same meaning as diffusion welding include diffusion bonding, solid-state welding, solid-state bonding, pressure welding/bonding, isostatic bonding, hot press bonding, forge welding and hot pressure welding.

Because the process welds in the solid state, below the melting point(s) of the metals being joined, it can be used for joining many types of metal combinations, as the metallurgical problems associated with fusion welding processes can be avoided.

Diffusion welding occurs in the solid state when properly prepared and cleaned surfaces are placed into contact under pre-determined conditions of pressure loading, time and, in some cases, elevated temperature.

All the parameters are closely controlled. The applied force or pressure is set to be above the amount required to ensure a uniform surface contact

142

Fig. 9.23 *Diffusion welding of steel using induction heating.*

(*Photograph courtesy of TWI, Cambridge, UK*)

between the components but below a level that could allow macroscopic deformation to take place.

If heating is employed, the temperature is generally kept to well below the melting point of the material(s). A filler metal can be used as an insert in order to lower the temperature, pressure or time required for welding. A filler metal insert can also enable welding to be carried out in a less expensive atmosphere, or even eliminate the need for a chamber.

Heating can be applied by a furnace or resistance methods. Figure 9.23 shows the diffusion welding of steel using induction heating.

Some considerations when making solid-phase welds

In principle, with solid-phase welding, all that is necessary is to bring two metal surfaces into contact and, if they are perfectly plain (within machining capabilities), with surface films removed, they will weld together.

Figure 9.24 shows how pressure will spread and crack a surface film, helping it to break up. Absorbed gases are not generally a hindrance to welding, but oil and grease are troublesome unless boiled off the surface by using high temperatures, which however must still be below the melting point(s) of the material(s) being welded.

Oxide films are also troublesome, but can usually be broken up by heavy deformation, as they are brittle compared with the metal. Alternatively, the oxide films may be removed by mechanical or chemical cleaning but it is difficult to prevent further oxidation before welding.

The amount of deformation that takes place is very important in solid-phase welding, particularly for cold pressure welding. Figure 9.25 illustrates the percentage deformation for bar and plate.

Surface oxide, sulphide, carbonate, absorbed gases, oil and grease Films can vary in thickness from around 10 – 1000 Ångström
$(\mathring{A} = 10^{-10}\,metre, \frac{1}{10000}\,micron, \mu)$

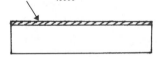

A ductile metal will spread and crack the surface film under pressure (point of minimum deformation in press-cold welding)

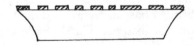

Fig. 9.24 *Surface films*

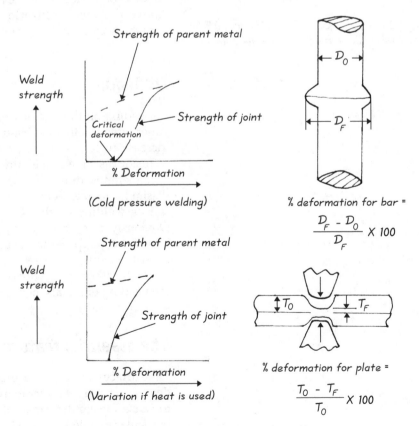

Fig. 9.25 *Showing percentage deformation for cold pressure welding*

143

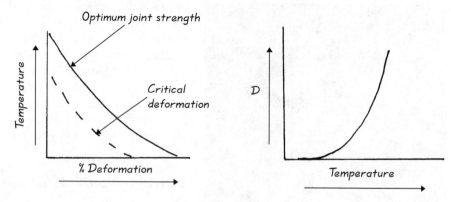

Fig. 9.26 *Effect of temperature on welding*

Fig. 9.27 *Showing exponential increase in diffusion rate with increased temperature*

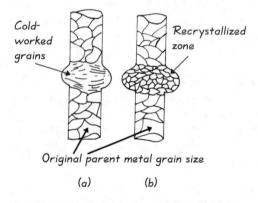

Fig. 9.28 *Comparison of bar welds made (a) below and (b) above the recrystallization temperature*

Increasing the temperature of welding (Figures 9.26 and 9.27) reduces both the critical deformation level and the amount of deformation required for optimum joint strength.

Recrystallization

This may occur in welding to an extent where grain growth occurs across the interface (Figure 9.28). This is not necessarily an essential feature of a good weld. Welds that are made below the recrystallization temperature will not have as good ductility as the parent metal but may well have higher strength. Above the recrystallization temperature, joint ductility improves and approaches that of the parent metal.

Diffusion

Depending on the size of atoms, diffusion can take place in different ways, either interstitially or substitutionally, for example. As mentioned earlier, an increase in temperature produces an exponential increase in the rate of diffusion, as shown in Figure 9.27.

Sometimes, diffusion at a junction may not be desirable. For example, if dissimilar metals were being joined, a brittle phase or compound may occur during welding (with a Cu–Al junction, for instance, $CuAl_2$ could form). Obviously, in such a situation, the more rapidly the joint is made and the lower the temperature used in welding, the lower will be the risk of embrittlement.

On the other hand, with other materials, diffusion is used to help promote the welding of metals, hence the diffusion welding process.

The essential features of cold pressure welding

The surfaces to be joined must be specially prepared. This usually entails degreasing and rotary scratch brushing. A pressure is then required, sufficient to cause plastic deformation in excess of the critical deformation, for the particular metal(s) involved.

Pressure can be applied by a number of means, such as dead weight loading, a press, or the use of differential thermal expansion.

A determination of whether the process is diffusion welding or diffusion brazing is that, if a filler metal is employed and it does not melt or alloy with the base metal to form a liquid phase, then the process is classed as diffusion welding.

Diffusion takes place through the metals using a mechanism involving vacant lattice sites or along grain boundaries. The use of vacant lattice sites is the more important at elevated temperatures.

The process of diffusion can be described by the following expression:

$$D = D_0 e^{-E/RT} \quad \text{(naperian)}$$

where D = the diffusion coefficient at temperature T (diffusion rate)

D_0 = a constant of the same dimensions as D (a constant of proportionality)

E = the activation energy for diffusion – the energy that is necessary to effect a motion of atoms from one site to another

R = the gas constant (Boltzmann's constant)

T = the temperature in absolute units.

The above expression illustrates the temperature dependence of the process, in that the diffusion process will vary exponentially with temperature (Figure 9.27).

Less energy is needed to cause the motion of a foreign atom than one of the same material in which the diffusion is taking place. Use is made of this fact when a thin film of filler metal is placed between the parts to be joined by diffusion welding. During the process, this layer will diffuse into both components and the finished joint may show no trace of the layer.

Because special equipment is usually required in order to apply heat and a compressive force simultaneously within a vacuum or chamber containing a protective atmosphere, the process is usually installed into production lines where moderate quantities of the component will be required.

In some industries, it has been found that scratch brushing is a better method of cleaning the surfaces than shot blasting, emerying or pickling, however, any such treatment gives an improvement when compared with attempting to weld without treatment. In each case, the material chosen for the bristles of the brush must be of a type that does not leave a metal deposit on the surface.

It is thought that scratch brushing is effective because of the way it breaks up the initial oxide film, by 'rolling' it into small areas.

Joint strength is impaired if welding is not carried out as soon as possible after preparation. Anodized aluminium (with a specially prepared thin oxide film) can be welded without the need for scratch brushing.

Amount of deformation required when welding cold

This varies from metal to metal, but some typical examples are:

Aluminium	approx. 60 per cent
Copper	approx. 80 per cent
Silver	approx. 94 per cent

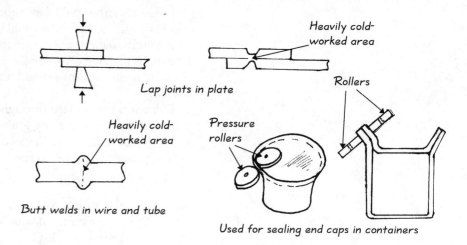

Lap joints in plate

Heavily cold-worked area

Rollers

Heavily cold-worked area

Pressure rollers

Butt welds in wire and tube

Used for sealing end caps in containers

Fig. 9.29 *Some typical applications of cold pressure welding*

Typical applications

The process is used for joining aluminium and copper. It is also occasionally used for joining other metals.

It is employed for sheathing cable, sealing aluminium containers (Figure 9.29), such as electrical condensers and the like, and in manufacturing detonators. The butt welding of wire and bar in copper and aluminium and similar types of joint also utilizes the technique.

Resistance welding processes: Part II – Electro-slag, consumable guide and electro-gas

Electro-slag welding

The **electro-slag** (or electric-slag) welding process (ESW, ISO 72) can in many ways be regarded as a development of the submerged-arc welding process.

Prior to 1900, graphite had been used to form moulds on each side of a space between vertical plates, in order to contain molten metal formed by the use of graphite electrodes. The molten metal then fused the edges of the plates to form a weld.

Copper and ceramic moulds have also been used to form welds in conjunction with conventional manual metal arcs, oxy-fuel gas torches and thermit mixtures.

In the early 1950s, scientists from the E.O. Paton Electric Welding Institute in Kiev announced the development of welding machines that employed an electrically conductive slag to make large single-pass vertical welds. The process started with an arc on a run-on plate. This arc became extinguished under a molten slag bath and the process then welded by resistance heating.

Subsequent work at the Bratislava Institute of Welding in Czechoslovakia became available to engineers in Belgium in 1958 and the process was introduced into the USA in 1959.

The ESAB company of Sweden purchased licences for the process and, in 1959, the equipment went into production in Britain and West Germany during this year.

Figure 10.1 shows the basic principle of the process. An electrode wire is fed by drive rolls through a curved electrode guide tube. It is common to use

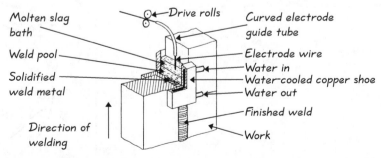

Fig. 10.1 *Non-consumable guide method of electro-slag welding. Some machines use more than one electrode. (See also Figure 10.2)*

more than one electrode wire (see Figure 10.2), because if one wire feed jams, the others can be speeded up to compensate, without scrapping the weld.

Welds are made in the vertical position, with plates aligned having a square edge preparation and a gap of usually 25–27.5 mm (1–1½ inch). Run-on and run-off plates are used. The minimum length of weld is about 75 mm (3 inches).

Two water-cooled copper shoes are fitted and held across each face of the weld operation. One, two or three electrodes (or more on some machines) are fed into the box formed by the shoes and the sides of the weld preparation.

Using run-on plates (material of the same thickness as the actual job, tacked on at the start and removed on completion of the weld), an arc is struck under a small amount of powdered flux. Initially the process is the same as for submerged-arc welding, but then more powdered electro-slag flux is added until the flux bath is deep enough to maintain electro-slag welding conditions. When these conditions are achieved, the arc extinguishes, and heat is generated by the resistance heating effect of the molten flux or slag, generally this being in excess of 2000°C (3623°F) and so it readily melts the electrode wire at the end and also the edges of the plates being welded.

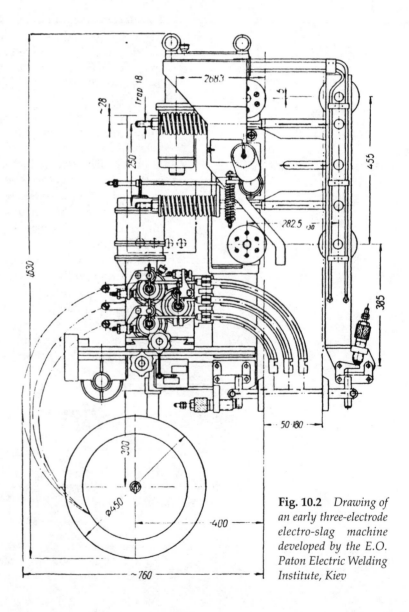

Fig. 10.2 *Drawing of an early three-electrode electro-slag machine developed by the E.O. Paton Electric Welding Institute, Kiev*

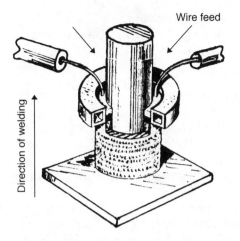

(a) Hardfacing the surface of a shaft using a circular shoe

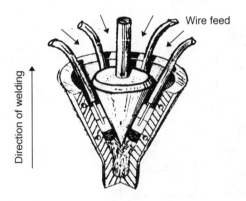

(b) Hardfacing a cone-shaped component using a cone-shaped water-cooled shoe

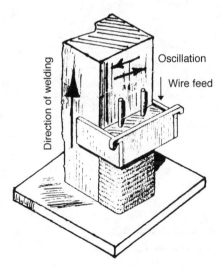

(c) Hardfacing a flat surface using a composite moulding shoe

Fig. 10.3 *Examples of hardfacing components using electro-slag welding process (after E.O. Paton Electric Welding Institute, Kiev)*

Sometimes, to produce a more uniform weld, the wire electrodes can be oscillated.

On some machines, the drive for the shoes and the flux addition is manual, while others are completely automatic, just requiring the operator to start and stop the process, check that the travel speed is set correctly, and maintain a good level of flux in the flux hopper.

Important parameters to be controlled are weld width, weld depth and penetration. The ratio of weld width-to-weld depth should be 0.8 or larger ($W/D \geq 0.8$ where W = width of weld and D = depth of weld). This ratio is increased by increasing weld voltage, decreasing welding current, or decreasing wire feed rate.

Fluxes

The fluxes used are similar to those in submerged-arc welding but usually of a higher calcium fluoride content. They are designed to have low conductivity at welding temperature. The depth of flux/slag in the pool should be from 37.5 to 62.5 mm ($1\frac{1}{2}$ to $2\frac{1}{2}$ inches). As mentioned, some machines have flux dispensers, but the amount to be added once the weld is started is usually so small that additions by hand, using a small scoop, are often preferred.

Sometimes the flux can be added via flux-cored wire, again similar to the method used in other automatic welding processes. Wires can be fairly large, starting at around 3 mm (1/8 inch) diameter.

The type of joint that can be made includes butts and fillets in the vertical position, together with special hardsurfacing applications, some examples of which are illustrated in Figure 10.3.

Longitudinal seam welds on thick-walled boiler shells and other types of vessel can be prepared for as shown in Figure 10.4. (a) illustrates one type of rolled shell preparation to allow the snug fitting of copper shoes and (b) shows how start and finish tabs are positioned, together with clamps or 'strongbacks' that are shaped to hold the work in position while still allowing the passage of the copper shoe mould.

When prepared, the whole vessel is then placed vertically for welding. Some large machines are equipped with a platform arrangement (Figure 10.5),

Clamps or "strongbacks", tacked in position to hold work but also allow passage of shoe mould

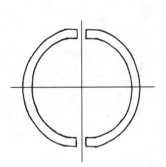

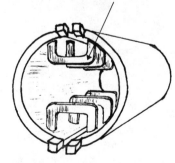

(a) End view of a boiler shell formed in preparation for electro-slag welding

(b) Finishing tabs (run-off plates) (starting tabs at other end – (run-on plates))

Fig. 10.4 *Preparation for welding longitudinal seams of a thick-walled boiler shell*

which allows the operator to travel with the welding head as the weld progresses up the seam. This allows close control of the process, enabling the operator to make any ongoing minor adjustments.

The welding of circumferential seams

The process can also be employed for the welding of circumferential seams and various methods are used in production. Welding is carried out by a fixed-head machine mounted on a column; such a machine is shown in Figure 10.6. The cylinder being welded is slowly rotated, at the pre-determined welding speed, using a large manipulator.

Great care must be taken in the arrangements for starting and finishing the weld, so that no discontinuity or defect is formed. Figure 10.7a shows one method of fixing an internal shoe mould using a 'slitter bar', which goes through the preparation gap. This has to be removed for the completion of the weld (Figure 10.7b) and an 'exit route' mould built up out of blocks of copper. At this point, rotation stops and the welding head is slowly raised up

Fig. 10.5 *Boiler-shell longitudinal seams being welded in the vertical position using electro-slag machines with platforms to allow full operator control*

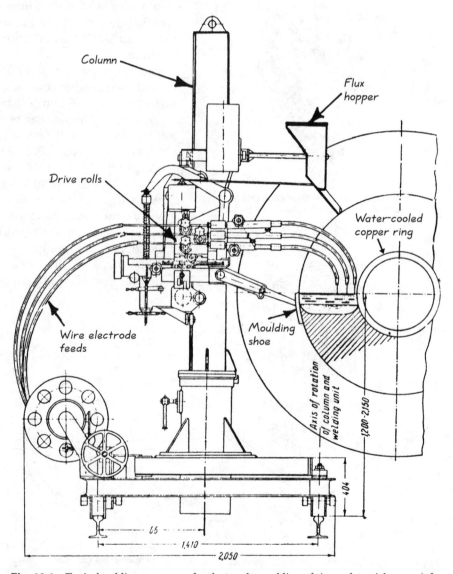

Fig. 10.6 *Typical welding apparatus for electro-slag welding of circumferential seams (after E.O. Paton Welding Institute, Kiev)*

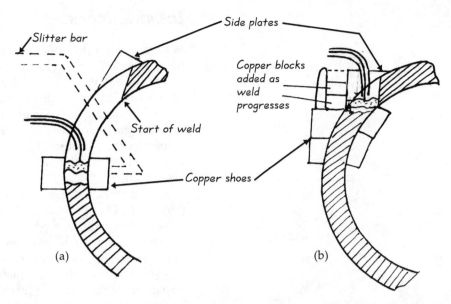

Fig. 10.7 *Method of completing a circumferential weld (after Jackson[1]): (a) slitter bar is removed and internal shoe clamped; (b) remaining shoes are clamped as weld is nearing completion, and an exit-route mould is built up by progressively adding blocks of copper*

the column to close the weld. Any excess metal can be removed by flame cutting.

Arms, or slitter bars can also be used on longitudinal seam welds. With circumferential welds, however, the practice of fitting internal shoes to close the weld requires the somewhat dangerous practice of an operator going inside the vessel while the welding operation is taking place.

Alternative methods include the use of:

(a) a water-cooled copper backing ring that can be split for removal;
(b) several copper shoes that can be fitted round the entire inner circumference;
(c) an integral steel backing ring, which can be either left in position if permitted, or removed by machining

Advantages of the electro-slag process

1. Plates can be left with a square edge as no further plate preparation, other than cleaning, is required.
2. High welding speeds are possible.
3. The thickness of plate that can be welded in one pass is almost unlimited.
4. Gases have plenty of time to escape from the weld and so there is usually a reduced chance of porosity.
5. If the carbon content is kept low and the manganese/sulphur ratio greater than 45, then there should be little danger of hot cracking.
6. The need for pre-heat is greatly reduced and because of the slower rate of heating and cooling, there is a reduced risk of cold cracking. (Care should be taken, however, when water-cooled copper shoes and moist sealing material are employed, as an atmosphere of steam can be produced, affecting the extreme weld surfaces. In these conditions, microcracks and porosity can result.)
7. There is little risk of slag inclusion, if welding can continue without interruption.
8. There is very little angular distortion and a favourable residual stress pattern in as-deposited electro-slag welds.

151

Disadvantages

Most as-welded electro-slag weldments have a very coarse as-cast weld metal structure and coarse grains in the heat affected zone (HAZ) – in steel welds this can be a Widmanstätten structure. Consequently, electro-slag welds may not be suitable for certain critical applications.

For ferritic steels, a normalizing treatment is generally employed, in order to refine the weld metal and HAZ structure. This can involve heating the whole structure to around 1100°C (2012°F) and removing all traces of the as-cast structure, which tends to even out the properties of the weld metal and base metal.

A second heat-treatment is then often applied at around 900–950°C (1652–1742°F) to provide grain refinement.

Such treatments are expensive and in some cases, because of the size of structures, they would be impossible to carry out. It is for these reasons that various methods have and are being investigated, in order to produce an as-welded structure with a sufficiently refined grain structure to give satisfactory impact properties without further heat-treatment.

The work of Cuddy and others[2] has shown that additions of titanium give promising results in terms of HAZ microstructure control. Additions of titanium and nitrogen were shown to cause the formation of finely dispersed titanium nitride particles, which in turn promoted the formation of equi-axed ferrite in the HAZ. The quantity of low-temperature microconstituents was also observed to reduce with increasing titanium content.

Applications of the process

The process is still mainly employed for the welding of thick-walled pressure vessels used in the power-generating, chemical petroleum and marine industries, shipbuilding, and the fabrication of large machine and press frames.

The process can also be used to cast components, such as large nozzles, on to thick-walled vessels.

The consumable guide method (variation)

The **consumable guide method** of electro-slag welding is illustrated in Figure 10.8. With this method the filler metal is supplied by both an electrode and the guide. The guide can be a simple tube, or for very large welds, plates can be added to each side of the tube. These are known as 'wing-type' guides, as shown in Figure 10.8. A flux coating may be provided on the outside of the consumable guide in order to insulate it and to help keep the slag-bath replenished. This system can be used to weld sections of virtually unlimited thickness. Electrodes can be oscillated or stationary. When using stationary electrodes, each electrode will weld approximately 63 mm (2.5 inches) of plate thickness, while one oscillating electrode can weld up to 130 mm (5 inches). The use of two oscillating electrodes increases the welding thickness capability up to 300 mm (12 inches), while three oscillating electrodes increase the use of consumable guide welding to thicknesses approaching 450 mm (18 inches). If oscillation is not employed, then additional stationary electrodes will increase the welding thickness capability.

Wing-type guide tubes are used when simple, round guide tubes cannot heat the entire cross-sectional area, or for irregularly shaped weldments. They are made by tack-welding low carbon steel plate or bar to the sides of a round guide tube.

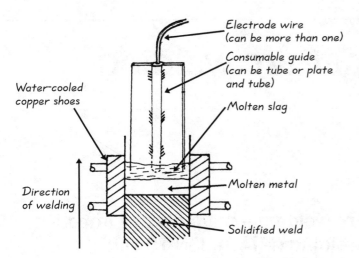

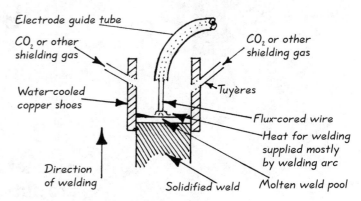

Fig. 10.8 *Electro-slag welding, consumable guide method (wing type)*

Fig. 10.9 *The electro-gas welding method*

Electro-gas welding (variation), ISO 73

Electro-gas welding (EGW) is a variation of electro-slag welding. It still employs water-cooled copper shoe moulds and carries out welding in the vertical position but combines features of the carbon dioxide (CO_2)-shielded flux-cored electrode process.

The flux-cored electrode is fed into the weld pool down a guide tube, the welding heat being supplied mostly by a welding arc, established between the end of the flux-cored wire electrode and the weld pool surface (Figure 10.9).

Carbon dioxide or another shielding gas is supplied through holes (tuyères), located near the tops of the copper shoes.

Again, oscillation of the electrode can be employed. A constant wire feed speed is employed, with a self-adjusting arc system.

Advantages

The process is similar to electro-slag but covers the plate thickness range from 12.5 mm ($\frac{1}{2}$ inch) up to 50 mm (2 inches), which is not covered very adequately by the electro-slag process.

Generally, weld properties are satisfactory and metals that can be welded include mild, low alloy and stainless steels.

The process has been used extensively in different parts of the world for welding the vertical seams on storage tanks.

References

1. M.D. Jackson, *Welding methods and metallurgy*. Charles Griffin and Co. Ltd, 1967
2. L.J. Cuddy, J.S. Lally and L.F. Porter, Toughness and micro-structures in the heat affected zones of ship steels containing titanium. In *USS Research Report*, 1983

Further reading

A.S. Bahrani and B. Crossland, *Institution of Mechanical Engineers, Symposium on High Rate Forming*, December 1964

R.D. Thomas, Jr. and S. Liu, Interpretive report on electroslag, electrogas and related welding processes. *Welding Research Council Bulletin 338*, New York, November 1988 [ISSN 0043–2326]

Magnet arc welding

Magnet arc welding – magnetically impelled arc butt welding (MIAB), ISO 18 and magnetically impelled arc fusion welding (MIAF) (rotating arc welding, ISO 185)

There are two types of MIAB machine (Figure 11.1, a and b). Figure 11.1a shows the most-used method, where the arc is drawn between the adjacent ends of the weldments.

An alternative technique is shown in Figure 11.1b, where the arc is held between a non-consumable auxiliary electrode and the ends of the parts to be welded. In both cases, the arc is rotated around the weld line by a force (*F*) which results from the interaction of the arc current (*I*) and the magnetic field (*B*). In both systems it is usual to employ carbon dioxide or other inert gas shielding.

The first method (a) – where the arc is drawn between the adjacent ends of the weldments in order to raise them to a high enough temperature to allow forging to form a solid-state weld – is the one, to date, that has been exploited by industry.

The alternative method (b) – using the auxiliary electrode – is suitable for the arc fusion edge welding of thin-wall pipes or tubes and certain pressed sheet fabrications.

This alternative method is known as **magnetically impelled arc fusion welding** or MIAF.

MIAB welding has an advantage of speed over other more conventional welding techniques. It also has the capability of welding thin sections and parts made of hollow section and not possessing rotational symmetry.

The technique enables workpieces to be accurately joined, so that in most cases, no further machining is required before assembly. In addition to this precision and the repeatable high quality of the welded joint, MIAB welding offers an extremely favourable basis for parameter monitoring and the keeping of quality assurance records.

Fig. 11.1 *Showing the basic differences between magnetically impelled arc butt welding (MIAB) and magnetically impelled arc fusion welding (MIAF) of mild steel tubing (after Johnson et al.[1]). MIAB welding is shown at (a), where the arc rotates about the interface. The heated abutting surfaces are then pressed together by a forging cylinder to form a solid-state weld. At (b), in the MIAF process, the arc is drawn between an auxiliary electrode and the weldments. The directions of current (I) and applied magnetic field (B) are indicated by the direction of the resultant force (F) on the arc*

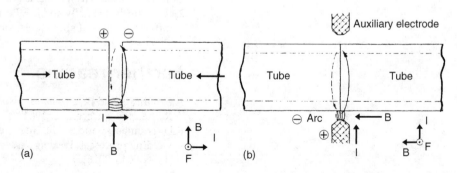

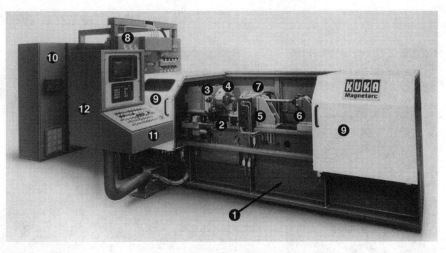

Fig. 11.2 *A standard Magnetarc machine*
(Photograph courtesy of KUKA Schweissanlagen and Roboter GmbH and Thompson Welding Systems Ltd, UK)

Key

1	Machine bed	8	Hydraulic power pack
2	Main carriage	9	Safety doors, electrically interlocked
3	Tool holder	10	Welding and magnetic coil current
4	Workpiece support as three-jaw chuck		source and programmable logic machine
5	Workpiece support as self-centring block		controller (PLC)
6	Workpiece stop	11	Operator terminal
7	Magnetic coil system	12	Parameter monitoring system

Standard Magnetarc machines

Unlike other welding techniques, Magnetarc (MIAB) welding requires only one simple, linear motion for the joining process. Low wear, minimal maintenance and a long service life of the machine are therefore inherent characteristics. A Magnetarc machine can be quickly and easily retooled and is thus capable of welding a broad range of different workpiece shapes and diameters.

The basis of the machine is formed by a rigid, self-supporting **machine bed**. A special foundation design is not required.

The **main carriage** is mounted, free of backlash, on a slideway and serves to support the tool holder. The carriage slideway is covered to protect it from dirt and damage.

With its ground, standardized short mounting taper, the **tool holder** can be fitted with commercially available chucks and enables the chucks to be changed with ease and precision. A setting device permits the individual fine adjustment of the chucks and thus ensures highly accurate welding results.

For the **support of the workpieces** a three-jaw chuck is provided on the carriage, while the stationary workpiece is held in a self-centring clamping block. The clamping insets can be adapted to the workpieces and are readily exchangeable when the machine is retooled.

The main carriage is driven hydraulically. The **hydraulic power pack** is fixed to the machine bed to form a single unit. The clamping tools are also actuated hydraulically.

The easily programmable **machine control system** is installed in a separate cabinet. The swivelling operator terminal, which also accommodates the PCD system, is located in an ergonomically favourable position next to the machine.

The **magnetic coil system** is selected to suit the workpiece and required weld. Two-piece external coils, internal coils and combined systems are available. The welding and magnetic coil control is enclosed in a separate cabinet. Coarse and fine regulating switches are provided for presetting the welding current. The coil current is thyristor-controlled.

The magnetarc welding process

Fig. 11.3 *Magnetarc welding process – phase 1*

Fig. 11.4 *Magnetarc welding process – phase 2*

Fig. 11.5 *Magnetarc welding process – phase 3*

Fig. 11.6 *Magnetarc welding process – phase 4*

(All photographs courtesy of KUKA Schweissanlagen and Roboter GmbH)

Simple sequence of operations (magnetically impelled arc butt welding (MIAB)

MIAB welding is a pressure welding technique using a magnetically impelled arc in a gas-shielded atmosphere. The sequence of the joining process can be divided into the following phases.

Phase 1 (Figure 11.3): The clamped workpieces are brought into contact. Installed in the area of the joint is an internal or external magnetic coil system, which does not come into contact with the workpieces (an internal coil was used in the application pictured in the sequence of photographs shown here).
Phase 2 (Figure 11.4): After the shielding gas, magnetic coil and welding current have been switched on, the workpieces are retracted to produce a defined gap, thereby striking the arc.
Phase 3 (Figure 11.5): The magnetic field of the coil causes the arc to rotate about the interface. The abutting faces are thus uniformly heated and melted.
Phase 4 (Figure 11.6): The melted abutting faces are pressed together by a forging cylinder; the materials are joined forming a porosity-free weld. The shielding gas, magnetic coil and welding current are switched off.

Some typical examples of MIAB welded components for the automotive industry

Figure 11.7 shows some examples of ready-to-fit MIAB-welded vehicle axle assemblies.

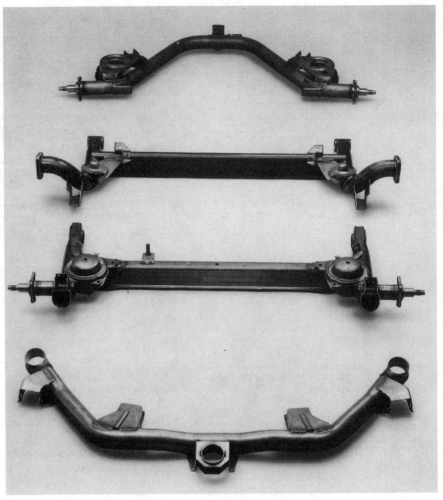

Fig. 11.7 *Some examples of ready-to-fit Magnetarc (MIAB) welded vehicle axle assemblies* (*Photograph courtesy of KUKA Schweissanlagen and Roboter GmbH*)

In conventional axle manufacture, MIG (metallic inert gas shielded) or MAG (metal-arc gas shielded) welding is generally employed for securing the flange plates to the axle tubes. With this method, before the wheel carriers can be bolted on, the welded assembly must be checked, straightened or finish-machined to obtain the required tolerances to cater for camber and caster angles. It is only after these operations that the wheel carriers can be attached to produce the completed assembly.

The production costs of this method are substantially reduced and the process much simplified by the use of MIAB welding, since this welding process allows the finish-machined wheel carrier to be welded directly to the axle-tube assembly. As there is minimum distortion, the angular alignments and the overall length can be maintained within the required tolerances.

When air-hardening carbon steels are used for wheel carriers, because MIAB welding is a solid-state process, brittle or stress cracking is much less likely to occur.

Using MIAB welding means that axles are ready for fitting immediately after welding, eliminating the need to straighten or further machine the assembly. Quality control systems allow the stringent safety requirements to be met, as the whole process sequence is electronically controlled and monitored. All significant parameters for acceptance are documented and

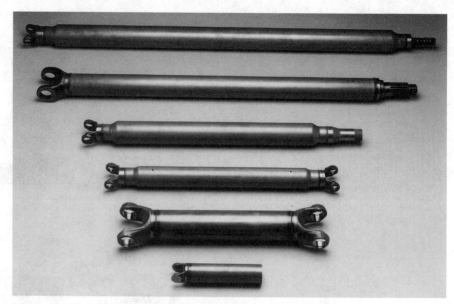

Fig. 11.8 *Some examples of Magnetarc (MIAB) welded vehicle drive shafts*
(Photograph courtesy of KUKA Schweissanlagen and Roboter GmbH)

recorded on a printout for each assembly. The data can be linked by number identification on the printout and the component itself.

Special technical data sheets can be obtained from the manufacturers of MIAB machines and these explain the precise machine and control options available.

Figure 11.8 shows a number of different MIAB welded drive shafts, also for the automotive industry. Large amounts of testing, both non-destructive and destructive, including fatigue strength testing, have proved the high quality of MIAB welds in this application.

The range of applications is increasing steadily and these are just two examples in every-day use.

Reference

1. K.I. Johnson, A.W. Carter, W.O. Dinsdale, P.L. Threadgill and J.A. Wright, The MIAB welding of mild steel tubing. In *Welding Institute Research Bulletin*, Vol. 19, October 1978. The Welding Institute, Cambridge, UK

Friction welding

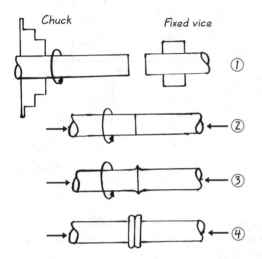

Fig. 12.1 *The basic steps in conventional friction welding:* ① *one workpiece is held in a chuck and rotated, while the other is held stationary in a fixed vice;* ② *when the pre-determined rotational speed is reached, an axial force pushes the components together;* ③ *friction produces heat and plastic displacement of material at the fraying surfaces (interface), and upsetting starts;* ④ *the weld cycle is completed as rotation stops and a final upset force is applied, producing a collar of plastically deformed metal (known as the 'flash') around the metal*

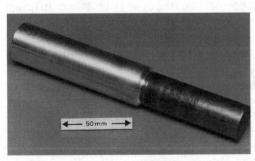

Fig. 12.2 *Friction welding is ideal for joining many dissimilar metal combinations. The photograph shows a bar of stainless steel on the left, welded to a bar of copper*

(Photograph courtesy of TWI, Cambridge, UK)

The process of **friction welding** (FRW), ISO 42, was used to join plastics in 1945 and a paper was published in Russia on the joining of metals using friction by Chudikov and Vill in 1956. It was not until the 1960s, however, that the development of production machines started in earnest.

Friction welding requires either rotating or linear movement of one component, which is brought into contact with the part or other component to which it is to be joined. This produces heat and the plastic displacement of material at the faying surfaces (interface), and upsetting starts (Figure 12.1). To complete the weld cycle, rotation stops and a final upset force is applied (Figure 12.1).

Such a joint is regarded as a **forge weld** or **solid-state weld**, as it is made at a temperature below the melting point(s) of the material(s). Under normal circumstances, no melting should take place and because of this fact, dissimilar metals can often be joined (Figure 12.2).

The typical friction weld shows plastically deformed material around the weld as a 'flash' (Figure 12.1, ④), which can either be removed as a post-weld operation or left intact, depending on the design and specification requirements of the product. Other characteristics of friction welds are a narrow heat affected zone and the absence of a fusion zone.

There are two main methods of supplying the mechanical energy for rotational friction welding: **direct** or **continuous drive**, which is sometimes known as conventional friction welding and uses a continuous power or energy input maintained for a pre-set period; and **inertia friction welding**, where the energy for welding is provided by kinetic energy stored in a flywheel or fluid storage system.

Direct drive welding

With this system, one of the workpieces, which is usually circular or tubular in cross-section, is held in a chuck which is driven by a motor unit, while the other workpieces is held stationary (Figure 12.3 – see also Figure 12.5 for variations in the arrangement).

The workpiece in the chuck is rotated at a pre-determined constant speed. The parts to be joined are moved into contact and a friction welding force is applied. This produces heat at the faying surfaces (weld interface), as they rub together. The friction phase continues for a pre-determined time, determined by joint size and type of material. It must be long enough to generate adequate heat in the weld zone for the forge phase. It can also be determined by a pre-set amount of upset having taken place, as this also helps to ensure that any surface irregularities are removed from the faying surfaces. At this stage, the drive to the chuck is switched off and the time taken to stop will be determined by a combination of the following factors:

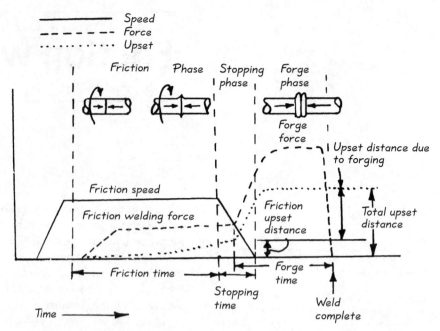

Fig. 12.3 *Showing the relationship of welding parameter characteristics with direct-drive friction welding*

(a) the efficiency of any braking system employed;
(b) the programme for powered deceleration where appropriate;
(c) any kinetic (stored) energy in the workpiece and machine transmission;
(d) size of workpieces, weld interface area;
(e) the type of material being joined
(f) the forge force

The friction welding force is maintained and usually increased as the 'forge force' for a pre-determined time after the workpiece has stopped rotating. The relationship of welding parameter characteristics with direct-drive friction welding is shown in Figure 12.3.

Inertia drive welding

In this process, one of the workpieces is placed in a chuck connected to a flywheel, which is accelerated to a pre-determined speed of rotation, in order to store the energy required for the weld. The other workpiece is held so that it cannot rotate. The drive is disengaged and the workpieces are pushed together by the friction welding force. The energy stored in the flywheel is given off as friction heat at the faying surfaces, causing its speed to decrease. An increase in the friction welding force (forge force) can be applied before workpiece rotation ceases and is usually maintained for a pre-determined time after rotation has stopped. Figure 12.4 shows the relationship between the welding parameter characteristics for inertia friction welding. By using the flywheel to store energy, an inertia machine can 'discharge' energy very rapidly, generally resulting in shorter welding times with narrower heat affected zones and less flash. Usually, less drive power is required with an inertia machine, when compared with a direct-drive machine welding the same cross-sectional area.

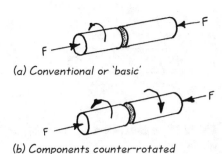

(a) Conventional or 'basic'

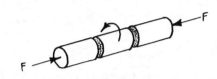

(b) Components counter-rotated

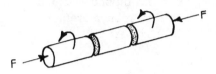

(c) Single rotating component in centre

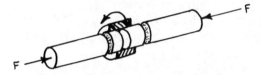

(d) Centre stationary, both outside components rotating

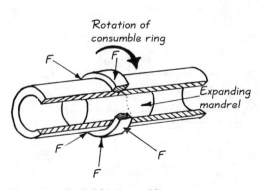

(e) Central chuck drive holding two components

Fig. 12.5 *Variations in the arrangement of the relative motion of workpieces in order to complete single or double friction welds in components with a circular or tubular cross-section*

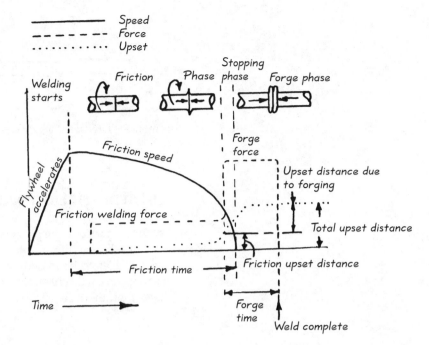

Fig. 12.4 *Showing the relationship of welding parameter characteristics with inertia friction welding*

Variation in arrangements for friction welding

The method of friction welding that is most used requires the rotation of one of the workpieces while the other is held stationary, and the rotating workpiece is invariably circular or tubular in cross-section (Figure 12.5a).

For the production of welds where a high interface speed is required, a machine can be employed that rotates each of the workpieces in opposite directions (Figure 12.5b). For the joining of long sections of pipe, or other awkwardly shaped workpieces that would be difficult to rotate, the two stationary workpieces can be pushed up against a central rotating section, as shown in Figure 12.5c. Figure 12.5d shows another method of making two welds at once, but this time with the centre section stationary and the two outer workpieces rotating. Figure 12.5e illustrates a system where two components can be welded at once, back to back, by means of a central chuck drive that is capable of holding two components. The other workpieces are then pushed against these rotating components from both sides, forming two welds at once and increasing production.

Radial friction welding

This is an interesting variation of the basic friction welding process. By rotating a consumable ring around the joint (Figure 12.6), the technique can be used for welding components that would be impossible or impracticable to rotate, such as long pipes or tubes. The consumable ring heats up both itself and the joint area as it is rotated; it is then compressed, while an expanding mandrel inside the bore supports the pipe walls and prevents the upset metal from the weld penetrating into the bore of the pipe.

Fig. 12.6 *Radial friction welding*

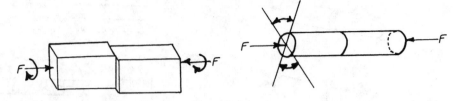

Fig. 12.7 *Orbital friction welding*   **Fig. 12.8** *Angular reciprocating (or angular oscillating)*

Orbital friction welding

With this process, the parts being joined do not need to be circular or tubular in cross-section (Figure 12.7). One component orbits around the other, so that the surfaces to be joined are touching. Neither component actually rotates around its own central axis or, if components are rotated around their own axes, then their axes are displaced relative to each other.

Angular reciprocating (angular oscillating)

Here, the moving workpiece rotates through a pre-determined angle, being less than one full rotation, and then reverses back to the starting position. This cyclic, reversing rotational motion is repeated and then a forging force applied (Figure 12.8).

Linear reciprocating (linear oscillating)

This system employs the moving of one component in a linear oscillating motion, being relative to and in contact with the mating face of the component to which it is to be joined (Figure 12.9). The technique is suitable for joining components whose symmetry would not allow rotation.

Friction surfacing

This process, illustrated in Figures 12.10 and 12.11, employs a consumable cylinder rotating in contact with the surface to which the surfacing material

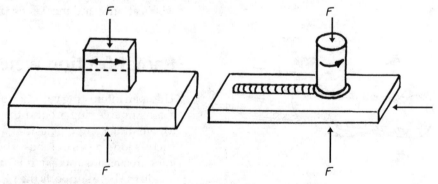

Fig. 12.9 *Linear reciprocating (or linear oscillating)*   **Fig. 12.10** *Friction surfacing*

Fig. 12.11 *Friction surfacing – applying stainless steel overlay to a carbon steel shaft*
(*Photograph courtesy of TWI, Cambridge, UK*)

is being deposited in the solid-state mode. The surface is moved in a direction perpendicular to the direction of rotation, and force is applied in order to keep the surfaces in contact.

Friction stir welding

This is a process currently under development at The Welding Institute (UK), which can friction-weld plate without any relative movement of the workpiece. It employs a hard, round 'stirring' tool of a small diameter which is attached to the bottom of a cylindrical device of a larger diameter. A rotary motion of approximately 1000 revs/min as the tool is moved along the joint is enough to cause coalescence. The only downward pressure is the force required to push the 'stirring' tool into the metal. The process shows great promise for the joining of aluminium.

Underwater friction welding

Small, hand-held friction welding machines have proved very effective for the attachment of studs underwater (see also Chapter 21 on underwater welding).

Although these process variations exist as research machines or machines designed for specific applications, by far the largest proportion of friction welders in use throughout the world are rotary machines, butt welding components made from at least one circular or tubular section. Figures 12.16 to 12.18 show typical examples of friction welded components.

The process of the welding cycle can be described as taking place in three stages (Figure 12.12):

Stage 1.   The cold parts are subjected to dry friction.
Stage 2.   Local seizures occur, gradually increasing until each seizure is followed by rupturing. The end of the second stage occurs at maximum torque.
Stage 3.   In this phase, the torque may drop, holding steady at some minimum level; the process of seizure and rupture gives way to plastic deformation and a steady state is reached. The greatest percentage of heat (85 per cent) is generated at this stage.

Metal in the plastic state is extruded from the interface to form a **flash** or **collar** around the joint. This causes a shortening of the work, known as the **upset**. This process also helps to ensure that any surface irregularities are removed. Low melting point constituents in the material, such as sulphur in 'free cutting steels', can cause the formation of low melting point films, which have little strength and can result in weld failure.

Sometimes, any elongated inclusions in the material(s) being joined can turn spherical, as a result of hot working, and then, as they are forced into the 'flash', they can become re-elongated. A transverse cut through a friction weld can sometimes show a 'swirl' (a distinct line or spiral from the centre to the outside circumference), indicating the path taken by surface material at the centre to the outside of the components.

A peak occurs in the torque curve at the end of the process; this is the result of speed reduction during braking and forging.

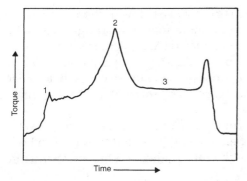

Fig. 12.12 *A graph of torque versus time shows that the friction welding process takes place in three stages*

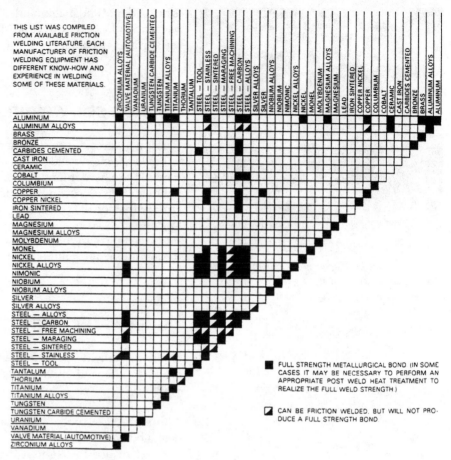

Fig. 12.13 *Material combinations weldable by friction welding*
(Courtesy of Thompson Welding Systems Ltd, UK)

Because friction welding is a solid-state process (no melting takes place), it enables many combinations of dissimilar metals to be joined together, as shown in Figure 12.13.

The duration of the rotational friction stage and the frictional characteristics (type of material, composition and metallurgical condition will also have an effect), mainly determined by the rotational speed and axial pressure at the faying surfaces, will control the peak temperature and amount of heat generated. The section size and thermal conductivity of the materials being joined will help to determine the temperature gradient across a friction weld.

For any given material, conditions giving rise to low frictional resistance at the faying surfaces will tend to promote longer welding times, with correspondingly wider heat affected zones (HAZs). A high coefficient of friction, however, will have the opposite effect, giving a shorter welding time and narrower HAZs. It is therefore possible to create a whole range of thermal cycles, which can be represented as thermal gradients (Figure 12.14).

Although, in practice, it is fairly common to carry out a series of test welds in order to determine parameters before going into production, it can be of benefit to consider the effect of the thermal gradient on the microstructure from a theoretical aspect.

Figure 12.14a illustrates the welding of two low carbon steel components in the normalized condition and the effects of the thermal cycle. The point T_1

on the thermal gradient represents arrest point 1 (A_1) on the iron/carbon equilibrium diagram, where the transformation from ferrite and pearlite to austenite commences (see Chapter 15 on metallurgy). At a point between this temperature and the interface temperature T_2, the A_3 temperature will be exceeded and the structure will become fully austenitic. As the weld cools down, the HAZs transform back to ferrite and pearlite. The high temperatures and cooling rates faster than those for normalizing tend to give small increases in hardness values.

The welding of two hardenable steel components in the hardened and tempered condition is represented in Figure 12.14b. Firstly, a small decrease in hardness can take place as a result of over-tempering causing softening at the edges of the HAZs. An increase in hardness then takes place, increasing along with the thermal gradient as the material becomes austenitic. On cooling, this can form martensite in materials with a high hardenability.

Figure 12.14c illustrates the possible effects when welding a material that has been precipitation hardened or age hardened, such as the alloy of aluminium and copper, Duralumin, or the aluminium–zinc–magnesium alloys. Here, increased temperatures along the thermal gradient will cause increasing thermal damage as the precipitate is taken back into solution. Strength can be regained after welding, in most cases, by heat treatment or allowing the

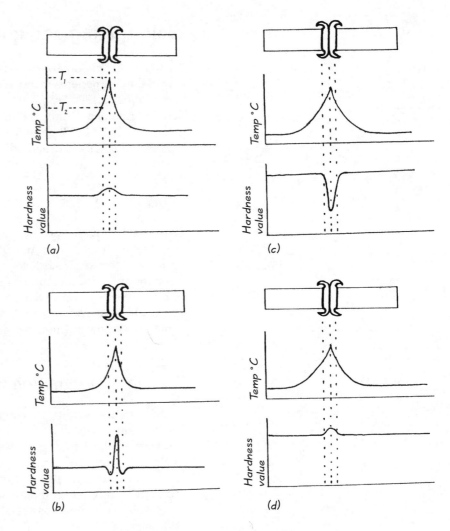

Fig. 12.14 *The thermal effects of friction welding on different metals*

natural ageing process to take place (in the case of Duralumin, heating and quenching – see Chapter 18).

Figure 12.14d shows the expected effects when a simple solid-solution alloy such as copper–nickel is friction welded. No phase transformations should occur during the welding operation but grain refinement giving a small increase in hardness is likely.

Figure 12.15 illustrates the relationship between thermal gradient and HAZ shape, showing the effect of welding conditions on these characteristics.

Figures 12.16, 12.17 and 12.18 show some typical examples of friction welded components, Figure 12.18 also includes some advantages of the friction welding process. Some limitations of the process are:

1. the cost of capital equipment is high;
2. in a number of applications, at least one of the workpieces must have an axis of symmetry and be capable of being rotated about that axis;
3. dry bearing and materials that will not forge cannot be welded by this method;
4. free-machining alloys can form low-strength films at the interface and are therefore difficult to weld;
5. if both components are longer than 1 m (3 feet) in length, then a special machine would be required.

Weld testing and quality control

These factors will obviously vary, according to the requirements of the end-user of the components manufactured by the friction welding process. British Standard 6223: 1990 *Specification for friction welding of joints in metals*, and the new International/European document *Friction welding*, being prepared by CEN/TC 121/WG 12, provide good sources of reference.

All weld testing is usually divided into two distinct areas:

Procedure tests

These tests are very specific, particularly in automated or semi-automated processes, such as friction welding, where there is no welding skill required from the operator. Such tests, which can involve mechanical testing and metallographic examination as well as NDT (non-destructive testing), and field trials serve to verify that the welding machine, when programmed with a particular set of welding conditions, will produce a weld quality of a standard to meet specified end-user requirements.

Testing of production weldments

Modern friction welding machines usually have in-built monitoring systems for continuous checking of the repetition of values within the welding cycle that were established in the development of the weld procedure. Checks on friction pressure, forge pressure, all times and component length are common. Some machines can also produce documentary evidence in the form of printouts traceable to component number.

As well as the above, as a further safeguard, destructive tests can be taken at various stages during production runs, in order to check all other variables

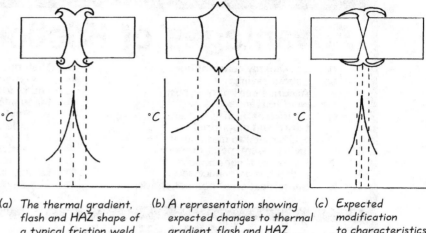

(a) The thermal gradient, flash and HAZ shape of a typical friction weld

(b) A representation showing expected changes to thermal gradient, flash and HAZ caused by a very long weld cycle

(c) Expected modification to characteristics caused by a very short weld cycle

Short weld cycles are produced by low rotational speed and high friction pressure. This can result in welds with higher peak hardness values (particularly in hardenable steel components), narrow HAZs and fine-grained microstructures.
Long weld cycles tend to produce wide HAZs having coarse microstructures. In extreme cases they can tend to cause overheating and a reduction in peak hardness values.

Fig. 12.15 *The effect of welding conditions on HAZ shape and thermal gradient (after Humphreys[1])*

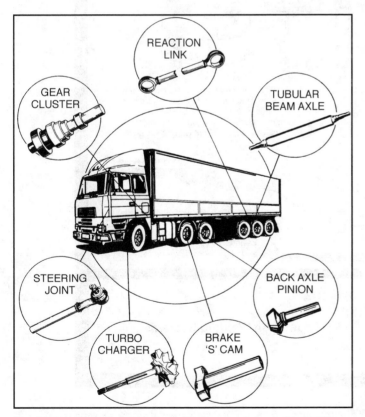

Fig. 12.16 *Examples of friction welded components used in HGVs*
(Courtesy of Thompson Welding Systems Ltd, UK)

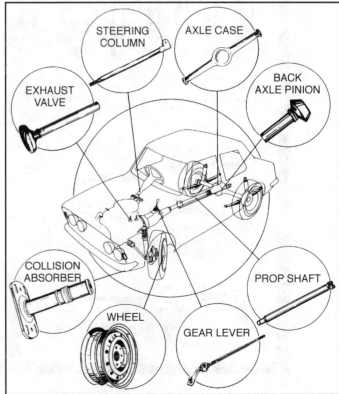

Fig. 12.17 *Examples of friction welding components used in automobiles*
(Courtesy of Thompson Welding Systems Ltd, UK)

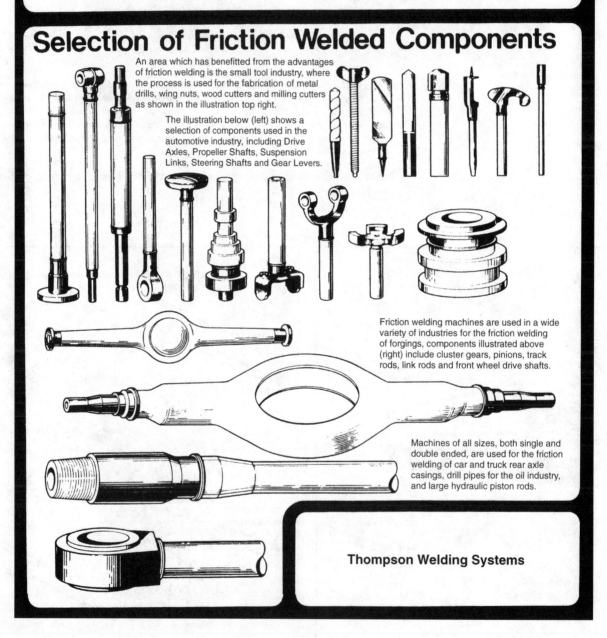

Advantages of Friction Welding

- Low energy consumption.
- Labour saving.
- Accurate control over post weld tolerances.
- Consistent quality is maintained and monitored.
- 100% metal to metal joints giving parent metal properties.
- No external consumables or protective gases necessary.
- Simplification of component design.

- High production rates.
- Low metal consumption and reduced machining.
- Manual loading or full automation optional.
- Unskilled labour can be used.
- Expensive material can be joined to cheaper material.
- Dissimilar material combinations.
- Simple clean mechanical operation.

Selection of Friction Welded Components

An area which has benefitted from the advantages of friction welding is the small tool industry, where the process is used for the fabrication of metal drills, wing nuts, wood cutters and milling cutters as shown in the illustration top right.

The illustration below (left) shows a selection of components used in the automotive industry, including Drive Axles, Propeller Shafts, Suspension Links, Steering Shafts and Gear Levers.

Friction welding machines are used in a wide variety of industries for the friction welding of forgings, components illustrated above (right) include cluster gears, pinions, track rods, link rods and front wheel drive shafts.

Machines of all sizes, both single and double ended, are used for the friction welding of car and truck rear axle casings, drill pipes for the oil industry, and large hydraulic piston rods.

Thompson Welding Systems

Fig. 12.18 *Advantages of friction welding and examples of friction welded components in the small tool industry and other applications* (Courtesy of Thompson Welding Systems Ltd, UK)

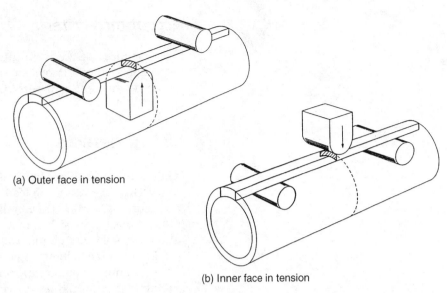

(a) Outer face in tension

(b) Inner face in tension

Fig. 12.19 *Typical bend tests for conventional friction welds in tube or pipe (see also the table beside Figure 12.20 for the recommended diameter of the former)*

and settings. The frequency of such tests will, of course, depend on the value of the component being tested and the cost of the testing procedure.

It is usual to take destructive test samples at the beginning and end of each shift or production run. Examples of bend testing arrangements for both tubular and solid-section components are given in Figures 12.19 and 12.20. Sections for macro- and micro-examination are taken parallel to the axis of the weldment in such a way that the critical section or sections of the weld can be examined.

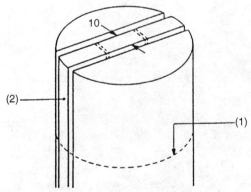

Dimensions in millimetres.

NOTE: (1) is the position of weld interface
(upset metal removed).
(2) is the test specimen for bending.

Diameter of former for bend testing selected materials

Material	Diameter of former
Carbon steel (0.25% C max.) Commercially pure aluminium Copper Titanium Austenitic stainless steel	$3t$
Carbon steel (over (0.25% C) Low alloy steel Brasses and bronzes Al–1% Mn alloys	$4t$
All other combinations (similar or dissimilar)	By agreement between the contracting parties

t is the specimen thickness or wall thickness.

Fig. 12.20 *Preparation of bend test specimens from a joint between solid components (after 'Recommendations for Friction Welding Butt Joints in Metals for High Duty Applications', TWI, Cambridge, UK, 1985)*

Hardness survey

Two series of hardness indentations should be taken, in lines parallel to the central axis of the weld. They should include parent metal(s) and heat affected zones on each side of the weld interface.

Acceptance criteria

If one or more of the weld imperfections listed is revealed by any of the inspection methods, then it should be established by reference to the product standard whether the imperfection(s) is sufficient to cause rejection.

List of defects

(a) Lack-of-bond: discontinuities present at the weld interface.
(b) Overheating: condition leading to detrimental grain structures.
(c) Inclusions: particles of metallic or non-metallic material in the weld zone.
(c) Cold-cracking: crack(s) in hardened weld zone formed after cooling.
(e) Intermetallic compound: can be formed along plane of interface in some dissimilar metal joints, giving low strength and ductility.
(f) Misalignment: one component offset with respect to the other.
(g) Bulging: deformation of weld region and adjacent parent material in pipes.

Safety

It can be seen from Figure 12.21 that friction welding machines for conventional applications resemble large lathes, in that one component is held in a chuck and rotated. Pressure is also used within the process and so any guarding or safety systems should be in accordance with those used for presses. Hazards to be protected against include rotating components, high noise levels and

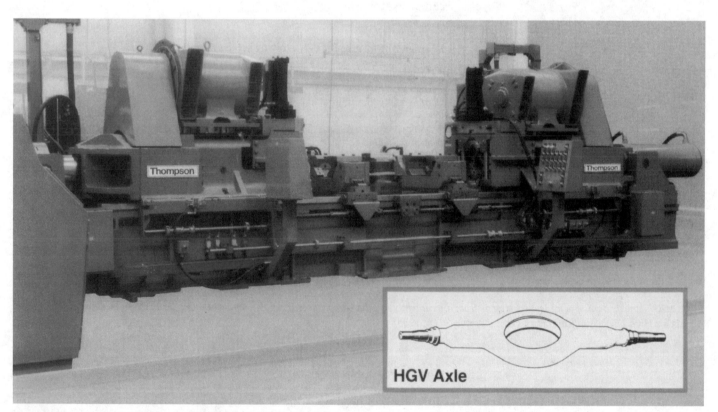

Fig. 12.21 *Type '12.5' friction welder used for welding HGV axles*
(*Photograph courtesy of Thompson Friction Welding Ltd, UK*)

flying particles. It is usual to have a guard that has to be swung in place before the machine can be operated. Eye and ear protection should be worn, together with normal machine shop overalls.

Care should be taken in removing the completed work from the machine. Gloves should be worn, for high temperatures are reached during welding, and components can remain hot for some time after welding has been completed.

References

1. B.A. Humphreys, *A Practical Guide to the Friction Welding Process*. Thompson Friction Welding Ltd

Further reading

AWS Welding Handbook, Vol. 2, 8th edn, Chapter 23 [ISBN 0–87171–354–3]
British Standard BS 6223: 1990 *Specification for friction welding of joints in metals*
Recommendations for Friction Welding Butt Joints in Metals for High Duty Applications. The Welding Institute, Cambridge, UK, 1975
ANSI/AWS C6.1–89, *Recommended practices for friction welding* (available from American Welding Society)

Explosion welding, ultrasonic welding, joining of plastics and Thermit welding

Explosion welding (EXW) and explosive welding, ISO 441

This process is used for cladding and welding tubes to tube plates. It has also been employed for the butt welding of pipes on pipeline construction.

Welding is produced by explosively forcing one plate (or component) against the one to which it is to be joined at an appropriate angle of incidence, known as the **angle of impact**.

The welding process has developed from the use of explosive forming, when it was found that the formed item was occasionally welded to the metal former. Bahrani and Crossland[1] published a paper on the subject in 1964.

Early experiments showed that 'jetting' was essential to the formation of a sound weld. The formation of jets was well known, since the principle had been used in the nose cones of armour piercing shells.

There are two methods used for the cladding of plates:

1. the inclined gap method (Figure 13.1a) and
2. the parallel gap method (Figure 13.1b).

The lower plate on which welding is to be carried out is placed on a steel anvil and is called the **target plate**; the plate that is to be joined to the target plate is called the **prime component** or **flyer plate**. In order to protect the flyer plate from surface damage and to control the explosive force, rubber, PVC or

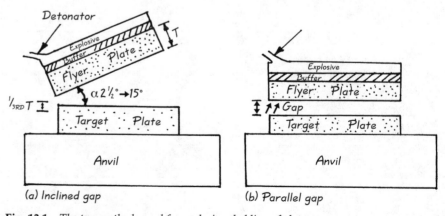

(a) Inclined gap   (b) Parallel gap

Fig. 13.1 *The two methods used for explosive cladding of plates*

similar material is placed in sheet form between the flyer plate and the explosive. This protective sheet is known as the **buffer** or **attenuator**.

With the inclined gap technique, various detonation velocities of explosives can be used, whereas with the parallel gap method the detonation velocity should be equal to or less than the velocity of sound in the metal being welded.

Theory of the welding operation

The angle between the 'flyer' and the target plates is selected so that a 'jet' is formed (Figures 13.2 and 13.3).

The 'jet' consists of a thin layer of metal stripped from the surfaces of both plates (Figure 13.3). This exposes the uncontaminated metal surfaces which are then welded in the high-pressure zone, known as the **stagnation point.**

Typically, the weld has a wavy interface, being essentially solid state with small pockets of melted jet material (composed of the two metals being joined), found on the front and back slopes of the waves.

It is thought also that some welding takes place by friction, as a result of the differences in velocity between the two plates: that is, flyer plate velocity $= V/\tan \beta$, target plate velocity $= V/\mathrm{Sin}\,\beta$, (where V = velocity imported to flyer plate, β is shown in Figure 13.3). If U = detonation velocity of explosive, then

$$\sin 0 = \frac{V}{U} \quad \text{(if plates parallel, } \frac{V}{\sin \beta} = U)$$

The smaller the angle β, the higher the jet velocity relative to the parent plate.

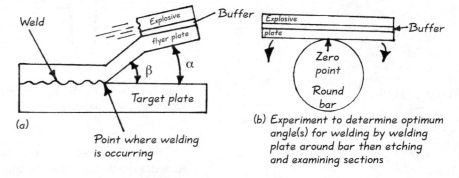

Fig. 13.2 *(a) Showing β-angle formation as inclined gap weld is being formed. (b) Experiment for determining optimum angles (after Bahrani and Crossland[2])*

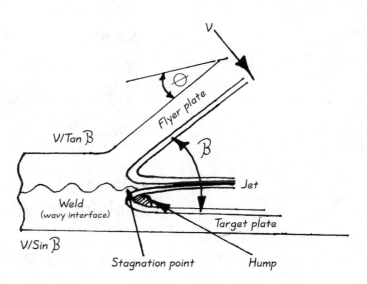

Fig. 13.3 *Illustrating jet formation*

$$\text{Velocity of jet} = V \left[\frac{1}{\sin \beta} + \frac{1}{\tan \beta} \right]$$

but, as the angle decreases, the mass of the jet diminishes:

$$\text{Mass of jet} = \frac{M}{2} [1 - \cos \beta]$$

There is the angle below which jetting does not occur.

Bahrani and Crossland[1] state that a 'hump' (Figure 13.3) should form in the metal ahead of the stagnation point, to ensure efficient removal of any dirty plate surface by the jet.

Applications

The technique is used for cladding plates, tube-to-tube plate welding, and tube closures with a plug (Figure 13.4).

The process has also been used for girth joints in pipe, where one pipe fits inside the other and a charge is placed both inside and out.

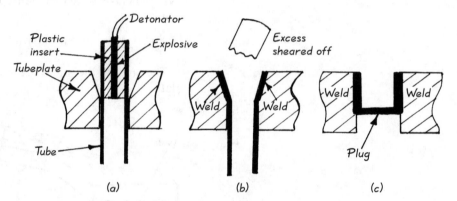

Fig. 13.4 *(a and b) Method of joining tube to tube-plate (YIMPACT welding) – Yorkshire Imperial Metals' technique. (c) Explosively welded plug for closing a tube*

The range of metals that can be welded is extensive but care has to be taken to avoid cracking when welding metals of low ductility.

Safety

Obvious safety considerations include correct procedures for handling and working with explosives. It is essential to clear the area involved and warn all within hearing distance of the impending loud retort. Various licences are required in most countries to permit the storage, handling and use of explosive substances.

Ultrasonic welding (USW), ISO 41 and welding/joining of plastics

This is another solid-state welding process, in which the weld is produced when the workpieces are clamped together between an anvil and a high-

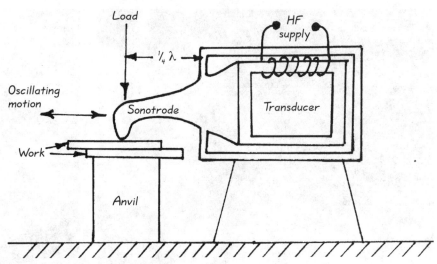

Fig. 13.5 *Ultrasonic welding (direct method)*

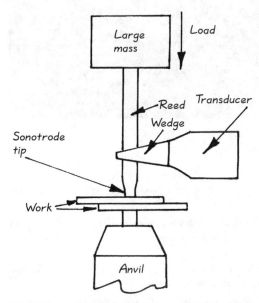

Fig. 13.6 *Ultrasonic welding coupled through a resonant bar (wedge-reed ultrasonic spot welding system)*

Fig. 13.7 *Sketch of a typical wedge-reed ultrasonic spot welding machine with movable head*

frequency vibrating probe (sonotrode). There are two main methods, the direct coupled, in which the transducer and sonotrode are connected directly (Figure 13.5), and the wedge-reed method, where the transducer is coupled through a resonant bar (Figure 13.6).

The sonotrode induces lateral vibration and local movement between the faying surfaces. This tends to disrupt any surface oxide films present and also raises the temperature, extending an area of plastic flow, and a solid-phase type of pressure weld is formed. Although the process raises the temperature at the point of welding, it is thought that in most instances, the maximum temperature is reached in less than half the material's melting point.

When welds have been studied, they show similarities to friction welds. The energy needed to produce an ultrasonic weld can be directly related to the hardness of the workpieces being joined and the thickness of the component actually in contact with the sonotrode tip. The analysis of much data has led to the following empirical relationship, which is accurate to a first approximation (AWS):

$$E = K(HT)^{3/2}$$

where: E = electrical energy
K = a constant for a given welding system
H = Vickers' hardness number
T = thickness of the sheet in contact with the sonotrode tip, mm (inches).

Variations of the process

Spot welding: Here, individual elliptical welds are produced on an ultrasonic spot welding machine (Figure 13.7) by the introduction of vibratory energy into the workpieces as they are clamped together under pressure between the sonotrode tip and the anvil. Such elliptical 'spots' can be overlapped if required, to produce a continuous weld.

Ring welding: In this variation, the sonotrode tip is shaped to give a weld of a particular shape, either circular, square, rectangular or oval. Shaping usually involves hollowing out of the sonotrode tip to the desired form.

Line welding: In this system a linear sonotrode tip can be employed to give a narrow linear weld, when it is oscillated parallel to the plane of the weld

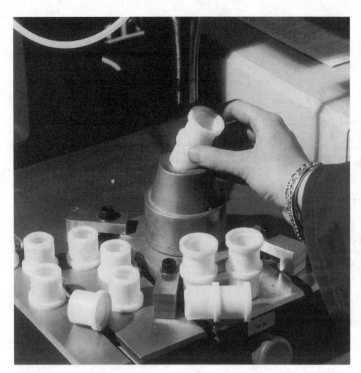

Fig. 13.8 *Ultrasonic welding of plastic components*
(*Photograph courtesy of TWI, Cambridge, UK*)

Fig. 13.9 *Microminiature welding*
(*Photograph courtesy of TWI, Cambridge, UK*)

interface and perpendicular to the weld line and the direction of the applied static force.

Continuous seam welding: The work is passed between a rotating wheel-shaped sonotrode and a roller-type or flat anvil. Such a system produces a continuous ultrasonic seam weld.

The process is suitable for joining materials other than metals and Figure 13.8 shows the ultrasonic welding of plastic components.

The largest growth area for ultrasonic welding is, however, in the field of microminiature welding and microjoining (Figures 13.9 and 13.10). In microelectric applications, the process is capable of welding very fine wires to the electrical components either direct or by the use of substrates (coatings).

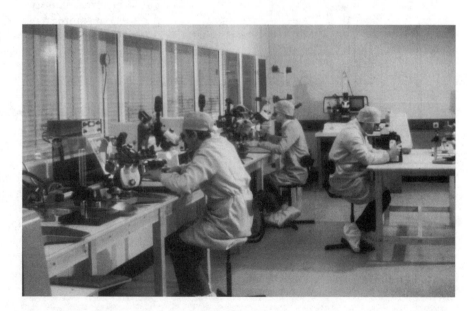

Fig. 13.10 *Microjoining laboratory clean room*
(*Photograph courtesy of TWI, Cambridge, UK*)

176

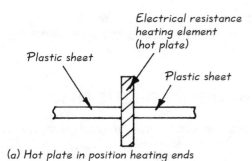

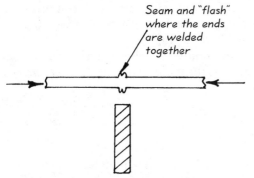

(a) Hot plate in position heating ends

Seam and "flash" where the ends are welded together

(b) Hot plate removed, ends pushed together to form weld

Fig. 13.11 *Hot plate welding of plastic sheet*

Thermosonic welding

As a direct development of ultrasonic welding and thermocompression welding (a deformation welding process where clean metal surfaces are exposed for welding by the mechanical disruption of surface films – usually carried out at temperatures ranging from 100 to 350°C (215 to 660°F)), **thermosonic welding** has emerged.

Thermosonic welding/bonding involve the use of ultrasonic welding and heated substrates. Interface temperatures of 100–200°C (215–400°F) are normally used. This has now become the most popular method for joining wire.

The heated tool welding of plastics

The general method for welding sheet is shown in Figure 13.11. Here, the edges of the plastic sheet are softened by contact with an electrically heated resistant strip. While the plastic is soft, the heated strip is removed and the two pieces of sheet are brought into contact. They are then allowed to cool down, while being held in position under pressure.

Accurate jigs and clamping devices are required in order to hold the sheets in alignment and also to provide adequate bonding pressure without buckling or flexing the material.

Welding pipe and tubing

Various procedures have been developed for the heated tool welding of pipe. Generally, for diameters above 50 mm (2 inches) some type of jig is required to support and align the tube or pipe during welding.

For some materials, the use of radiant heat rather than direct contact with a hot plate decreases oxidation and material degradation. However, when pipe ends to be joined are heated by direct contact with a hot plate, the oxidized surface that is formed is usually left on the plate and fresh, fused material remains at the ends for making the welded joint.

The heated tool or hot plate method of welding plastic pipe is shown in Figure 13.12.

Fig. 13.12 *Heated tool welding of plastic pipe*
(Photograph courtesy of TWI, Cambridge, UK)

Regardless of the method used to heat the ends, experience would indicate that unless a flash of extruded plastic is readily apparent, the strength of the joint will be questionable.

The hot-air welding of thermoplastic sheet

In this process (also known as hot-gas welding), the nitrogen or compressed air is heated, usually by an electric heating element in the torch, and the heated gas exits through a nozzle and is used for welding.

The thermoplastic sheets/components to be joined are suitably prepared and cleaned, and a thermoplastic welding rod of the same composition as the parent material is either fed in separately, as in oxy-acetylene welding, or through the torch/gun as shown in Figure 13.13.

The welding rod and weld bead are simultaneously heated by the hot-gas stream issuing from the welding gun. As the mating surfaces soften and fuse, they form a homogeneous bond.

Adhesive bonding

The use of adhesives for the joining of metals and other materials is increasing.

The term **adhesive** is a general one, which includes substances such as glue, cement, paste and gluey substances from plants such as mucilage.

Synthetic organic polymers are usually used to join metal assemblies.

Adhesives are available in a number of forms:

1. a range of thin to thick liquids;
2. pastes;
3. gum-resin type mastics;
4. powders
5. solids
6. films (supported or unsupported).

Consideration of the method used for the application of a particular type of adhesive should include an evaluation of the following points:

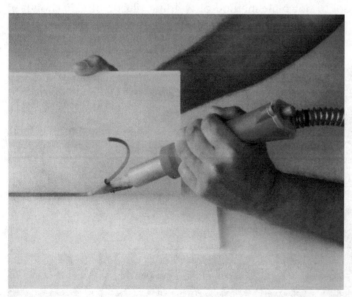

Fig. 13.13 *Hot-air welding of thermoplastic*
(*Photograph courtesy of AWS, Miami, USA*)

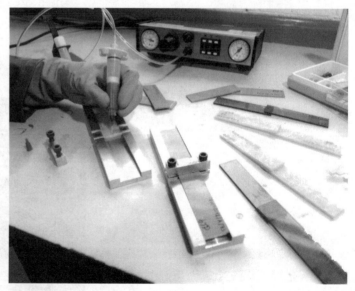

Fig. 13.14 *Manufacture of adhesively bonded lap-shear (plastic) specimens using pneumatically driven dispensing equipment*
(*Photograph courtesy of TWI, Cambridge, UK*)

1. the available forms in which the selected adhesive may be obtained;
2. method(s) available for applying the specific form of adhesive;
3. the requirement of the specific joint design(s) in question;
4. the rate of production required;
5. the cost of any equipment necessary.

Many applications use adhesive bonding in conjunction with other joining processes, for example spot or seam resistance welds, rivetting and bolting.

Very light but strong structures can be made by joining thin sheets together with adhesives and/or other methods. This is known as 'sandwich construction'.

Figure 13.14 shows the manufacture of adhesively bonded plastic lap specimens for testing purposes.

Thermit welding (TW), ISO 71 and its variation, Thermit pressure welding

The Thermit welding process employs a chemical reaction to provide the heat for welding. In one type of welding, molten metal runs around the stationary components, which are held in a mould, and a weld is formed as the molten metal solidifies. This is known as **Thermit fusion welding**. The other system uses the molten Thermit steel to heat up the ends of the components being joined to welding temperature and then pushes them together. This second method is known as **Thermit pressure welding.**

Thermit fusion welding is the process that is used mostly at present.

Thermit welding is based on aluminium exothermic reactions (an exothermic reaction is one that gives off heat). The reduction of metal oxides by means of finely divided aluminium was demonstrated by Sainte-Claire-Deville and by Wohler in the early Nineteenth Century. It was not until the 1890s, however, when Herault and Hall used the electric furnace method for producing cheaper aluminium, that commercial applications of the alumino-thermic method could be considered.

In 1894 Vautin discovered that extremely high temperature (approximately 3000°C) (5457.6°F) could be obtained by igniting mixtures of finely divided aluminium with iron oxide.

Dr Hans Goldschmidt advanced the process on a commercial scale by incorporating the important development of starting the reaction with a fuse. The earlier experiments had required the heating of the whole reaction charge in order to reach ignition temperature.

This method is known as the **Goldschmidt** or **Thermit process**. (The name Thermit is derived from the word thermite, which denotes a mixture of powdered aluminium and metallic oxide.)

Although the Goldschmidt principle is applied in the extraction of certain metals from their oxides, Thermit mixtures can be used for the welding of iron and steel or in incendiary bombs.

The Thermit welding process

The basis of the welding process is the chemical reaction between the finely powdered aluminium and the iron oxide.

The chemical equation describing this reaction is as follows:

$$Fe_2O_3 + 2Al \rightarrow Al_2O_3 + 2Fe + heat\ (4187\ J)$$

The approximate proportions of the mixture by weight are three parts of aluminium to ten parts of iron oxide.

The charge is placed in a crucible, which consists of a refractory magnesia-lined conical steel pot covered with a steel lid having a large hole in the centre. At the base of the crucible is the tapping device. Modern tapping thimbles are automatic and melt on completion of the reaction.

The reaction is started by an igniter, having a flash temperature of 200°C (392°F). These igniters must always be stored and transported separately from the Thermit charges.

Once started, the reaction is rapid, and continues until all the oxygen from the iron oxide has been transferred to the aluminium. This causes the temperature to rise to approximately 3000°C (5432°F) (neglecting heat losses). The time taken for this reaction to complete is usually about 30 seconds.

An important factor is the difference in specific gravity between the molten iron and the molten alumina, causing them to separate and the alumina slag to float to the surface. At this stage the metal will have at least 600°C (1137.6°F) superheat, and this allows it to be used for welding purposes.

The hot liquid Thermit steel is cast into a mould, welding the ends of the workpiece by flowing between and around them.

Moulds can either be pre-formed to the shape of the desired weld, as for railway lines, or one-off moulds can be manufactured for repair welding using the 'lost-wax' process.

It is usual to pre-heat the mould and the ends of the work to be joined, before pouring in the hot metal. This ensures that the mould is dry and that full advantage can be taken of the superheat for welding purposes.

The first patent for Thermit welding was granted in 1897. By the early 1900s the process was well established in Germany, France and the United Kingdom. At this time it was mainly used for smelting, rail welding and repairs to heavy machine parts. These still remain the main applications of the process.

In 1909 Dr L.A. Groth published a work illustrating the applications of the Thermit process to large marine repairs.

Figure 13.15 shows the general mould and crucible arrangements for Thermit welding.

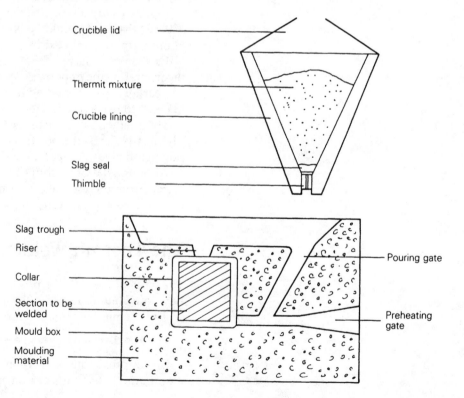

Fig. 13.15 *General arrangement for Thermit welding ('Thermit' is a registered trade name)*

180

Long-life crucibles

In the UK, Europe and Southern Africa, crucible linings were made from dead burned magnesite refractory and could be used for eight to ten reactions before needing replacement.

Increased fuel costs have made the dead burning of the magnesite very expensive and the Thermit company now use a refractory made from a special form of alumina. These long-life crucibles are pressed into a thin mild steel shell and give a life of approximately 35 reactions. The whole crucible is then discarded and replaced. The magnesite-type lining required the removal of the slag coating after each reaction to prevent residual Al levels in the steel becoming too high. The long-life lining is totally inert and only requires the removal of the slag once during the life of the crucible. This helps to increase the life of the lining because mechanical damage is not inflicted by frequent slag removal.

Automatic tapping thimbles

In conjunction with the development of the long-life crucible, automatic tapping thimbles are employed, which release the molten steel into the mould at the appropriate time (a **thimble** is a plug fitted into the crucible tapping hole; automatic tapping thimbles melt at the desired tapping temperature).

This relieves the welder of the responsibility of deciding when to release the steel and also makes the process much safer, because there is no need for welders to stand near the crucible while it contains superheated molten steel.

Figures 13.16 and 13.17 show the process in use welding railway lines.

Fig. 13.16 *Welding railway lines by the Thermit process. Here the reaction has been started*

Fig. 13.17 *Molten Thermit steel pouring into the mould to form the weld*

Reference

1. A.S. Bahrani and B. Crossland, Explosive welding and cladding. In *Inst. Mech. Eng. Proc*, Vol. 179, Part 1, No. 7, pp 264–81, 1964–65

Further reading

AWS Welding Handbook, Vol. 2, 8th edn [ISBN 0–87171–354–3]

Some notes on brazing processes (General ISO Category 9)

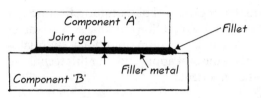

Fig. 14.1 *A typical brazed joint filled by capillary action*

There are many brazing and soldering processes and a list complete with ISO numbers is given on page 129 of *Basic Welding*. Soldering uses filler metals that melt below 450°C (840°F). **Brazing** is the joining of metals without melting them, using a filler metal which has a melting point above 450°C (840°F) but below that of the parent metal, and which fills the joint by **capillarity** (capillary action) (Figures 14.1 and 14.9 – see also Chapter 17).

Advantages

1. Low distortion
2. Can be carried out, in most instances, without affecting parent metal properties. Because of this, post-brazing heat-treatments are rarely required.
3. Ease of automation allows for semi-skilled or unskilled labour to be used.
4. If correctly designed, the properties of the joints can be made to suit most applications.
5. A wide range of brazing techniques (mainly variations of heating method) is available.

Some typical examples of situations where brazing might be used

1. Where the parts to be joined cannot be conveniently welded, owing to the lack of compatibility between the parent metals.
2. On application where a soft soldered joint would not be strong enough.
3. Where distortion might be a problem.
4. In a situation where parent metal properties would be affected by the heat of fusion welding.
5. Where the joint would be inaccessible by other processes.

Brazing techniques

1. Torch heating methods (flame brazing), ISO 912

Air and natural/coal gas torches can be used up to 800°C (1472°F) for silver soldering/brazing, but are not suitable for AlSi alloys and have too slow a heating rate for large components in alloys and metals of high conductivity, such as copper and aluminium.

Natural/coal gas used with oxygen gives a higher-temperature flame and slightly higher heating rates.

Oxy-acetylene torches give a more concentrated heat source and are therefore suitable for carrying out brazing operations on copper and aluminium. However, care must be taken not to overheat thin-gauge components.

2. Induction heating, ISO 916

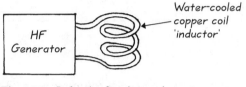

Fig. 14.2 *Induction brazing equipment*

The inductor (Figure 14.2) is placed close to, but not in contact with, the parts to be brazed together. In most cases the coil surrounds the component. A high-frequency (HF) current in the inductor induces a heating current in the work. The brazing cycle can be precisely controlled using timing equipment built into the HF generator.

Advantages of this process are that it gives a rapid, uniform heating rate (there is, therefore, little distortion and discoloration) and the system can be used with an inert atmosphere or even under vacuum conditions. The quality production of identical components is the best application for this technique. It is mostly used for brazing steel components.

3. Furnace brazing methods, ISO 913

These can be either batch or continuous, controlled-atmosphere or vacuum. Figure 14.3 shows the basic principle of a controlled-atmosphere 'retort-type' batch furnace and Figure 14.4 the continuous brazing furnace system.

When it is required to braze under vacuum or closely controlled atmosphere conditions, the work is sealed in a retort which is heated in a furnace. By using a moving hearth conveyor system, a high output of brazed components can be obtained.

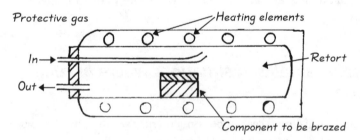

Fig. 14.3 *Sketch of controlled-atmosphere retort-type furnace*

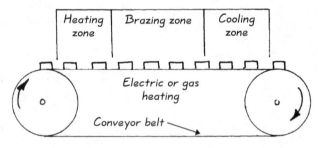

Fig. 14.4 *Principle of a continuous brazing furnace*

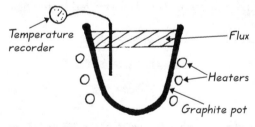

Fig. 14.5 *Dip brazing*

4. Dip brazing, ISO 914

Figure 14.5 illustrates the equipment used with this method.

The work is pre-assembled, pre-heated and dipped into the pot slowly to allow the flux to penetrate the joint. It is then held under the molten filler metal for a pre-arranged time before being lifted out. The assemblies must be carefully wired up to prevent them falling apart in the bath.

The flux has to be replaced frequently, because of the effects of oxidation.

5. Salt bath brazing (salt bath soldering, ISO 945)

This process is similar to dip brazing but has just molten flux in the crucible. In the technique, the filler metal is pre-placed in the joint and the work is heated by immersion in the molten salt bath, which brings it up to brazing temperature.

This method can be applied to components of intricate shape, where many joints are to be made simultaneously, and the components may be made from low carbon steel, copper and its alloys, or aluminium and its alloys. Sodium nitrate can be used for steel and copper but salts with a fluxing action such as chlorides and fluorides are used for aluminium.

6. Resistance brazing, ISO 918

With this process, heating is achieved by passing a high current at low voltage through the joint in which the filler is pre-placed (Figure 14.6).

Standard resistance welding machines may be used. The heating is extremely rapid and localized. One of the main applications of the technique is for making copper-to-copper electrical connections.

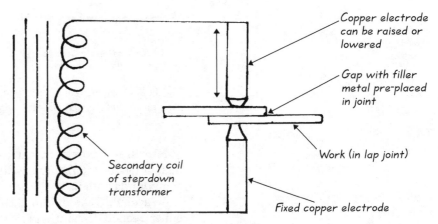

Fig. 14.6 *Resistance brazing method*

7. Exothermic brazing, ISO 93 – other brazing processes

This system (Figure 14.7) uses a chemical reaction to provide the heat for brazing. It is used for joining small tubes and pipes.

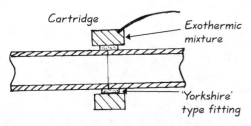

Fig. 14.7 *Method for exothermic brazing*

8. Arc braze welding, ISO 72

With this method, either a twin-carbon or a single carbon-arc torch is employed and the filler metal is supplied in the form of a flux-coated rod (see Chapter 6 on carbon-arc welding).

Principles involved in brazing

The nature of the bond is complex, varying from true metallic bonding to Van der Waals' force (an attractive force existing between atoms or molecules of all substances, arising as a result of electrons in neighbouring atoms or molecules moving in sympathy with one another).

The most important consideration as regards strength is the continuity of the bond, which can vary from 0 to 100 per cent, as it is dependent on the ability of the brass metal to **wet** the surface of the gap.

Wetting

In general, liquid braze metals will not wet, clean, unfilmed surfaces unless:

(a) the liquid metal is intersoluble with the parent metal;
(b) the liquid and solid metals react to form an intermetallic compound or compounds (see also 'Fluxes' below and Figure 14.8).

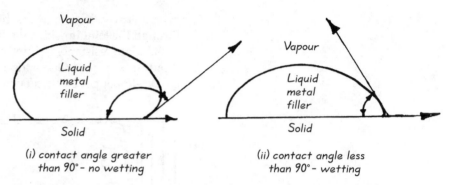

(i) contact angle greater than 90° – no wetting

(ii) contact angle less than 90° – wetting

Fig. 14.8 *Showing that a contact angle of less than 90 degrees measured between the solid parent plate and the liquid filler metal usually indicates a positive wetting characteristic*

From this, it is evident that in order to achieve a satisfactory bond, the filler metal must be intersoluble with the base metal or able to form intermetallic compounds with it.

To achieve clean unfilmed surfaces at high temperatures, the surfaces are either cleaned first and kept clean, or the surfaces are cleaned during the actual brazing operation.

Fluxes

Fluxes are the most common method of ensuring good wetting (achieving the spreading and adhering of a thin continuous layer of liquid filler metal and / or flux on a solid surface of base metal), generally achieved by dissolving oxides. However, in some cases, other properties can be attributed to the flux, such as the deposition of metals on to the surface of the parent metal and reaction with the surface, thus 'preparing' it chemically.

In soldering, with zinc chloride flux, zinc is deposited on iron surfaces, giving a 'tinning' effect.

The flux also has a blanketing effect on the surface, keeping out oxygen.

Capillary flow and joint clearance

The clearances (or distances between plates), determined by the formula shown in Figure 14.9, give good agreement with values found satisfactory in practice. Clearances are usually set slightly larger than those predicted, because too small a clearance can have the affect of altering the braze metal by causing a high level of alloying with the base metal, and this can cause the braze metal to solidify prematurely.

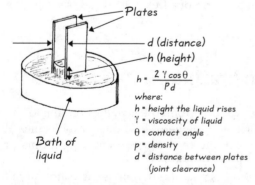

$$h = \frac{2\,\gamma\cos\theta}{p\,d}$$

where:
h = height the liquid rises
γ = viscoscity of liquid
θ = contact angle
p = density
d = distance between plates (joint clearance)

Fig. 14.9 *Principle of capillary flow and joint clearance*

Filler metals

These are covered by the specification standards of each country, being British/European, American and/or combinations that fit into international categories.

Generally they are made from pure metals or eutectic compositions. This is a deliberate choice to ensure good flow in the joint and achieve rapid solidification over a narrow temperature range.

Non-eutectic alloys are sometimes specially selected for 'wide-gap' brazing.

A typical filler rod suitable for brazing copper or mild steel is the silicon-bronze type which contains Cu 57–63 per cent, Si 0.20–0.50 per cent, Sn optional up to 0.50 per cent max., and the balance Zn. Such a rod would have a melting point in the range of 875–895°C (1632–1668°F).

One type of silver solder typically contains: Ag 43 per cent, Cu 37 per cent and Zn 20 per cent, having a melting point of around 698–788°C (1303–1468°F).

Depending on the make and standard, filler metal manufacturers state the exact compositions and melting points on the packaging and will always give help and advice on the correct use of their products.

Use of fluxes

There are two main types of flux:

1. **borax** based – for temperatures above 750°C (1400°F);
2. **fluoride** based – for temperatures below 750°C (1400°F) (particularly for silver brazing alloys).

It is always important to remember the following two points:

1. a flux will not clean to the extent that work need not be initially cleaned by some other means;
2. fluxes are often corrosive and should be removed after brazing. Borax-based fluxes are less corrosive than fluoride types.

Preparation for brazing

Cleaning

Mechanical cleaning by abrasion will usually be necessary on large components, however it is usually less effective and more costly than chemical cleaning when large numbers of small components are involved in the production process.

Chemical cleaning

The use of chemicals will, of course, be subject to the appropriate safety measures being taken in accordance with regulations and the control of hazardous substances.

Degreasing is usually the first operation, using one of the following:

1. solvent cleaning by the use of petroleum solvents or chlorinated hydrocarbons;
2. vapour degreasing using stabilized trichloroethylene, carbon tetrachloride or acetone.

Scale and oxide removal can then take place by acid cleaning or pickling. Salt bath pickling can also be used.

Typical examples are: iron and steel – 10 per cent sulphuric acid; brass – 10 per cent sulphuric acid for 10 minutes' maximum immersion; stainless steel – 7 per cent nitric acid, and 21 per cent sulphuric acid, in water.

Cleaning after brazing

Borax-type fluxes dissolve to some extent in boiling water, but pickling in cold 5 per cent sulphuric acid is more effective, followed by a water rinse.

Fluoride type fluxes are treated in a similar manner and wire brushing eases removal.

The fluxes used for brazing aluminium and its alloys are particularly corrosive and should be removed as follows:

1. wash in boiling water for 10 minutes;
2. rinse in hot water;
3. wash in 10 per cent nitric acid at 65°C (160°F) for 20 minutes;
4. rinse in hot water;
5. inspect and, if free from visible signs of flux, wash in 10 per cent nitric acid and 2 per cent sodium dichromate solution;
6. rinse in hot water, drain and dry.

Joint design for brazing

Type of joint

Where possible, lap-type joints should be employed, to give a good surface area for the braze, since the brazed metal and joint can withstand loading better in shear.

The examples shown in Figure 14.10 indicate that a good joint design is one that has a large area of contact and, for this reason, it is usually required

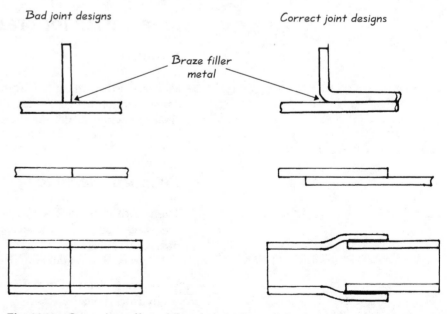

Fig. 14.10 *Comparison of brazed joint designs. Correct joint designs give increased surface areas for brazing to take place*

that the joint length should be three times the thickness: $L = 3 \times t$, where $t =$ thickness of thinnest section.

Components may have the filler metal pre-placed, but venting is essential with components and joint configurations in which air cannot easily escape.

Joints where the gap varies owing to poor location and fit-up can tend to unsoundness where the filler metal is thickest.

The minimum quantity of filler metal to fill the joint completely is optimal, and a brazed joint when completed should only have a small fillet at the points of entry and exit. Excessive heating rates, if localized, can result in faulty brazing and when high carbon steels are being brazed, decarburization of the surface should be avoided.

Brazing procedure

A typical brazing procedure should contain the following:

1. The Brazing Procedure, Number and any title of component/component name.
2. Joint detail(s), with measurements and sketch showing dimensions.
3. Sizes and thickness of base metals.
4. Composition (chemical) of base metals.
5. Pre-cleaning treatment(s) required.
6. Filler metal specification – size and type.
7. Flux type.
8. Brazing process.
9. Post-braze heat-treatment.
10. Post-braze cleaning.
11. Inspection.
12. Date.

15

Introduction to welding metallurgy

Fig. 15.1 *If every atom in your hand was the size of your fingernail, then your hand would be big enough to hold the world!*

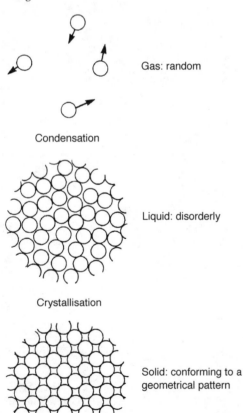

Gas: random

Condensation

Liquid: disorderly

Crystallisation

Solid: conforming to a geometrical pattern

Fig. 15.2 *The arrangement of atoms in the three states of matter*

Atoms and molecules

Everything is made from atoms and the atom is the smallest part of an element that can exist chemically.

It is very difficult to comprehend just how small individual atoms are. A copper atom, for example, measures approximately 1.92 Angstrom units, that is 1.92×10^{-8} cm.

Another approximation that can give some indication of relative size, is to say that if one atom was the size of your fingernail, then your hand would be big enough to hold the world (Figure 15.1)!

The atom is not solid, it is mainly space, rather like a miniature solar system. It consists of a number of negatively charged particles called **electrons**, which surround a small, dense **nucleus** of positive charge (see *Basic Welding*, page 115).

How the atoms are packed together determines what we call the three states of matter – that is, whether a substance is solid, liquid or a gas (Figure 15.2). We can change a substance from one state to another by either adding or taking away heat, for example, ice to water to steam by adding heat, and steam to water to ice by taking heat away.

Atoms build up into a fixed pattern to form crystals as a liquid solidifies. There are therefore two types of substances: those with no fixed crystalline pattern, where the atoms are completely disordered, such as liquids (known as **amorphous** substances); and those with a fixed crystalline pattern, such as solids (known as **crystalline**).

Many crystalline formations, such as the ones produced by minerals, are large enough to be seen with the naked eye. The crystalline structure of metals is, however, so small that it can only be seen if viewed under high magnification using a metallurgical microscope. Crystals consist of plane faces (or facets) arranged in a symmetrical pattern. The geometrical pattern which the vast number of atoms build up within the crystal determines the geometrical shape of the crystal, and this varies from one substance to another (Figure 15.3).

Close packing of atoms in a fixed geometrical pattern makes a solid rigid and more difficult to distort. Therefore, if a solid is stressed (below its elastic limit) it will distort, but any distortion will be temporary and the solid will return to its original shape once the stress has been removed.

Some substances that we consider as solids are in fact amorphous. Pitch is a very viscous liquid and if we place a ball of pitch into a container it will, over time, spread to cover the base of the container. Some metals have amorphous tendencies and lead can, over time, creep or flow. Creep or flow in metals that have amorphous tendencies will take place faster with increased temperature.

When a pure liquid starts to solidify, it does so at a set temperature, known as the **freezing point**. Figure 15.4 shows a cooling curve for a pure metal. It

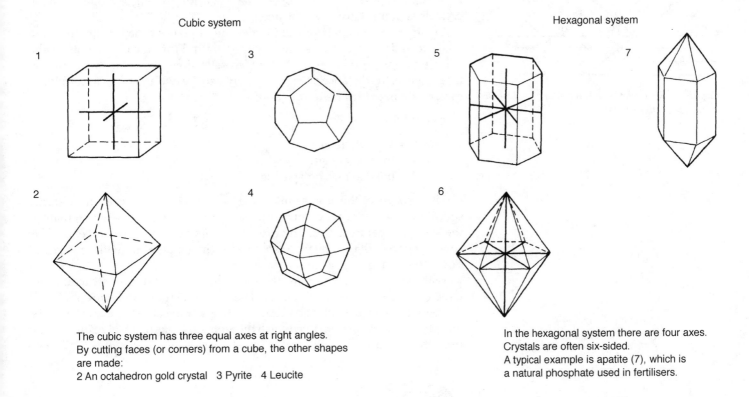

Cubic system

Hexagonal system

The cubic system has three equal axes at right angles. By cutting faces (or corners) from a cube, the other shapes are made:
2 An octahedron gold crystal   3 Pyrite   4 Leucite

In the hexagonal system there are four axes. Crystals are often six-sided. A typical example is apatite (7), which is a natural phosphate used in fertilisers.

Fig. 15.3 *The shapes of crystals are based on their axes of symmetry for the rows of atoms (called space lattices). The faces of the crystals are therefore always aligned with the axes of symmetry*

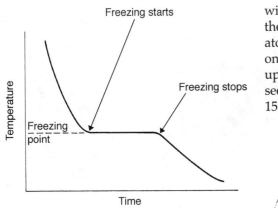

Fig. 15.4 *Typical cooling curve for a pure metal*

will be noticed that the temperature drops steadily over a period of time, once the heat is taken away, until the freezing point is reached. At this point, the atoms begin to arrange themselves into set geometrical patterns, depending on the material. These atomic patterns are called **space lattices** and they build up to form the crystals of the solid. These space lattices are far too small to be seen, requiring a hypothetical magnification of around 35 million times (Figure 15.5).

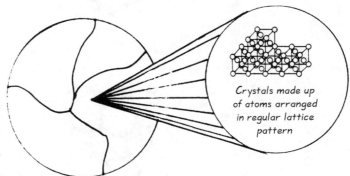

Crystals made up of atoms arranged in regular lattice pattern

Fig. 15.5 *Showing a regular pattern built up from space lattices. This is what we would expect to see if a high enough magnification were possible*

Material surface magnified 1000 times (after polishing and etching) in order to show grain boundaries

Hypothetical magnification of around 35 million times (if such a magnification were possible, the arrangement of atoms could be seen)

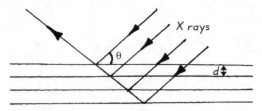

Fig. 15.6 *Reflections of X-rays from atomic planes in a crystal (after Bragg)*

The method developed by Sir William and Sir Lawrence Bragg, using X-rays to study crystals, has proved that the arrangement of atoms in metal crystals is in this lattice-type structure.

Using the X-ray spectrometer, the different planes or layers of atoms are made to reflect X-rays of known wave-length. From measurements of the glancing angle, θ (Figure 15.6) at which reflection takes place without any interference, the distance or 'spacing' between the two adjacent atomic planes can be calculated from the equation:

$$\eta\lambda = 2d \sin \theta$$

where: η = any whole number
 λ = the wavelength
 d = the distance or 'spacing'.

This process of the atoms building up to form the crystals as the material solidifies gives off heat (latent heat), and so the cooling 'curve' will remain at the same temperature until the process is complete and freezing ends. The cooling curve will then begin to show a loss in temperature with time again, down to room temperature.

With only a few exceptions, there are three main types of structure into which metallic elements crystallize. These are illustrated in Figure 15.7.

These three space lattice patterns show a graphical representation of the orderly geometric pattern into which the atoms arrange themselves on cooling from the liquid to the solid state. The shapes of these structures tend to be

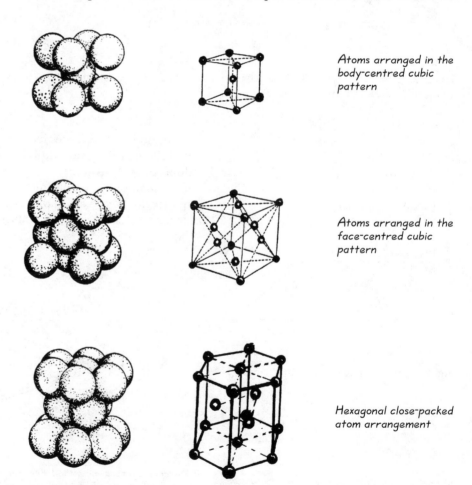

Atoms arranged in the body-centred cubic pattern

Atoms arranged in the face-centred cubic pattern

Hexagonal close-packed atom arrangement

Fig. 15.7 *The three main types of space-lattice structure into which metallic materials tend to arrange themselves*

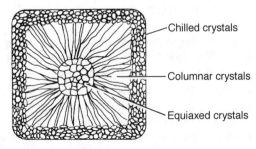

A metallic dendrite fir-tree crystal

Dendritic crystals taking shape and growing at many centres (nuclei) throughout the melt

Cooling

Grain boundaries forming as crystal growth stops when arms meet

Grain boundaries

In a pure metal, when the process is complete, all that can be seen through a microscope is the grain boundary outline. If the dendrite can be seen, it indicates impurities trapped between the arms, thus highlighting the shape

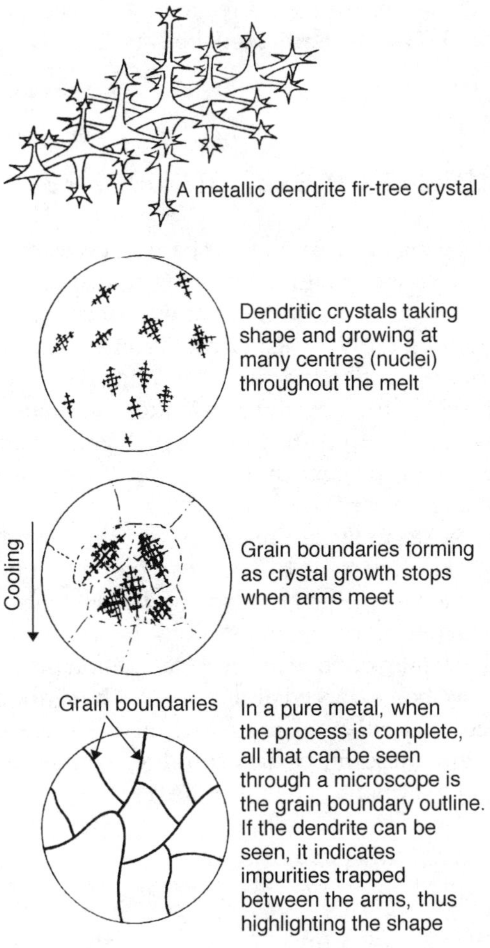

Fig. 15.8 *The stages of crystallization as a metal solidifies*

Chilled crystals

Columnar crystals

Equiaxed crystals

Fig. 15.9 *An ingot of a pure metal showing the crystal structure*

illustrated more easily if the atoms are shown to be held in position by imaginary lines.

The first type, the **body-centred**, has nine atoms, one at each corner of the cube and one at the centre. This pattern is found in such metals as iron, molybdenum, chromium, tungsten and vanadium. The atoms in an 0.83 per cent carbon steel are in the body-centred cubic form at temperatures below 723°C (1333°F). Above this temperature (the **critical temperature**), the atoms assume the second pattern, the **face-centred cubic structure**.

As its name implies, the face-centred cubic structure has an atom in the centre of each face. Other typical metals that have this lattice pattern are aluminium, copper, nickel, lead, platinum, gold and silver.

The third type is called the **close-packed hexagonal** and the structure is more tightly packed than the other two. Cadmium, bismuth, magnesium, cobolt, titanium and zinc are among the metals having this type of crystalline structure.

Generally, metals with the face-centred lattice structure are ductile, plastic and workable, and metals with the body-centred structure have higher strength with lower cold working properties. Metals with the close-packed hexagonal lattice lack plasticity and cannot be cold worked.

In a pure metal, these individual crystals will start to form at many centres (or nuclei) throughout the cooling material. The process starts with one unit of the crystal lattice and rapidly builds up with the addition of atoms and lattice structures into a crystal framework called a 'dendrite' (see Figure 15.8).

From the dendrite, arms of space lattices start to grow in other directions, giving an appearance similar to a fir-tree with the growth of its twigs and branches, and the dendrite is sometimes called a fir-tree crystal because of this.

As the temperature continues to fall, the dendrite increases in size until its arms come into contact with other similar structures, at which point growth stops in that direction and solidification begins to take place between the arms of the dendrite. This solidification continues until the crystal is formed (Figure 15.8) without a trace of the original dendrite, unless impurities have been trapped between the dendrite arms, or if the metal has cooled so rapidly that metal has not fed completely into the spaces between the dendrites, causing shrinkage cavities and allowing the outline of the dendrite to be visible.

Effect of the rate of cooling on grain growth

The rate at which a metal cools will affect the number of nuclei formed, with slow cooling promoting relatively few nuclei. The result is that the crystals (or grains) will be large enough to be visible without the aid of a microscope. On the other hand, fast cooling will promote many nuclei and a small fine grain structure.

In a large ingot of cast metal or a large weld, the crystal sizes can vary considerably from the outer edges to the centre (see Figure 15.9). This is due to variation in the temperature gradient, because as an ingot (or large weld) solidifies, heat will be transferred from the metal to the mould (or heavy parent material).

When the metal first makes contact with a cold mould, there will be a chilling effect and the formation of many small crystals. As the mould warms up, the direction of cooling will be mainly inwards and extremely elongated columnar crystals are formed. As heat is lost still further, the metal in the centre begins to form its own nuclei and a third type of crystal is formed, showing no preference for directional growth. These centre crystals are much larger than the chilled crystals at the outer edges, as a result of a slower rate of cooling. Such central crystals are known as **equi-axed**, because they have no preference for directional cooling.

Effect of grain structure on metal properties

The deformation of grains can have a big effect on the mechanical properties of a metal. Deformation can take place when a metal is being cold worked, during operations such as hammering, pressing, rolling, drawing or bending. An example of how crystals can become greatly elongated and strain hardened is shown in Figure 15.10, where a plate is undergoing a cold rolling operation.

This undesirable structure can be rectified by the application of enough heat to produce the growth of new equi-axed crystal grains which will return the structure to an unstrained state, having the same properties that it possessed before the cold working that had taken place (Figure 15.11).

The temperature at which these new grains form is called the **recrystallization temperature**. It is important to hold the metal at this temperature just long enough for the new grains to form, and then control the cooling rate so that the structure consists of the refined equi-axed crystals. If the metal is subjected to excessive heating above its recrystallization point, grain growth can take place. These weaker enlarged grains can also be formed if too slow a cooling rate is employed. Such enlarged grains can cause a decrease in ductility and tensile strength, but these properties could, of course, be restored by correct recrystallization treatment (see Figure 15.11).

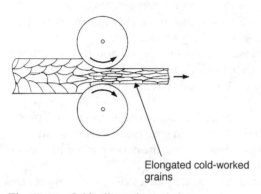

Fig. 15.10 *Cold rolling of a steel plate*

Elongated cold-worked grains

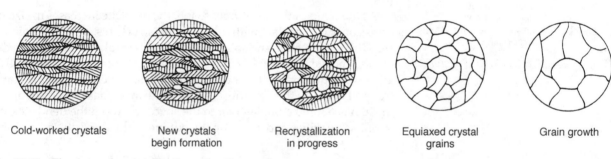

| Cold-worked crystals | New crystals begin formation | Recrystallization in progress | Equiaxed crystal grains | Grain growth |

Fig. 15.11 *The stages of recrystallization and grain growth*

Alloying and thermal equilibrium diagrams

Thermal equilibrium diagrams are very important in metallurgy and welding engineering, as they indicate the type of structure we should expect to find in an alloy of two elements at a specific temperature and composition **under equilibrium conditions**.

In practice, a true equilibrium condition is not usually achieved. However, the diagrams still provide a guide to understanding the structural changes that take place within alloys and the consequences of these changes.

Before looking at how these diagrams are made, it is important first to understand the concepts involved in the process of alloying and have an appreciation of the mechanism of solution, precipitation and saturation.

Many substances can be dissolved in water and as the temperature rises, so will the amount capable of being dissolved, making what is known as a **solution**. The water is the **solvent** and the substance dissolved the **solute**.

If sugar is added to water, the water will dissolve only a small amount and then **saturation** will be achieved (no more sugar will dissolve – it will only settle to the bottom of the container), but if **heat** is now applied to the solution, the sugar will gradually be absorbed as the temperature rises and even more sugar may be added and dissolved. At this point the solution has reached a **supersaturated** state, because as it cools down, a large amount of sugar will be **precipitated** out (come out of the solution). This precipitation will continue down to room temperature and the point where the solution reaches **equilibrium** once more, having the original amount of sugar in solution with the water.

With metals, solution and precipitation can take place while in the **solid state** and therefore, the solution of one metal in another can be described as a **solid solution**.

The rate of cooling in metals is extremely important, because in the solid state, precipitation does not occur as easily in the liquid state and in most cases, it is the precipitation that the metallurgist/welding engineer is trying to control during the heat-treatment and/or welding operations.

Pure metals have their uses in applications where electrical conductivity, ductility or corrosion resistance are required.

However, by alloying metals, material with quite different characteristics can often be formed.

Useful alloys are usually formed when one metal is capable of being dissolved in another. Often a material having advantages of one or more of the following characteristics can be formed: increased hardness, increased strength, improved corrosion resistance, reduction in weight and lower costs.

Sometimes the added material does not dissolve, as in the case of 'free cutting steels', which contain additions of lead or in some cases sulphur.

In this instance, the additions do not dissolve, but remain suspended in the material, causing a 'discontinuous chip' formation, giving easier machining.

Depending on the materials, one of three conditions can result when alloys are made and solidification takes place:

1. The solubility ends and the two metals that were mutually soluble as liquids become totally insoluble in the solid state, separating out as particles of the two pure metals.
2. The solubility in the liquid state is completely or partially retained in the solid state to form a solid solution.
3. The two metals react chemically, to form an intermetallic compound.

Any one of the above is referred to as a **phase**, being either:

(a) pure metal
(b) solid solution or
(c) intermetallic compound of two or more metals provided it exists within the structure of the alloy as a separate structure.

In a solid **binary** (an alloy made from two metals), not more than two phases can exist.

Eutectic

The word 'eutectic' is derived from the Greek *eutektikos*, which means 'able to be easily melted'.

When two metals which are completely soluble in the liquid state, but completely **insoluble** in the solid state, solidify, they do so by crystallizing

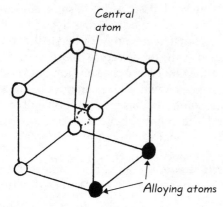

Central atom

Alloying atoms

(a) Substitutional system

If each lattice structure replacement is the same throughout the structure, then the system is known as 'ordered'.

If the replacement is random throughout the structure, then the system is known as 'disordered'.

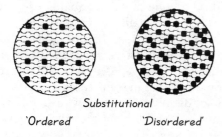

Substitutional

'Ordered'   'Disordered'

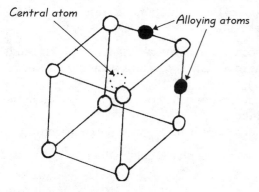

Central atom

Alloying atoms

(b) Interstitial system

The alloying atom inserts itself in the space between the atoms of the lattice structure.

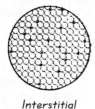

Interstitial

Fig. 15.12 *The ways in which solid solutions can occur: (a) substitutional; (b) interstitial systems*

out as alternate layers of the two metals, giving a laminated structure similar in appearance to plywood.

This structure, formed from the liquid to the solid state, is referred to as the **eutectic**. In alloys of lead and tin, for example, the eutectic composition (alloy of the lowest melting point) is 63 per cent Sn/37 per cent Pb, melting at 181°C (356°F).

Eutectoid

This is a similar structure to the eutectic but it is formed in the solid metal as it cools down towards room temperature (that is, it is precipitated at constant temperature from solid solutions). Pearlite is an example of the eutectoid structure formed under equilibrium conditions in steel (see discussion of the iron–carbon equilibrium diagram in the section on 'Metallurgy of a weld in iron', later in this chapter). However, if a steel is quenched from above the eutectoid temperature of 723°C (1335°F) it can prevent the formation of a pearlite structure, and martensite, troosite or sorbite structures can result, depending on the severity of the quench.

Solid-solution structures

As mentioned previously, all solids form from atoms arranging themselves into a set 'lattice' formation, the patterns of which are many and varied, including cubic, hexagonal and diamond shapes.

Generally, metals tend to arrange themselves into one of the three shapes shown in Figure 15.7.

The space or distance between the atoms varies, depending on the metal. Lead, for example, is extremely dense and the gap between atoms is therefore smaller than for some other metals.

When solid-solution alloys are formed, this means that two metals that were mutually soluble in the liquid state have remained dissolved in each other after crystallization.

If such a structure is viewed through a microscope, the parent metals would not be seen as separate entities. Only crystals consisting of a homogeneous solid solution of one metal in the other would be visible and, if the metals were completely soluble in each other, the visible crystals would be of one type.

For one metal to be able to dissolve into another and form a solid solution its atoms must be able to fit into the crystal lattice of the other metal, and there are two main systems by which this operation takes place.

The first system is known as **substitutional** (from the Latin *substituere, substitutum* – the replacing of one quantity by another). In this system the atoms of the alloying material actually replace some of the atoms in the original space lattice structure (Figure 15.12a). If the replacement is the same throughout the structure (same atom or atoms replaced in each lattice) then the system is known as 'ordered'. If, however, the replacement is random throughout the structure, then the system is known as 'disordered'.

The other system is known as **interstitial** – (from the Latin *intersitium,* between; *sistere, statum,* to stand). Interstitial solid solutions can only be formed if the atoms of the added elements are small enough to fit between (into the interstices or spaces) the atoms of the crystal lattice of the parent material (Figure 15.12b).

In substitutional alloys the high-melting-point metal will freeze first, forming the backbone or core of the dendrite (Figure 15.13(i)). As the temperature drops further, the alloyed solutions and then the pure solute metal freezes (Figure 15.13(ii)). By annealing the alloy, the high melting point metal is allowed to diffuse slowly throughout the grain (or crystal), giving a more homogeneous solid solution (Figure 15.13(iii)).

In an interstitial solid solution, diffusion can take place relatively easily, as the solute atoms will be smaller and able to move through the crystal lattice structure of the parent material. Diffusion is speeded up with the substitutional system if vacancies (spaces empty of atoms) exist in the structure, allowing passage of the substitutional atoms.

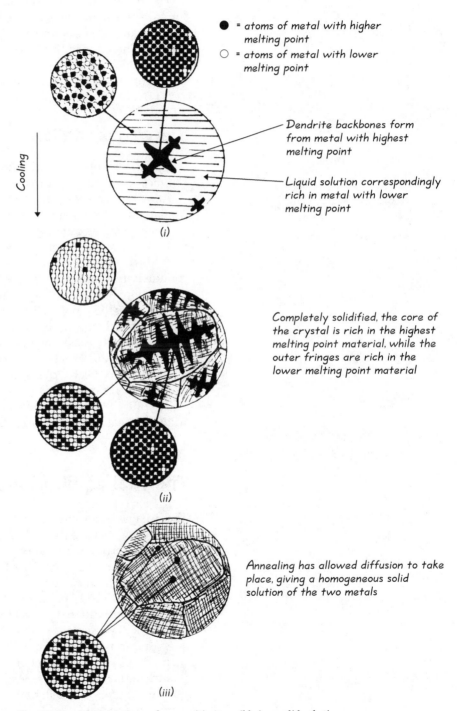

● = atoms of metal with higher melting point
○ = atoms of metal with lower melting point

Dendrite backbones form from metal with highest melting point

Liquid solution correspondingly rich in metal with lower melting point

Cooling

(i)

Completely solidified, the core of the crystal is rich in the highest melting point material, while the outer fringes are rich in the lower melting point material

(ii)

Annealing has allowed diffusion to take place, giving a homogeneous solid solution of the two metals

(iii)

Fig. 15.13 *The variations of composition possible in a solid solution*

Intermetallic compounds

Many metals will combine chemically to form compounds, producing metals that can differ considerably from the parent materials in colour and other properties.

The melting points are usually definite and often higher than the melting points of their constituent elements.

Examples are: $CuAl_2$, Cu_3Sn, Cu_2Zn_3, Cu_2Sb, Fe_3C, etc. Compounds are usually very hard and brittle, and alloys composed **entirely** of these would be useless for mechanical purposes, except perhaps as 'bearing metals'.

The intermetallic compound is not, therefore, generally used alone but can form many useful alloys in conjunction with other metallic solutions.

For example, Duralumin is an alloy that contains approximately 4 per cent copper. The copper first forms a solution with aluminium and then an intermetallic compound ($CuAl_2$). When correctly alloyed, Duralumin possesses the strength and hardness of mild steel but is so much lighter.

The effects of alloying

Alloying elements can form either substitutional or interstitial solid solutions, chemical compounds, or remain suspended in the solid alloy as a mixture.

Alloying elements often improve strength and hardness, they are also often used to improve a material's corrosion resistance.

Some elements can be added to an alloy to help in deoxidizing and/or to make the alloy flow more easily in the liquid state.

Adding alloying elements to a metal lowers the thermal and electrical conductivity. Copper is a good example of this, as it has the very high rate of conduction; however, if zinc is added, forming brass, the conduction rate is reduced. It is therefore, easier to maintain the heat input required for welding with less heat being lost by conduction. For this reason, the welding of copper alloys is usually easier than the welding of pure copper.

Solid solutions usually form strong, tough and fairly ductile materials and, by using the correct alloying, mechanical and thermal treatments a range of useful alloys is produced.

By applying this knowledge to welding engineering, the metallurgy of the weld and heat affected zone (HAZ) can be better understood and controlled.

Thermal equilibrium diagrams

These diagrams are graphs obtained by mixing many percentage combinations of the elements used in producing a particular alloy. The cooling curve for a particular percentage is then recorded. A series of cooling curves, each related to a particular percentage of alloy, build up to form a thermal equilibrium diagram for the alloy concerned (Figure 15.14(I) and (II)).

In Figure 15.14(I), points A_1, A_2, A_3, A_4, A_5 and A_6 show the temperatures at which solidification starts. The higher-melting-point element starts to solidify with a small percentage of the lower-melting-point element in solution with it. As the temperature drops still further, so more solution freezes, until at B_1, B_2, B_3, B_4, B_5 and B_6, the last of the liquid freezes off.

Such a diagram as this is formed when the two metals are soluble in each other in all proportions in both the liquid and solid states, as is the case with copper and nickel or gold and silver.

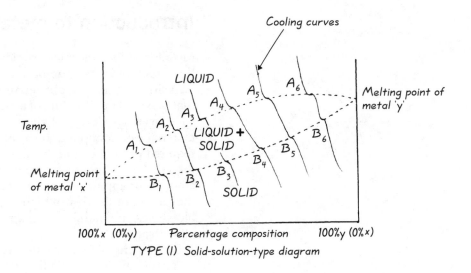

TYPE (I) Solid-solution-type diagram

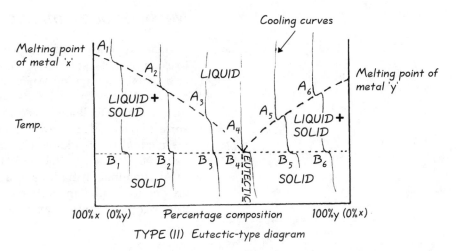

TYPE (II) Eutectic-type diagram

Fig. 15.14 *Showing the construction of thermal equilibrium diagrams using cooling curves; Type (I) solid-solution type; Type (II) eutectic type*

If the metals are completely insoluble in the solid state a eutectic-type of equilibrium diagram would represent the system (Figure 15.14(II)).

Here, the eutectic part of the structure of any percentage of alloy in the series solidifies at the same temperature and is the part of the alloy with the lowest melting point. The freezing points B_1, B_2, B_3, B_4, B_5 and B_6 are all at the same temperature. At the eutectic point A_4, B_4 (Figure 15.14(II)), where the alloy percentage is at the eutectic composition, because of insolubility, the alloy will freeze off in a laminated structure. Cadmium and bismuth are insoluble in the solid state and will form a diagram similar to this.

Thermal equilibrium diagrams play a major part in the understanding of the structural changes that can take place during and after operations that have employed fusion welding.

Later in the chapter we will be looking at a thermal equilibrium diagram that is of major importance to welders and welding engineers – the thermal equilibrium diagram for steel (the iron–carbon equilibrium diagram).

We will also be looking at various metallurgical structures connected with the diagram and welds in steel. Because certain of these structures can only be viewed through a metallurgical microscope, the next section is included in order to give an introduction to metallography and the metallurgical microscope.

Introduction to metallography

Metallography is the microscopic study of the structure of metals and relating this structure to their composition and physical properties.

A chemical analysis can determine the percentages of the various elements that make up the structure of a metal, including any possible impurities, but such an analysis may not give an indication of the physical properties of the material. For example, a chemical analysis would not show if a metal had the required properties after it had been heat treated, cold worked, hot rolled or forged. Also, although a chemical analysis may give the type and average amounts of impurities present, it might give little indication of the form that they are in or the distribution across the material. Such defects as porosity, blowholes, internal shrinkage cavities and cracks also would not show up at all with just a chemical analysis.

Metallographic examination, on the other hand will reveal more about the nature of any defects and the internal structures of the material.

Methods of examining metal structures

The naked eye, a magnifying glass and a microscope are all methods employed to examine metal structures. If the naked eye only is used, then this is known as a visual examination. An examination using a magnifying glass, in order to try and detect larger defects, such as cracks and blowholes, using up to 10× magnification, is known as a **macro-examination** or **macroscopy**. When a surface is cleaned and an acid used to etch it prior to using a magnifying glass, the method is often known as **'macro-etch examination'** (see *Basic Welding* pages 68–70, and also Table 15.1).

An examination made using a microscope with higher magnifications than are possible with a magnifying glass (above 10× magnification) is known as **microscopy** or a **'micro-examination'**.

Table 15.1 *Etching reagents suitable for preparing specimens for macroscopic examination*

Type of etchant	Composition	Comments and uses
Reagents suitable for steel		
Diluted hydrochloric	50 ml hydrochloric acid, with up to 50 ml water	Used boiling for 5–15 minutes, it can reveal flow lines, weld structure, small cracks and other defects such as porosity. It is also useful for determining the depth of hardness in tool steels
Nital (with increased acid content for **macro** use)	10 ml nitric acid, 90 ml alcohol	Suitable as a cold swabbing reagent on larger areas
Reagents suitable for copper and its alloys		
Acid and ferric chloride	25 g ferric chloride, 25 ml hydrochloric acid, 100 ml water	Useful for showing up dendritic structure
Reagent suitable for aluminium and its alloys		
Dilute hydrofluoric acid	20 ml hydrofluoric acid, 80 ml water	Degrease specimen, wash off in hot water and then swab with etching reagent

Table 15.2 *Micro-etching reagents*

Type of etchant	Composition	Comments and uses
For iron, steels and cast iron		
Nital	2 ml nitric acid, 98 ml alcohol (industrial methylated spirits)	This produces the best general etching reagent for irons and steels. It will etch pearlite, martensite and troosite. It will also attack the grain boundaries of ferrite. For use on pure iron and wrought iron, the concentration may be increased to 5 ml of nitric acid. Nital is also satisfactory for use on ferritic grey cast irons and blackheart malleable iron. Etching with nital must be very light if pearlite is to be resolved
Picral	4 g picric acid, 96 ml alcohol	Most suitable for etching pearlite and speroidized structures but will not attack ferrite grain boundaries. Picral is the most suitable etching reagent for the cast irons, with the exception of alloy and completely ferritic types
Mixed acids and glycerol	10 ml nitric acid, 20 ml hydrochloric acid, 20 ml glycerol, 10 ml hydrogen peroxide	Recommended for iron–chromium and nickel–chromium austenitic steels. Also suitable for high chromium–carbon steels, high speed steels and other austenitic steels. The specimen should be warmed in boiling water prior to immersion
Acid ammonium persulphate	10 ml hydrochloric acid, 10 g ammonium persulphate, 80 ml water	This is a particularly suitable reagent for use on stainless steels. It must, however, be freshly prepared for each application to give best results
For copper and its alloys		
Ammoniacal ammonium persulphate	20 ml ammonium hydroxide, 10 g ammonium persulphate, 80 ml water	Provides a good etch on pure copper, brasses and bronzes, revealing the grain boundaries readily when freshly made-up
Ammonia–hydrogen peroxide	50 ml ammonium hydroxide, 20–50 ml hydrogen peroxide (3% solution), 50 ml water	This is the best general-purpose etchant for copper, brasses and bronzes. Again, it should be freshly made-up, as the hydrogen peroxide deteriorates
For aluminium and its alloys		
Dilute hydrofluoric acid	0.5 ml hydrofluoric acid, 99.5 ml water	This is a good general etchant for use by swabbing on to the specimen.
Caustic soda solution	1 g sodium hydroxide, 99 ml water	Again, a good general etchant for swabbing the specimen
Mixed acid – Keller's reagent	1 ml hydrofluoric acid, 1.5 ml hydrochloric acid, 2.5 ml nitric acid, 95 ml water	Useful for etching Duralumin-type alloys by immersion of the specimen for 10–20 seconds

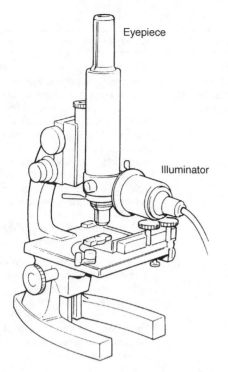

Fig. 15.15 *A typical single-tube and eyepiece metallurgical microscope*

It is generally considered that the first documented metallographic examination was carried out by Aloys Beck von Widmanstätten, when, in about 1808, as the director of the State Technical Museum in Vienna, he cut, polished and etched the surfaces of some meteorites to reveal a type of metallurgical structure (usually associated with prolonged heating and slow cooling) which is still named after him. (The Agram (Zagreb) iron meteorite was observed to fall in 1751 and had a characteristic mesh-like structure.)

It was not until 1861 that Professor H.C. Sorby of Sheffield developed the systematic examination of metals through the microscope, founding metallography as a branch of metallurgy. Sorby mainly studied iron and steel, illuminating his microscope by reflected light. He published photomicrographs and his findings to the Iron and Steel Institute in 1887.

A single-tube metallurgical microscope is shown in Figure 15.15, with examples of modern stereo microscopes in Figures 15.16 and 15.17.

If a rough, fractured surface were to be placed under a metallurgical microscope, we would be unable to see very much, as it would be impossible to bring the uneven surface into focus. A specimen for examination under a metallurgical microscope has to be ground flat and then polished, in some cases using a combination of both hand and machine polishing, with grades of increasingly fine abrasives until a mirror-like surface is obtained. It is important that during the actual removal of the specimen and the polishing process, methods are employed to keep the specimen cool, in order not to alter the grain structure in any way.

With rotary disk polishers, water is constantly dripped on to the disk while polishing. This has the effect of making a finer abrasive, taking away any polishing debris and keeping the specimen cool.

Such polishing processes spread a thin film of metal over the surface of the specimen. In order to see the grain structure and examine the weld/metal in detail, an etching reagent is used to remove this film.

Some typical etching reagents are given in Tables 15.1 and 15.2.

Full laboratory safety precautions should be employed when carrying out etching operations. Goggles must be worn, all skin protected from contact with the reagents and fume extraction be in operation.

A metallurgical microscope is different from a biological microscope. A biological specimen is sliced very thinly and placed on a glass slide under the microscope. Light is then shone through the specimen by adjusting a small mirror from beneath or by electrical illumination from underneath. Light will, of course, not shine through a metal specimen, so metal samples are viewed by reflected light from the polished and etched surface. This is achieved by having an illuminator built into the metallurgical microscope (Figure 15.18).

A magnification of 100 times is usually enough to show the grains or crystals of a pure metal (the word *crystal* and *grain* are both used to mean the same thing).

Sometimes, for ease of handling or if specimens are to be kept for some time, they can be mounted in cold setting plastic (Figure 15.19). Any colour of plastic can be used, but clear plastic is often preferred as an identification label can also be placed inside the plastic and other aspects of the specimen can be ascertained through clear material.

In general metallography, the magnifications most commonly employed range from 50× to 1000× and standard microscopes are usually capable of obtaining 5000× to 6000× magnification.

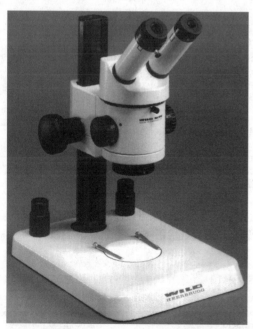

Fig. 15.16 *'Wild' stereo metallurgical microscope*
(*Photograph courtesy of Leica UK*)

All reagents should be made up by a qualified chemist or laboratory technician and used and handled in accordance with country and/or state health and safety regulations, for example, Control of Substances Hazardous to Health Regulations (COSHH) 1988, and the Health and Safety at Work etc. Act 1974 for the UK (see *Basic Welding*, pages 5 and 6).

Full body and eye protection must be employed, together with fume extraction.

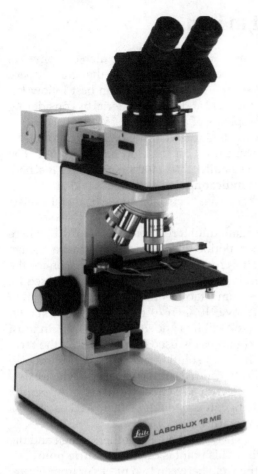

Fig. 15.17 *Leitz compound microscope for metallurgical use*
(Photograph courtesy of Leica UK)

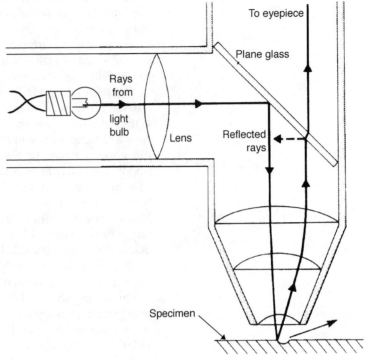

Fig. 15.18 *How the illuminator works in a metallurgical microscope*

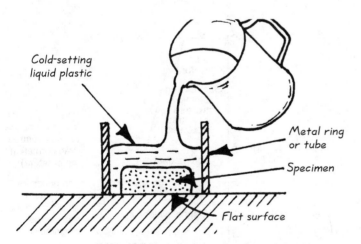

The inner surface of the ring can be coated with chalk to allow easy removal of plastic cylinder when set. Most plastics set in about 5 minutes, depending on amount of hardener added.

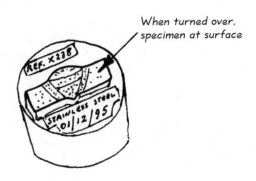

Produced cylinder of clear plastic with specimen at surface when turned over.

Identification labels can be positioned within the protection of the plastic.

Fig. 15.19 *Mounting specimens in plastic ready for microscopic examination. It is often easier to polish and etch specimens when they are mounted in this way. They are also easier to handle, store and identify*

Metallurgy of a weld in iron

Iron is a very interesting metal because it can exist in more than one physical form (being **allotropic**). At various specific temperatures, the atomic space lattice structure changes from one type to another. We can best follow the various changes that take place in the weld and parent material by considering a simple fusion weld between two pieces of iron (Figure 15.20).

At around 1530°C (2786°F) the molten iron weld pool will begin to freeze, the temperature remaining constant until all the iron has solidified. The temperature again begins to drop steadily until 1400°C (2552°F) is reached; here the temperature lags slightly, indicating that a change is taking place in the solid iron, but this particular change has been determined by metallurgists to be of little importance.

The temperature falls steadily again until it reaches 910°C (1670°F) where it again remains constant for a short period of time, indicating that a change is taking place in the solid iron. This change is important; the reason for the constant temperature at this point is due to heat being given off, caused by the atoms rearranging themselves from a face-centred cubic lattice crystal to a body-centred cubic formation. (Above 910°C (1670°F) the structure is known as gamma iron and below 910°C (1670°F) it is known as alpha iron. In most texts on metallurgy the symbol γ (gamma) is used to denote both the face-centred cubic form of iron and the solid-solution austenite. The symbol α (alpha) is used to denote both the body-centred cubic form of iron which exists below 910°C and the solid-solution ferrite. See Greek alphabet in Appendix 4.) The temperature at which a change in crystalline structure takes place is known as the **transformation temperature**.

On cooling still further, to 769°C (1417°F), there is another change and the iron regains its magnetic properties. This point is called the **Curie point**.

We can see from the sketch of the weld between two pieces of iron (Figure 15.20) that the single-pass weld builds up grains growing inwards and slightly

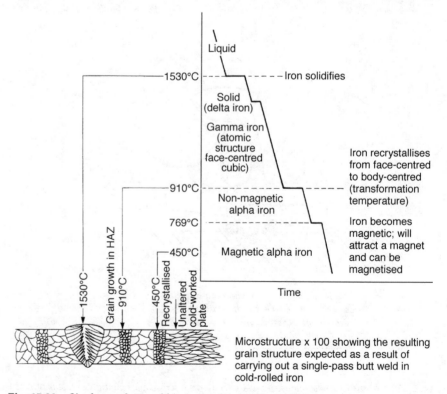

Fig. 15.20 *Single-pass butt weld in iron, related to the cooling curve*

upwards, the last of the columnar grains forming as the last portion of liquid iron freezes at the surface of the weld. The parent material that has been heated above 910°C will have undergone the crystalline structure change and the grains closest to the weld in the heat affected zone will be slightly weaker, because of an increase in size. The parent material that did not reach 910°C (1670°F) will have not been subjected to any atomic change, but if the parent material had been cold worked prior to welding, then recrystallization will have taken place in areas reaching above 450°C (842°F). Any iron not heated to over 450°C (842°F) at a distance further from the weld is still in the cold-rolled state.

When carbon is added to iron, the iron becomes known as steel. The first change that an addition of carbon will have is to cause the material to freeze over a temperature range, instead of freezing at a definite temperature. The more carbon in solution with the iron, the lower the freezing temperature range.

With the body-centred atomic structure, iron will only be able to hold a very small amount of carbon in solution, so little that it is generally ignored. However, in the face-centred cubic structure, iron is capable of holding large amounts of carbon in solid solution. This solid solution of carbon in iron, if viewed through a microscope at 1000 magnification would appear as pure metal rather than an alloy, with just the grain boundaries being visible.

It is easy to imagine liquid solutions, such as salt, dissolving in water to form brine, but more difficult to imagine that iron can hold dissolved carbon in a solid solution. The fact that it does is extremely important, as it means that iron–carbon alloys can be heat treated to give a wide range of properties.

Iron with a very small amount of carbon added, that is up to 0.006 per cent, and pure iron are known as ferrite. If carbon is added above 0.006 per cent, changes to the microstructure occur and the temperature of the alpha–gamma (body- to face-centred atomic structure) is gradually lowered from 910°C (1670°F) to 723°C (1333°F), when the carbon content reaches 0.83 per cent. The changes that occur with various carbon contents and the resulting microstructures can best be understood by looking at Figure 15.21 (the

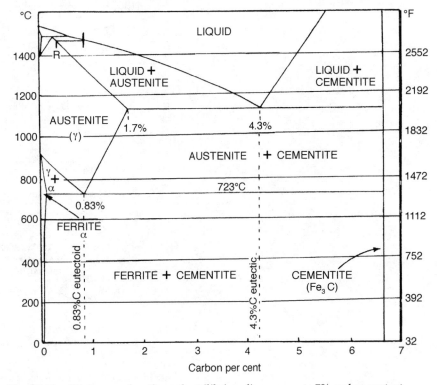

Fig. 15.21 *The iron–carbon thermal equilibrium diagram, up to 7% carbon content*

iron–carbon thermal equilibrium diagram). This diagram has been constructed by plotting a series of curves (heating curves in this instance, there being approximately 30°C (86°F) lowering of results when cooling curves are used) and noting when the internal structural change takes place for each carbon percentage. These change points are then joined up to give the equilibrium diagram.

Alloys of carbon and iron having a carbon content between 2 and $4\frac{1}{2}$ per cent are known as 'cast iron'. Increased carbon content improves hardness but decreases ductility. The welding of these materials is discussed in Chapter 17.

The part of the iron–carbon equilibrium that contains the steels is therefore the section up to 2 per cent carbon.

Figure 15.22 looks at this part of the diagram in more detail as it covers the ordinary steels that are most frequently welded. These microstructures will occur when cooling takes place slowly; rapid cooling, such as quenching, will have the effect of not allowing the steel time to carry out the internal changes, giving a different structure (this is discussed in the next section).

At 0 to 0.006 per cent (Figures 15.22a, 15.23a and 15.24) the carbon will not be visible through a microscope and the structure will appear as a pure metal, with just the grain boundaries visible (Figure 15.22a). As the level of carbon is increased, iron carbide (**cementite**) is formed by three atoms of iron combining with one atom of carbon (Fe_3C). The more cementite present, the greater the hardness and tensile strength of the steel. Steels containing up to 2.5 per cent carbon are known as **low carbon** or **mild steels**. Figure 15.22b shows the microstructure of a typical mild steel containing 0.25 per cent carbon. Here, in addition to the grains of ferrite, there are grains which contain alternate

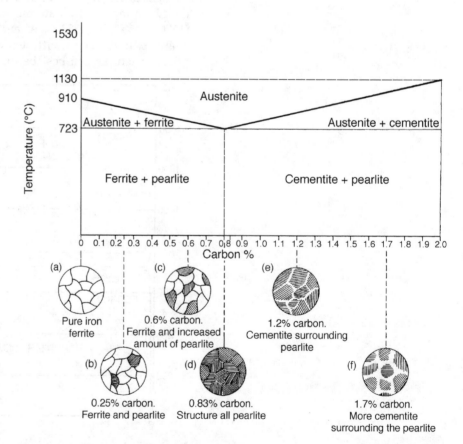

Fig. 15.22 *The iron–carbon equilibrium diagram, showing the microstructures that result from increasing the carbon content up to 2% (see also Figure 15.21)*

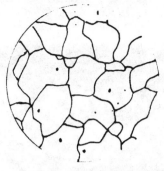

(a) Iron showing crystals
of ferrite x 100
(etched in 2% NITAL)

(b) Ferrite and pearlite
x750 to reveal lamellar
nature of pearlite
approx 0.6% carbon

(c) Cementite and pearlite
approx 1.2% carbon

(d) Typical structure of lamellar
pearlite x 1000 0.83% carbon

Fig. 15.23 *Typical microstructures produced by varying carbon content*

layers of black and white, giving an appearance similar to the edge of a piece of plywood. These grains are called **pearlite** and are made up of alternate layers of ferrite and cementite. The amount of carbon in pearlite is always 0.83 per cent. The pearlite will be distributed evenly through the microstructure, and increases the tensile strength and hardness of the steel but reduces the ductility.

Mild steel is generally easily welded by the majority of welding processes and all the welding data and general notes in this book refer to the welding of mild steel unless they specifically refer to the welding of other materials.

At 0.6 per cent carbon (Figures 15.22c and 15.23b) more pearlite is present, giving a further increase in tensile strength and hardness with a further decrease in ductility. The opposite of a ductile material is a brittle material, therefore, as a steel becomes less ductile it will become more brittle. The more brittle it becomes, the greater the risk of cracking when welding takes place. Steels within the range of 0.25–0.45 per cent carbon are classed as **medium carbon steels**. Above 0.6 per cent carbon they are classed as **high carbon steels**, with those above 0.8 per cent carbon being known as **tool steels**.

A steel containing 0.83 per cent carbon is known as a **eutectoid steel** (Figure 15.22d), as it is formed at the **eutectoid point** on the equilibrium diagram (the point where the lines for the upper transformation point, known as the upper critical point, and the lower transformation point, known as the lower critical point, meet). The transformation from austenite to pearlite takes place at one temperature and the whole microstructure will be pearlite (composed of alternate ferrite and cementite layers). Such steels are classed as eutectoid steels.

Increasing the carbon content to 1.2 per cent (Figures 15.22e and 15.23c) distributes excess cementite around the pearlite grains, as only 0.83 per cent carbon is required to form the pearlite, the remainder being distributed at the grain boundaries.

Steels containing less than 0.83 per cent carbon are called **hypoeutectoid** and the predominant ferrite will contribute to the metal's relative softness and ductility. Steels having more than 0.83 per cent carbon content are called **hypereutectoid**, the cementite producing the properties of hardness and brittleness.

Increasing the carbon content still further, to 1.7 per cent (Figure 15.22f) produces an even greater amount of cementite around the pearlite. (See also Figure 15.23 for typical microstructures produced by varying the carbon content.)

Some effects of heat-treatment and alloying on the structure of steel

As we have discussed, iron has a body-centred structure at below 910°C (1670°F) but undergoes an allotropic change when heated above this temperature. The rearrangement of atoms for a face-centred cubic lattice structure allows the iron to take a much greater amount of carbon into the solution, giving a condition at high temperature where all the carbon present in the steel has gone into solution with the iron to produce a new structure called **austenite** (named after the English metallurgist, Sir W.R. Austen). This solid solution of carbon in face-centred iron will not be retained as the steel cools to room temperature, for, as the iron–carbon equilibrium diagram (Figures 15.21 and 15.22) shows, depending on the carbon content, the resultant structures on cooling will be either ferrite and pearlite or cementite and pearlite. However, by the addition of high percentages of manganese or nickel, the

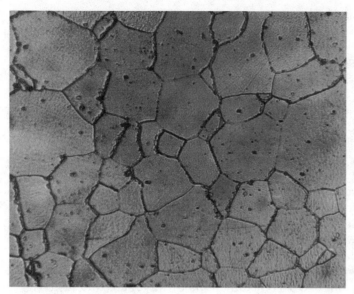

Fig. 15.24 *Ferrite under higher magnification reveals only grain boundaries*

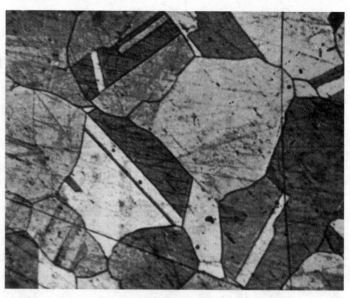

Fig. 15.25 *Austenitic structure formed by adding nickel, as seen through a microscope. The addition of alloying elements is one method of maintaining an austenitic structure at room temperature*

alloy steel formed will remain austenitic when cooled to room temperature (Figure 15.25).

Rapid cooling of a steel from above the transformation range, while it is in the austenitic state, can be carried out in a tank of water or oil. Such a treatment is known as **quenching** and can give a structure on cooling that is entirely different from that of a steel cooled slowly. Quenching obtains a different structure by not allowing the steel time to carry out the internal changes that would normally take place in the transformation range. The carbon atoms are trapped, causing a further distortion of the lattice arrangement, which is already distorted because it has not had time to arrange correctly. With a very rapid quench, lattice structures can be diamond shaped instead of cubic; this prevents rows of atoms sliding when a metal is stretched (tensile stress), forming a very hard but brittle structure. For very fast quenches, brine (salt water) can be used, as the particles of salt explode, bursting the steam envelope that may form around the component in an ordinary water quench, and so preventing further cooling. With brine, the bursting of the steam envelope as fast as it forms enables the water to be in constant contact. If the quench is fast enough, the hardest structure obtainable in steel can be formed, which is known as **martensite** (Figures 15.26 and 15.27).

It is for this reason that you usually have to let a weld cool down slowly in its own time. If you quench a weld as soon as you have completed it, the chances are that it will be hard and brittle and, if not already cracked, will easily crack in service.

Relationship between structure and rate of cooling

The iron–carbon thermal equilibrium diagram indicates the type of structure and properties expected in a steel that is cooled under equilibrium conditions.

However, as we have just mentioned, a steel that is cooled slowly can have a completely different structure to one that is cooled quickly.

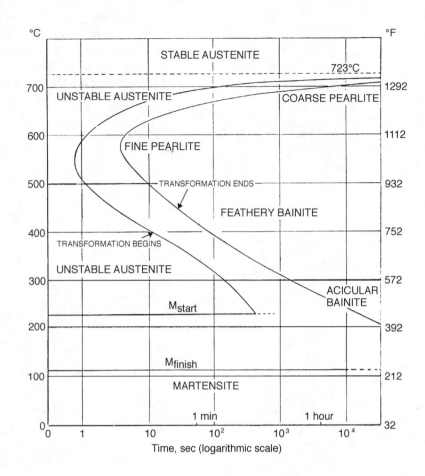

Fig. 15.26 *Time, temperature, transformation (T.T.T.) curves for a plain carbon steel of eutectoid composition.* M_{start} = *martensite transformation start;* M_{finish} = *martensite transformation finish. See also Figure 15.27*

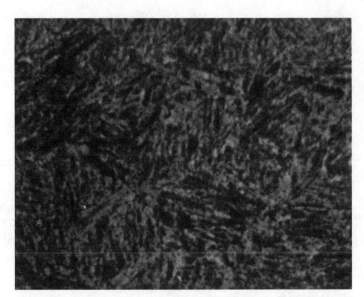

Fig. 15.27 *Martensite, ×750, showing the needle-like structure that is characteristic of etched sections of hardened steel (steel that has been subjected to a rapid quench)*

The relationship between the rate of cooling and the structure for a given steel can be studied by reference to sets of isothermal transformation curves called **T.T.T.** (time, temperature, transformation) curves.

These curves are constructed by heating a number of specimens of a particular composition of steel into the austenitic range and then quenching them into baths held at different temperatures. At precise time intervals, individual specimens are removed and further quenched in water. The microstructure of each specimen is then examined in order to determine how much transformation has taken place at the holding temperature.

Figure 15.26 shows the T.T.T. curves for a steel of eutectoid composition. It can be seen that if a steel of this composition is quenched in less than one second, then there is every possibility of a completely martensitic structure being formed. Less severe quenches result in structures that are not quite as hard and brittle as martensite. For example, a specimen quenched to 500°C (932°F) and held there for ten seconds before water quenching would form a structure composed of **bainite**. Under high magnification, bainite shows up as a laminated structure very similar to pearlite. For this reason it is generally regarded as an extremely fine type of pearlite. Even slower rates of cooling will produce normal pearlite.

Bainite is named after its discoverer. In 1930 E.C. Bain and E.S. Davenport, of the United States Steel Corporation, constructed isothermal-transformation diagrams using the microscopic examination of quenched specimens to obtain data.

16

The welding of mild, carbon and alloy steels (including the problems of cracking and crack prevention)

We have seen from the previous chapter that steels with a low carbon content, when cooled under equilibrium conditions, result in a structure of ferrite dispersed in pearlite, and steels of a higher carbon content produce resulting structures of cementite dispersed in pearlite. The amount of ferrite and pearlite in a steel at room temperature is therefore controlled directly by the carbon content.

As the carbon content increases above 0.25 per cent we need to take increasing precautions when welding operations are to be carried out, as although the carbon will increase the tensile strength and hardness, this is done at the expense of ductility. This is because of the amount and dispersion of the carbide (cementite) which is extremely hard and brittle.

The main precaution when welding steels of higher carbon content is to pre-heat and cool slowly after welding. This is essential in all cases, as if the work is cooled too quickly, the brittle effects of a rapid quench will result, causing cracking or the possibility of cracking. Pre-heating can be as simple as using the oxy-acetylene torch prior to welding, or may involve large, specially built furnaces. Slow cooling is sometimes obtained by a post-heat treatment, that is letting the component or fabrication cool down slowly in the furnace by gradually reducing the furnace temperature. It may be sufficient for small components to let them cool down in sand. In the case of large vessels, special electrical induction or resistance pads can be used (see *Basic Welding*, page 64). **Stress relieving** is another type of post-heating operation that removes internal stresses in the work caused by the prevention of full contraction when a welded fabrication cools.

Mild steel

Although mild steel is generally easily weldable by all the major welding processes, consideration should be given to process selection and methods of welding for particular applications. Figure 16.1 shows the various microstructures for a butt weld in mild steel plate under slow (equilibrium) cooling conditions. It can be seen that, with a single-pass run, grain size can be quite large. Grain size is dependent on prolonged heat imput and would therefore be larger for an oxy-acetylene weld than for an electric arc weld, with a faster heat imput. However, multi-pass welds with both processes will

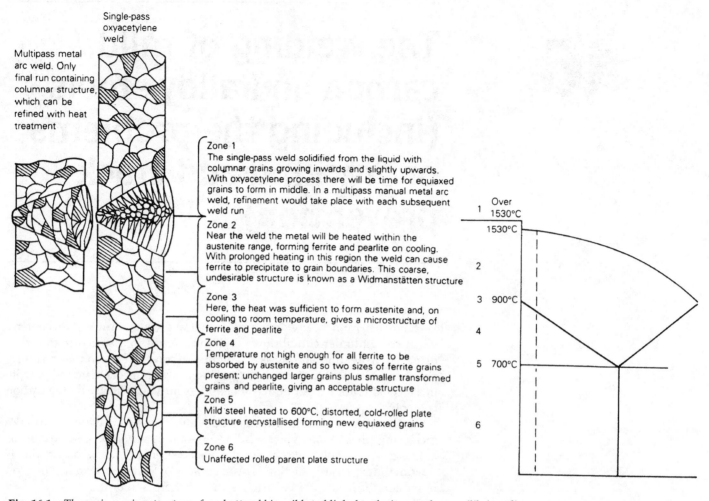

Multipass metal arc weld. Only final run containing columnar structure, which can be refined with heat treatment

Single-pass oxyacetylene weld

Zone 1
The single-pass weld solidified from the liquid with columnar grains growing inwards and slightly upwards. With oxyacetylene process there will be time for equiaxed grains to form in middle. In a multipass manual metal arc weld, refinement would take place with each subsequent weld run

Zone 2
Near the weld the metal will be heated within the austenite range, forming ferrite and pearlite on cooling. With prolonged heating in this region the weld can cause ferrite to precipitate to grain boundaries. This coarse, undesirable structure is known as a Widmanstätten structure

Zone 3
Here, the heat was sufficient to form austenite and, on cooling to room temperature, gives a microstructure of ferrite and pearlite

Zone 4
Temperature not high enough for all ferrite to be absorbed by austenite and so two sizes of ferrite grains present: unchanged larger grains plus smaller transformed grains and pearlite, giving an acceptable structure

Zone 5
Mild steel heated to 600°C, distorted, cold-rolled plate structure recrystallised forming new equiaxed grains

Zone 6
Unaffected rolled parent plate structure

1 Over 1530°C
1530°C
2
3 900°C
4
5 700°C
6

Fig. 16.1 *The various microstructures for a butt weld in mild steel linked to the iron–carbon equilibrium diagram*

have the effect of refining the lower deposits, because of reheating, leaving only the final capping run with the larger, weaker structure. Refinement of this last deposit would require a further post-weld heat-treatment.

Boiler steels and ship plate steels

These are special mild steels developed for use in the above applications. Boiler-quality steel limits the phosphorus and sulphur content to 0.05 per cent, while ship-quality allows up to 0.06 per cent. The carbon content in both types must be kept down to 0.26 per cent, to allow welding to take place. The majority of steels used for welding fall into the low or mild steel category with less than 3 per cent carbon. When any carbon steel is welded, a zone on each side of the weld will be heated to above the upper critical temperature and can, therefore, harden when cooling. The degree of hardness will increase with increased carbon content and the rate of cooling. For medium carbon steels with above 0.3 per cent carbon, and high carbon steels, having more than 0.6 per cent carbon, the risks will obviously be greater, as the tendency of the metal to crack is related to the hardness. Every attempt must therefore be employed to reduce the hazard. The first precaution should be to use a steel with the minimum carbon content necessary to obtain the properties required in the job. The cooling rate should also be controlled with pre- and post-heating, as well as employing a large-diameter electrode to give a higher heat input.

Pre-heating

In most cases, it will not be necessary to pre-heat steels having a carbon content up to 0.3 per cent. Above this percentage up to 0.5 per cent carbon, pre-heating to 100–350°C is recommended, the temperature being increased as the carbon content increases if mild steel electrodes are being used. Above 0.6 per cent carbon, steels are known as 'tool steels' and are not often welded by the arc process. If, however, welding is required on this type of material, it is advisable to use the buttering technique.

The use of buttering to prevent carbon pick-up

Cracking in a weld can sometimes be caused by 'carbon pick-up', this is, when carbon from the parent material diffuses into the weld metal. This has the effect of making the weld brittle and prone to cracking, usually along the centre of the weld. The 'buttering' technique can help overcome this problem by depositing a layer of weld metal on the faces before joining them together (see Figure 16.2). Some carbon will be picked up by this layer, but because the components are not joined together, the weld metal will not be highly stressed and therefore should not crack. The carbon content of the 'buttering' runs will be considerably lower than that of the parent material, so that when the joining weld is made, the carbon pick-up will be much less than if directly made with the parent material. The joining weld or welds should be made while the work is still warmed up from the buttering process. 'Buttering' can also be employed in joining certain combinations of dissimilar materials and in cast iron repair welding.

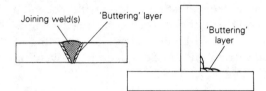

Fig. 16.2 *The use of 'buttering' on butt and fillet welds*

Low alloy steels – medium and high tensile

These steels contain small additions of alloying elements to improve their mechanical properties, such as tensile strength, ductility and toughness. Typical elements include manganese, silicon, nickel, chromium, molybdenum and vanadium. Such steels have similar welding properties to carbon steels and care must be taken to avoid hardening. Medium and high tensile steels denote low alloy steels possessing even higher physical properties. These steels generally have welding instructions issued by the supplier, which must be closely followed. A broad outline of the welding procedure is given below and care must be taken to avoid cracking, as detailed in the following pages. In general, the effect of the various alloying elements can be calculated as a carbon equivalent, (CE) and any pre- and post-heating requirements can be determined from this value. For example:

$$CE = C + \frac{Mn}{6} + \frac{Cr + Mo + V}{5} + \frac{Ni + Cu}{15}$$

Because the zones of parent metal each side of the weld will be heated above the 'critical' temperature and will therefore be subject to hardening on cooling, the cooling rate should always be controlled, particularly in the range from 400°C to 100°C (752°F to 212°F). Use a low hydrogen process (such as GMAW (MIG/MAG) or low hydrogen electrodes. To ensure that there is no moisture in electrode coatings, they should be baked for at least 30 minutes at 150–200°C (302–392°F) in an electrode drying oven (see Figure 16.3) if

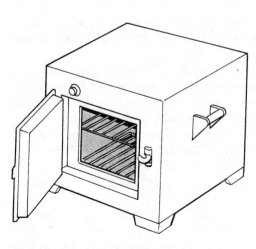

Fig. 16.3 *An electrode-drying oven*

available, otherwise an ordinary oven with temperature control will suffice. This is essential, as moisture will cause the formation of hydrogen. Avoid stray arcing, that is, catching the electrode on the work in an area that is not being welded, as stray arcing on these materials can create a stress-raiser and cause subsequent cracking. The same can apply if small tacks are used; always make strong tack welds and use a planned sequence of welding to reduce the chance of distortion and restraint on joints. On these materials never use small electrodes, low current or deposit welds with a shallow cross-section, as such actions can increase the risk of cracking.

Armour plate

This is a nickel–chromium–molybdenum steel which can contain between 0.25 and 0.35 per cent carbon. These steels must be welded with austenitic electrodes that have been designed for the purpose. All the previous precautions, including pre-heating, should be employed. The welding of these materials usually has to be carried out to a set government welding procedure.

Manganese steels

These fall into two groups: steels containing up to 2 per cent, and high-manganese steels containing 12 per cent or more. The first type can be classed as 'low-alloy' and, if the carbon content is below 0.25 per cent, normal procedures as for steel are generally satisfactory. Above 0.25 per cent carbon, pre-heating and the other special precautions for low alloy steels will be necessary. With high-manganese, as heating and slow cooling will cause brittleness, it is recommended to use the lowest current capable of making a satisfactory weld, in order to keep the heat input to an absolute minimum.

The problems of cracking in welds

Methods of controlled cracking and crack-susceptibility tests

The contraction forces occurring during a weld cooling will set up tensile stresses in the joint and may cause what is one of the most serious of weld defects – **cracking**.

Cracking can occur in the actual weld deposit, the heat affected zone, or in both regions. Cracks can either be of the large type, visible to the naked eye and known as **gross** cracks or **macrocracking**, or they can be of the type visible only through a microscope, known as **microcracking** or **microfissuring**.

Cracks that form above the solidus temperature are known as **supersolidus** cracks, or **hot cracking** to show that cracks occurred at elevated temperature. Cracks forming below the solidus are known as **subsolidus** cracks or **cold cracking**, a term that is often applied to cracking in low alloy steel welds at room temperature.

From a metallurgical point of view and therefore a welding engineering point of view, the distinction between cracks that have occurred at high temperature and those that have occurred at low temperature is very important, and therefore weld cracking is studied under these two headings.

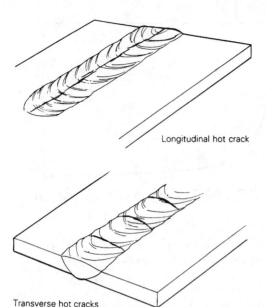

Longitudinal hot crack

Transverse hot cracks

Fig. 16.4 *Longitudinal and transverse hot cracking in welds*

Hot cracking

Two conditions must be present for cracking to occur during the weld thermal cycle:

1. the metal must lack ductility;
2. the tensile stress developed as a result of contraction must exceed the fracture stress.

The lack of ductility at high temperatures is usually due to the effects of low-melting-point films on the grain boundaries – such as sulphides in steels or a eutectic in certain crack-sensitive aluminium alloys. **Manganese** will tend to globularize sulphur and therefore will help to prevent weld cracking as a result of low-strength sulphide films forming in carbon and low alloy steel. Therefore the main element causing hot cracking is **sulphur**.

Hot cracking is likely to occur when sulphur-bearing steels are welded, particularly under restraint. Another cause is the weld metal picking up impurities from a dirty or contaminated surface.

The stresses which aid hot cracking are as a rule the shrinkage stresses associated with the cooling weld metal and these are, of course, greatest when the weld joint is restrained from moving. Restraint is therefore an important factor to consider when fusion welding and should be minimized or counteracted where possible.

A typical example is the welding of the centre disk into a heavy gear ring blank. It is necessary to heat the ring in order to prevent cracking. This heating is required not because the material is necessarily difficult to weld but in order to counteract the contraction stresses that would tend to crack the weld. Careful and uniform pre- and post-heating are therefore sometimes necessary even when welding fabrications in mild steel, merely to prevent expansion/contraction stresses and not as a metallurgical requirement.

Hot cracking (Figure 16.4) is usually longitudinal, occurring down the middle of the weld, however it sometimes takes place across the weld at roughly regular intervals which may range between 12 mm and 150 mm. This type of cracking is generally associated with contraction stresses in the weld metal, especially if the contraction rate is high or if it is greater in the weld metal than in the parent metal.

As mentioned earlier, manganese can help to prevent cracking occurring, owing to the formation of low-strength sulphide films. If the carbon content is less than 0.15 per cent and the ratio of the manganese content to the sulphur content (Mn/S) is greater than 15 in the weld metal, there is little danger of hot cracking (Figures 16.5 and 17.3).

Cold cracking

Parent metal cracks occur close to the weld and are usually associated with too high a rate of cooling coupled with the action of hydrogen. They form after the weld metal has cooled and can sometimes occur hours or days after the weld has been made. Because of this, they are called **cold cracks**.

Cold cracking is likely to occur in all ferritic steels when arc welded if the alloy content is above a certain level and if adequate precautions, mainly pre-heating, have not been employed. These cracks are not normally visible on the surface and are therefore difficult to detect. Because of the dangers that hidden cracks could cause, it is very important that suitable welding procedures are established and followed carefully.

When a steel is fusion welded by the arc process, some of the parent material will mix with the weld metal (dilution). If no filler metal is added, such as in

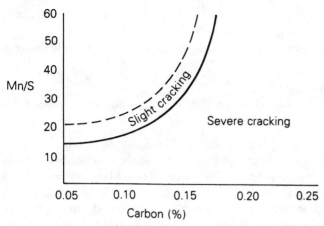

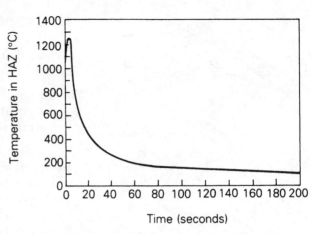

Fig. 16.5 *Effect of Mn/S ratio and carbon content on cracking possibilities. (Based on TWI and Russian work for manual metal-arc welding)*

Fig. 16.6 *A typical cooling curve for a fillet weld between two 6 mm thick low-carbon steel plates using a 3.25 mm electrode*

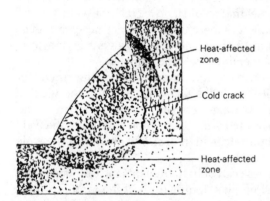

Fig. 16.7 *A typical cold crack in the heat affected zone of a fillet weld (magnification ×5)*

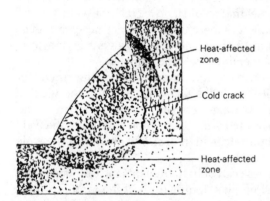

Fig. 16.8 *The solubility of hydrogen in steel*

an upturned edge weld, the weld will be classed as autogenous as it will consist entirely of melted parent material.

Because the material is heated to its melting point, there will be a temperature gradient from the weld through the material, which is some distance away. The whole area of the weld will cool down as soon as the arc has been removed and the hottest material will cool most rapidly from the higher temperatures. A typical cooling curve is shown in Figure 16.6.

The heat affected zone (HAZ) will contain material that has been affected structurally by the heating and cooling associated with the welding cycle. An HAZ containing a cold crack is shown in the photomacrograph of Figure 16.7. Part of the HAZ becomes heated to the austenitic condition and transforms to martensite if cooled rapidly. Cooling is mainly by conduction into the surrounding metal and, as shown in the cooling curve of Figure 16.6, it can be very rapid in welds that have not been pre-heated, particularly when heavy sections are being welded. This rapid cooling results in the formation of martensite, which is an extremely hard substance and is more easily formed when alloying elements are present in the material.

The existence of the hard and brittle martensite means that there is a structure present that is susceptible to crack formation. Any cracks that do form in the region are generally associated with **hydrogen**.

The graph of Figure 16.8 shows the solubility of hydrogen in steel at various temperatures. Hydrogen is much less soluble in the products of transformation and tends to get trapped in the HAZ.

Atoms of hydrogen tend to combine and form molecular hydrogen which builds up to exert tremendous pressure, causing cracks some time after welding. The extent of cracking will be aggravated by any residual stresses that may be present as a result of restraint.

Hydrogen must be eliminated as much as possible, and the rate of cooling should be slow enough to prevent the formation of martensite and to allow any trapped hydrogen to diffuse out of the weld and the HAZ. Low hydrogen electrodes and processes such as GMAW (MIG/MAGS) help to reduce the amount of hydrogen present.

The composition of a steel determines its response to heat-treatment and therefore its susceptibility to cracking. The element **carbon** has the biggest effect of all, and other alloying elements may be expressed in terms of percentage carbon. In this way, the susceptibility of a steel to hard zone cracking (or cold cracking) may be given by a single number known as the **carbon equivalent**. One simple formula for calculating the carbon equivalent is

$$\text{Carbon equivalent} = C + \frac{Mn}{6} + \frac{\text{All other elements}}{14}$$

(see also Figure 16.16)

A mild steel will generally have a carbon equivalent in the range of 0.21–0.40 and may be welded without any special precautions. If the carbon equivalent is above 0.40 it is strongly advised to use a low hydrogen process. It is generally accepted that steel up to 0.47 carbon equivalent can be welded without pre-heat using a low hydrogen process, provided the temperature is above freezing. Steels with carbon equivalents above 0.47 will require pre-heating in most cases (even when using a low-hydrogen process or electrode).

A simple workshop test to determine how an unknown steel must be welded

There are many tests to determine the pre-heat temperature that may be needed but the **clip test** provides a quick method on site or in the workshop. It is only suitable for sections down to 3/8ths of an inch (1 cm) thick. It was developed by Sheeham of the Portsmouth Naval Ship Yard.

To carry out this test, a clip or lug of low carbon steel, 2 or 3 inches (50 to 75 mm) square and $\frac{1}{2}$ inch (12.5 mm) thick, is fillet welded to the unknown steel using the same type of electrode and method to be employed on the welding job (with a large electrode and a slow speed, clips up to 1 inch (25 mm) thick can be used) (Figure 16.9). On completion, the weld is allowed to cool for five minutes, then the lug is broken by hitting it with a hammer. If the lug bends and then finally fails through the weld, the test indicates that the analysis of the steel, the temperature of the steel before welding and the welding procedure are such as to prevent serious underbead cracking.

A steel that requires pre-heating will fail the clip test by the weld pulling out of the parent metal (Figure 16.9c). The pulled-out metal is usually bright,

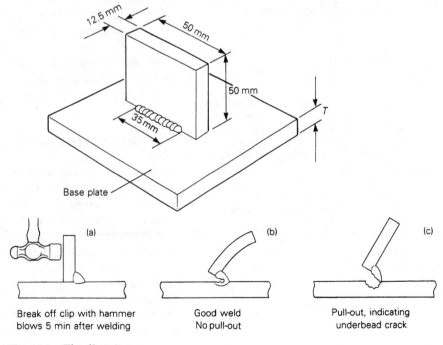

(a)	(b)	(c)
Break off clip with hammer blows 5 min after welding	Good weld No pull-out	Pull-out, indicating underbead crack

Fig. 16.9 *The clip test*

217

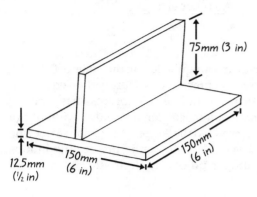

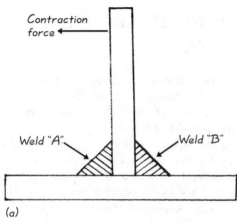

(a)

(b)

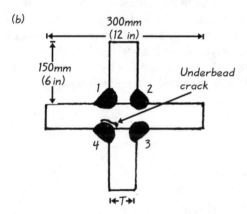

Severity can be increased as a cruciform test. Progressively higher degrees of restraint are placed on each weld with the maximum on the fourth and last weld to be deposited. The specimen is examined for external cracks after 48 hours at room temperature, followed by stress relief. It can then be sectioned for metallographic examination to determine if cracking is in the weld metal or base metal heat affected zones.

Fig. 16.10 *(a) The fillet-weld hot cracking test. (b) Cruciform test for plate cracking susceptibility*

indicating that the crack did not extend to the surface or that crack formation had occurred after the metal had cooled below the temperature at which temper colours occur.

If pull-out takes place, the clip test should be repeated using pre-heat, starting at 200°C (392°F), until satisfactory results are obtained. When the correct parameters are found, these should then be employed on the job.

If the pre-heat is too high for use on the job, then try the test using a low hydrogen electrode and reduced pre-heat. The 18/8 stainless steel electrodes will also make crack-free welds at much lower pre-heat temperatures than ordinary low carbon steel electrodes.

Several other welding tests have been developed in order to determine the susceptibility of a welded joint to cracking in either the weld metal or the base-metal heat affected zone.

In crack susceptibility tests, the stresses associated with the forces of contraction are increased by imposing a high degree of restraint, or an opposing force on the base metal or one of the base metal pieces.

One of the simplest weld metal cracking tests for use in the workshop is the double fillet weld joint, as shown in Figure 16.10a. This is known as **the fillet weld hot cracking test** and particularly tests the susceptibility of the weld metal to hot cracking, as opposed to the underbead HAZ cracking tested for with the clip test.

Weld 'A' is made first using a large-diameter electrode, usually 5 mm (6 SWG) at maximum current. Weld 'B' is then commenced within 10 seconds of completing weld 'A'.

The contraction stresses due to the cooling of weld 'A' have a direct effect on weld 'B' while it is cooling through the high temperature range where it could have low ductility.

When cool, any fracture can be broken open and examined for signs of hot cracking, such as blue-gold temper colours (see also Figure 16.10b for the cruciform test).

More reliable results can usually be obtained by rotating one plate with respect to the other while the test weld is being made. This can be done in a machine made for the purpose, such as the 'Murex' hot crack testing machine. With such a machine, one plate starts to rotate 5 seconds after the weld has been started. The plate then rotates up to 30 degrees in the next 30 seconds. The speed of rotation can be varied. Such machines can be controlled under laboratory conditions to give results valuable to the production of welding consumables.

Results are usually based on the crack length at a particular machine setting, with up to a 12.5 mm ($\frac{1}{2}$ inch) crack at the start being ignored. An average of six tests are taken to give a mean crack length.

The Houldcroft crack-susceptibility test

This test involves the making of a butt weld between two plates with saw cuts positioned as shown in Figure 16.11.

The test was designed for determining the hot cracking tendencies of sheet materials when welded by the gas tungsten arc (GTAW/TIG) process.

If the material is weld-crack sensitive, then a crack will initiate and propagate along the weld until it reaches a point where the degree of restraint is insufficient to make the crack continue. The length of the crack can then be used as a measure of the filler material's susceptibility to hot cracking.

The specimen is not usually clamped during welding but welding conditions must be carefully determined and held constant throughout the test. Factors such as an increase in weld width or welding speed can tend to increase the risk of cracking, as could variations in penetration.

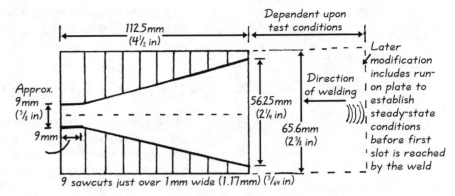

Fig. 16.11 *The Houldcroft weld-cracking test*

Steady-state thermal distribution is more readily established in the first part of the test by the addition of a run-on plate, as shown in Figure 16.11.

The Reeve cracking test

In this test (Figure 16.12), restraint is produced by bolting the 150 mm (6 inch) square and 175 mm × 250 mm (7 inch × 10 inch) rectangular plate down to a massive bedplate. Four fillet welds are then placed around the square in the sequence shown (1, 2, 3 and 4). The highest stresses will be generated in the area of the fourth fillet weld.

When the welds have all cooled down to room temperature, the plates are unbolted and sections removed, as shown in Figure 16.12 (three strips). These can then be metallographically examined for cracking.

Great care must be taken with this test to tighten down the bolts to a pre-determined torque setting, otherwise there can be variations in the amount of restraint offered.

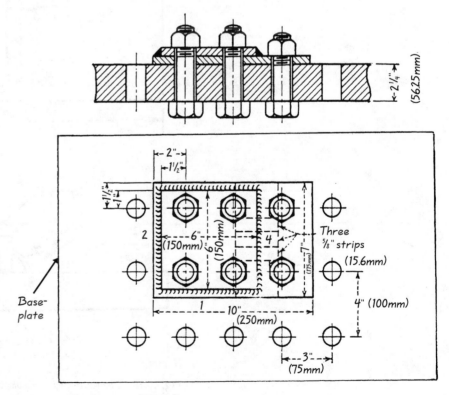

Fig. 16.12 *The Reeve cracking test*

It has been found that less variation in restraint occurs when restraining welds are employed, and this has lead to a greater use of the test discussed next.

The controlled thermal severity (CTS) test

The CTS cracking test is designed to assess the susceptibility of carbon and alloy steels to underbead cracking in the base-metal heat affected zone area. Because the rate of cooling is a major factor in producing this type cracking, the control of the cooling rate must be a requirement of any test proposed for evaluating the tendency of a material to crack in the heat affected zone.

The specimen is bolted together and then the anchor welds are deposited, as in Figure 16.13. Once the anchor welds are deposited, the bolt is not considered to have any influence on the amount of restraint upon the test welds.

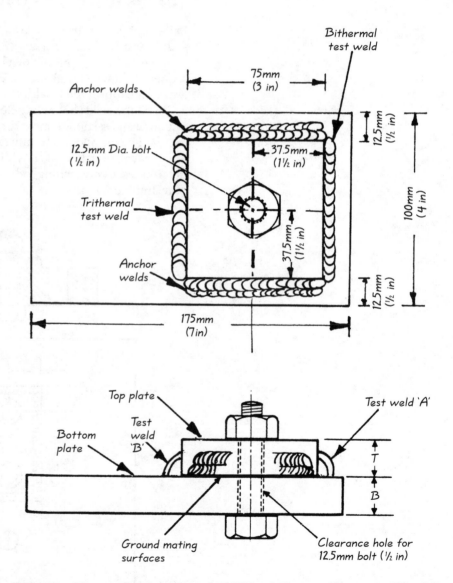

Fig. 16.13 *The controlled thermal severity (CTS) to indicate cracking susceptibility (after Cottrell)*

After the anchor welds have been deposited, the specimen is allowed to cool to room temperature (or any other temperature specified by requirements).

The greater the thickness of base metal, the faster heat will be conducted away from the weld, and a 'T' fillet configuration will dissipate heat 50 per cent more rapidly than would a butt joint between plates of the same thickness.

It can be seen, by referring to Figure 16.13, that test weld 'B' will be subjected to greater thermal severity than test weld 'A' for this reason.

Each CTS test, therefore, provides two levels of thermal severity and two cooling rates. The extent of any heat affected zone cracking is determined by the metallographic examination of three sections taken out of each test weld.

The temperature should be allowed to stabilize after depositing weld 'A', and before weld 'B' is deposited. The completed specimen should then be left for 24 hours before examination, to allow time for cracks to form.

The test is useful in establishing procedures for the welding of low alloy steels and for research concerning new steels, consumables and processes.

The Lehigh cracking test

This test employs slots cut into the sides and ends of the plate (Figure 16.14) in order to control the degree of restraint offered by the base plate on the weld. The amount of restraint can be expressed numerically by the width of the specimen between the slots.

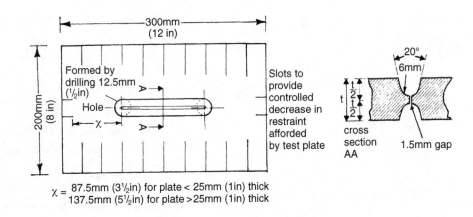

$$\chi = \begin{array}{l} 87.5\text{mm } (3\tfrac{1}{2}\text{in}) \text{ for plate} < 25\text{mm } (1\text{in}) \text{ thick} \\ 137.5\text{mm } (5\tfrac{1}{2}\text{in}) \text{ for plate} > 25\text{mm } (1\text{in}) \text{ thick} \end{array}$$

Fig. 16.14 *The Lehigh cracking test*

A specimen is first welded under full restraint, by using a plate with no saw cuts. If any specimen shows cracking, another specimen is prepared which imposes less restraint (deeper saw cuts), until a restraint level is reached (the threshold level) where no cracking occurs.

Tests can be conducted with conditions fixed and then with variations of heat input. The test appears to be sensitive enough to determine consumable characteristics and differentiate between types of electrodes, as well as evaluating the effect of base-metal dilution upon weld-metal cracking.

The test is suitable for use with shielded metal-arc, submerged-arc and gas metal-arc processes.

Results can be obtained by noting the percentage of the bead length that has cracked and also by metallographic examination of cut sections.

Patch tests

Several other weld cracking tests are designed around welding a circular plate into a solid sheet using different types of edge preparations, depending on thickness. The angle over which cracking occurs can then be measured.

The welding of a circular plate into a circular hole provides far greater stresses than would normally be encountered in a more conventional type of joint having the same mass of material, and cracking can occur in the weld metal, at the fusion line, or in the heat affected zone of the base metal. Figure 16.15 shows the dimensions for a typical patch or circular-groove welding test, however there are so many of these tests that further literature should be consulted if more information is required.

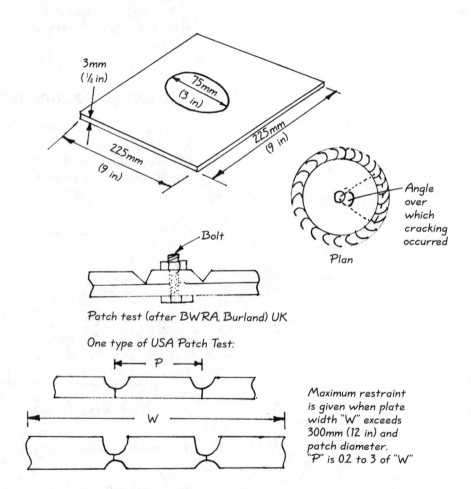

Fig. 16.15   *Examples of some types of circular patch cracking tests*

The rigid restraint cracking (RRC) test

A test which gives good correlation between the carbon equivalent and the crack susceptibility is the **rigid restraint cracking test** (Figure 16.16). In this test, two test plates are clamped into a heavy frame and welded together using a single, partial penetration weld run.

The restraint is constantly measured by strain gauges mounted on the frame, cracking being detected by a fall in the amount of strain being registered.

The amount of restraint can be altered by varying the restraining length, the rate of heat input, or the plate thickness.

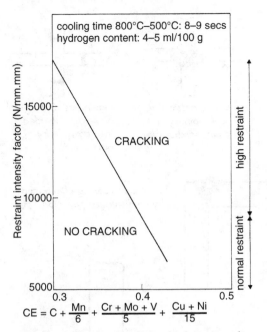

cooling time 800°C–500°C: 8–9 secs
hydrogen content: 4–5 ml/100 g

CRACKING

NO CRACKING

high restraint

normal restraint

Restraint intensity factor (N/mm.mm)

15000

10000

5000

0.3   0.4   0.5

$$CE = C + \frac{Mn}{6} + \frac{Cr + Mo + V}{5} + \frac{Cu + Ni}{15}$$

Fig. 16.16 *Showing the relationship between carbon equivalent, restraint intensity factor and cold cracking in the RRC test. (From Lancaster and after A.T. Fikkers and T. Muller, 1976 Metals Technology Conference, Sydney, Australia)*

The main variables are the cooling time from 800°C down to 500°C (1472°F down to 932°F), the type of steel, the restraint intensity, and the hydrogen content.

Lamellar tearing

Lamellar tearing is a type of cracking that can occur in the base metal of a welded structure. It is formed by the separation at the interface between inclusions or voids and metal caused by the through-thickness stresses induced by the weld-metal shrinkage as it contracts. Figure 16.17 shows types of joint particularly susceptible to lamellar tearing, when the localized stresses caused by weld-metal shrinkage may become many times greater than the yield point of the base material.

This can create a very dangerous situation when welding heavy fabrications. For example, lifting lugs have been known to pull out of fabrications. The welds are still completely intact, but a large section of base metal, to which the welds are fused, pulls out of the main fabrication.

A lamellar tear will occur only in the parent metal and although it can originate near to the toe or root of a weld, it can often originate outside the heat affected zone and may not propagate to the surface.

The typical tear takes place in three stages: firstly, separation voids are formed, then, as the stresses increase, these voids link together in a planar manner by necking, cleavage or coalescence. The third stage involves the joining together of these discontinuities by vertical shear walls. It is this joining together that gives the stepped appearance of a typical lamellar tear, as the planar discontinuities are formed at different levels.

It has been proved that silicate and sulphide inclusions can help initiate lamellar tearing, and the reduction of sulphur content in the steel is accepted as one control method. The use of a low hydrogen process is recommended, as although hydrogen is associated with the formation of cold cracking, in plain carbon steels it can also increase the susceptibility to lamellar tearing.

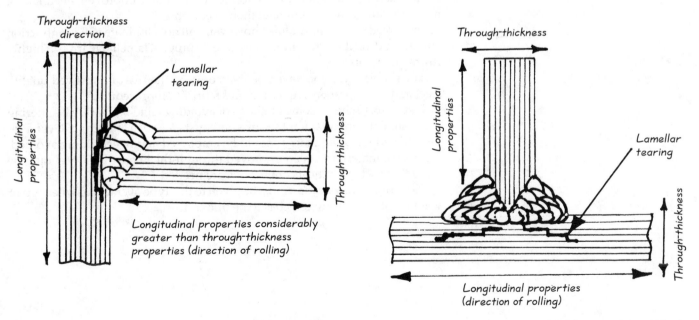

Fig. 16.17 *Showing types of joint that are particularly susceptible to lamellar tearing as the contraction forces of the welds exceed the yield point of the base material in the through-thickness direction of one plate*

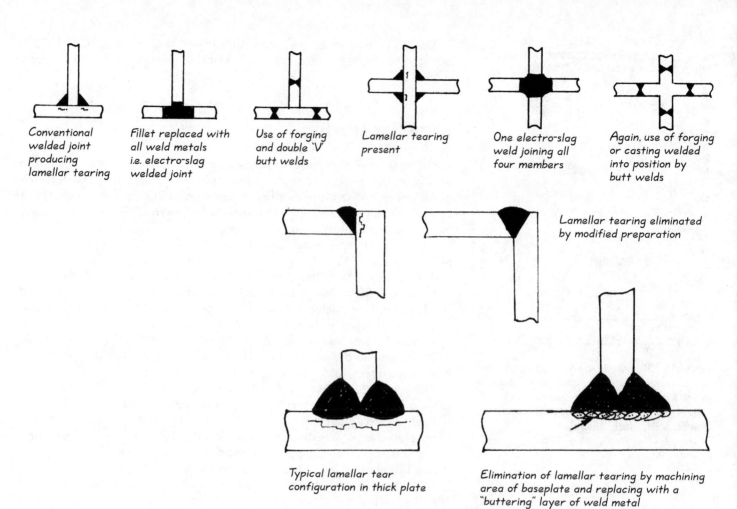

Fig. 16.18 *Some methods of avoiding lamellar tearing*

Actual laminations (pre-existing planes of weakness) can, in some cases, be detected by ultrasonic inspection techniques and improved steelmaking methods can virtually eliminate their occurrence.

Improved steel quality alone, however, will not eliminate weld contraction stresses and, on its own, may not necessarily prevent lamellar tearing in highly restrained joints.

The problem can, however, be overcome by the use of improved quality steel and a comprehensive design and fabrication procedure.

Figure 16.18 shows some methods of avoiding lamellar tearing by altering the design of particular joint set-ups and by the use of a 'buttering' layer. Pre-determined weld sequences should be employed and overwelding avoided. The use of material with an improved through-thickness ductility should be considered where critical connections are needed.

Particular consideration must be given to this type of cracking when repair welding of older structures has to be undertaken.

Joining dissimilar metals, coated or clad metals and cast iron

Difficulty is often encountered when dissimilar metals are required to be joined, for a number of reasons – dilution of one metal, different melting points, removal of oxide films, and so on. In some instances, conventional welding processes may not be able to join certain combinations. In such cases, it may be necessary to use a non-fusion technique such as brazing or bronze welding. For example, brass may be joined to steel by depositing bronze on to the steel and then joining the two surfaces with a bronze filler rod or electrode of the same composition. There is a difference between bronze welding and brazing.

Brazing was developed for applications when soldering was not strong enough, using a filler material with a melting point above 430°C (806°F) but below that of the metals being joined. It has many valuable applications but is limited by the fact that pre-heating must be general; a flux is used to help clean the oxide from the work surfaces and help the filler metal to flow more easily. This flux is usually applied by dipping the heated end of the rod into the tin of flux, although rods can be obtained with a flux core. Typical examples of the use of brazing are the manufacture of motorcycle and cycle frames. It is also frequently used for the joining of tungsten carbide tips to machine tools.

For motorcycle and cycle frames, one tube usually fits inside another, which it is to be joined to. The tube surfaces are pre-cleaned by wire brushing or the use of emery paper, and the smaller tube is dipped into a flux paste, made by mixing the flux with water, before being inserted into the larger-diameter tube. The joint area is then brought up to red-heat using an oxy-acetylene flame, and the flux-coated brass rod melted into the joint. In the case of brazing steel, the brass (copper and zinc) filler material melts at around 950°C (1742°F), which is much lower than the melting point of steel. As heat is applied to the joint area, the brass will flow around the joint between the tube surfaces by capillary attraction. (Capillary attraction is the name given to the act of a liquid travelling in a narrow gap; it is a phenomenon that does not occur if the gap is wider. For this reason, joints for brazing are made a fairly close fit. See also Chapter 14.) For brazing, any suitable heat source can be employed, such as a blowpipe, resistance heating or furnace. For bronze welding, a brass filler rod or electrode is employed and the heat is therefore obtained by an oxy-acetylene flame or an electric arc. The technique of oxy-acetylene bronze welding is explained in more detail in subsequent sections – for joining galvanized steel (Figure 17.1) and for joining cast iron (Figure 17.2).

In both brazing and bronze welding, the filler material melts at a lower temperature than the base material and so they are classed as non-fusion processes.

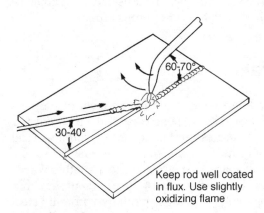

Keep rod well coated in flux. Use slightly oxidizing flame

Fig. 17.1 *Bronze welding of galvanized mild steel sheet*

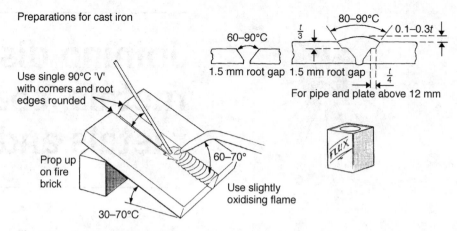

Preparations for cast iron

Use single 90°C 'V' with corners and root edges rounded

Prop up on fire brick

30–70°C

60–70°

Use slightly oxidising flame

60–90°C

1.5 mm root gap

80–90°C

$\frac{t}{3}$

0.1–0.3t

1.5 mm root gap

$\frac{t}{4}$

For pipe and plate above 12 mm

FLUX

Fig. 17.2 *Bronze welding of cast iron*

Awkward combinations of materials can sometimes be joined together by using what is known as a 'transition piece'. This is a piece of metal to which both materials can be readily welded. A novel method of manufacturing a transition piece is by the use of friction welding. **Friction welding** is capable of joining dissimilar metals together as it is a solid-state welding process (melting does not take place); it is therefore possible to weld materials such as aluminium to steel with this method. To make a transition plate that can be used within a fabrication, an aluminium tube can be welded to, say, a stainless steel tube of the same diameter by friction welding. The tube can then be cut and flattened into a plate and the respective sides welded by a conventional welding process, aluminium-to-aluminium and stainless steel-to-stainless steel.

Buttering is often used when joining dissimilar metals, again to reduce the amount of carbon pick-up by the weld. This is used in many cases, a typical example being the joining of mild steel to manganese steel. It is best to deposit a layer of manganese on to the mild steel before carrying out the joining weld. Pre-heat may also be necessary. Another method is to 'butter' the manganese steel surface with a layer of 18/8 stainless steel, and then join the latter to the low carbon steel either by a stainless steel electrode of similar composition or a mild steel electrode.

When joining cast iron to steel, one method is to deposit either a couple of layers of steel on the cast iron, or a layer of nickel-iron and then join using either a steel electrode or a nickel-iron electrode. For really difficult material, the cast iron can be bronze-welded to the steel and this can sometimes be made easier by coating the cast iron with a layer of bronze first.

Welding clad materials

A typical clad material is low carbon steel covered with a layer of stainless steel, to increase corrosion resistance at less expense than if the material as a whole were stainless steel. Such materials can be welded by stainless steel electrodes throughout, or by using mild steel electrodes to just below the cladding and then completing with stainless steel electrodes. Another way is to treat the clad side as the root and place the bulk of the preparation within the mild steel plate. The mild steel is then welded normally with mild steel electrodes, and the root chipped out to clean metal and welded from the clad side using a stainless steel electrode.

Welding galvanized steel plate

Galvanized plate is just steel plate with a coating of zinc, again to give protection from corrosion. The welding of this material gives off zinc fumes (see *Basic Welding*, page 14 for Health and Safety precautions) and should only be welded out-of-doors or with the use of a fume extractor. For prolonged welding a respirator should be worn. After welding using a mild steel electrode and technique or oxy-acetylene process, the weld area should either be regalvanized or at least painted to restore protection.

The technique of bronze (braze) welding mild steel (including galvanized steel)

Before bronze welding, ensure that the work surfaces are free from oil, grease and dirt. Oxides can be removed by grinding, filing or wire brushing. When welding galvanized steel, the zinc coating near the weld area can be protected to some extent by applying copper welding flux in paste form.

If pre-fluxed rods are not being used, then apply flux by heating the end of the rod and dipping it into powder, or by applying in paste form. A square-edge preparation with a gap of half the plate thickness can be used for up to 3.5 mm plate; above this thickness, change to 60–90° V preparation with a gap of around $\frac{1}{4}$ plate thickness. Set the plates with a slight tapered gap (wider at the end you are welding towards), and use a slightly oxidizing flame. The blowpipe and filler rod should be held as in Figure 17.1. To prevent overheating or oxidation, ensure that the flame impinges only on the melting filler and weld deposit.

When the pool is established, a technique of slightly withdrawing the flame allows partial freezing of the molten pool; the filler rod is then again introduced and the flame brought back. This careful procedure, when repeated, allows for heat control and the controlled progression of the molten pool. The weld area only needs to be heated to bright red for the brass to run freely and so care must be taken not to overheat. The finished weld surface should be bright in appearance and free from any porosity. Clean the flux residue from the completed weld. Fillet joints can be completed in a similar manner.

The welding of cast iron

Again, the bronze welding technique can be employed satisfactorily to many types of cast iron repair. It is generally used when a colour match is not important (that is, if the component is to be painted afterwards), or if a lower temperature process is necessary. The oxy-acetylene method is shown in Figure 17.2.

Tacks are made and then the surfaces to be joined coated with a layer of filler metal for a distance of about 20 mm ($\frac{3}{4}$ inch). The weld is then put in place by means of repeating 'tinning' and welding sequences.

'Cast iron' is a general term which is used to cover a wide range of alloys of carbon and iron with carbon content between 2 and $4\frac{1}{2}$ per cent. As was noted when discussing steels, increasing carbon content improves hardness with a decrease in ductility. From a welding point of view, we must always consider cast iron as being extremely hard and brittle; great care must therefore be taken in planning and carrying out welding operations on these materials.

Very broadly, the composition of cast iron will be 2–4 per cent carbon, 0.4–1 per cent manganese, 0.04–0.15 per cent sulphur, 0.4–1 per cent phosphorus and 1–3 per cent silicon. From the welding point of view, after the high carbon content, the sulphur is the most undesirable element because it will reduce the ductility of the material at red heat and increase the amount of shrinkage on cooling. In manufacture, therefore, attempts are made to keep the sulphur content as low as possible.

Cast iron welding is used mostly in the salvage or repair welding of cast iron components. Because of this fact, each welding job will present its own problems and require a planned procedure. Before welding can be undertaken on this material, then, we need to consider the different types of cast iron and the various welding methods available.

The various forms of cast iron are now considered in turn.

White cast iron

This is an extremely hard and brittle material, the carbon being combined as either cementite or martensite. A fracture in this material will have a white appearance which gives the metal its name. It is produced by rapid cooling, which will prevent the carbon from separating out, or by adding alloying elements that prevent the carbon from forming graphite. White cast iron is classed as very difficult to weld and is therefore not usually welded. In certain cases, welds can be made that are satisfactory, such as building up chilled iron rolls. Using coated manual metal-arc electrodes, the first couple of layers will be cracked and porous but will improve with a third and fourth layer. The coating will probably serve its purpose, but will be only moderately fused to the white iron. Oxy-acetylene welding of this material is also limited.

In all welding operations, care must be taken to avoid producing white cast iron, unless it is required for providing a hard surface. (When welding, fast cooling will produce white cast iron from grey cast iron, and the only way to restore it back to grey cast iron would be to remelt the whole casting!)

Grey cast iron

This is the most widely used type of cast iron, being employed for components where strength and wear resistance requirements are not too high and there is no shock loading in service. Grey cast iron can be found in use as flywheels, gear wheels, bedplates, machine frames, crankcases, cylinder blocks, brackets and pump bodies etc. Grey iron is so called because of the grey colour of a fracture surface in this material. When compared with white cast iron, grey cast iron is much softer and readily machinable. This is because all the carbon is not in the combined state, as with white cast iron, but a certain amount will be found in the form of free graphite as well as in the combined cementite form. The precipitation of the carbon in graphite form can be aided by the addition of silicon to the cast iron, however slow cooling is essential for the cementite to break down into free graphite and iron. This takes place between around 750°C (1382°F) and 650°C (1202°F) when the casting is cooling down and a definite expansion of the casting occurs between these temperatures. Too rapid a cooling rate, even when a casting contains silicon, will prevent the carbon from leaving the combination with iron, which will result in the formation of white cast iron. Manganese is added to cast iron to prevent the formation of sulphide films (see Figure 17.3). Both sulphur and manganese, as separate elements, will tend to prevent the breakdown from cementite to graphite but, as a compound of manganese sulphide, the effect is removed as

Low-strength sulphide films on grain boundaries will cause metal to tear apart under construction, as they remain liquid until a low temperature is reached

Manganese sulphide globules. Sulphur is removed from grain boundaries

Fig. 17.3 *Showing how addition of manganese prevents grain boundary films*

the sulphur is in globular form and not able to form boundary films.

The small amount of phosphorus present in cast iron will have no adverse effect on the carbon in grey cast iron. It will, however, improve the fluidity of the cast iron.

Grey cast iron can be welded by the electric arc or oxy-acetylene processes and the respective techniques are covered in the following pages. Large castings can also be welded, after pre-heating, by either the carbon-arc process using cast iron rods or the Thermit process (see Chapter 13).

Malleable cast iron

This type of cast iron is made by the prolonged heating of white cast iron in the presence of iron oxide. This has the effect of decarburizing the surface to a depth of about 3 mm (depending on the length of time it is heated). The result is to make the surface soft and ductile, and the general strength and ductility of the whole casting is considerably increased. It can be distinguished from ordinary cast iron by removing a chip by means of a hammer and chisel. A chip from a malleable iron casting will curl off in a similar manner to mild steel, instead of breaking in a brittle manner like a brittle cast iron chip.

As fusion welding of malleable iron would destroy its malleability and ductility, non-fusion welding is recommended. Oxy-acetylene welding using bronze filler rods will give the best results, as this will involve no reduction in the malleability of the metal and yet provide a joint as strong as the casting.

Painting the completed job after welding can conceal the difference in colour match.

Spheroidal cast iron

This is a special type of cast iron to which magnesium has been added in order to change the flake-graphite structure into spheroidal or nodular form. This has the effect of approximately doubling the strength and also increasing the ductility and shock resistance. It is readily welded by the normal methods discussed in the following pages. A typical example is the use of a 55 per cent nickel and 45 per cent iron electrode for electric arc welding.

Choice of welding methods

In general, if a full pre-heat at 650°C (1202°F) or 700°C (1292°F) for a casting of complicated shape, can be given, then the oxy-acetylene method, using the leftward technique and a cast iron filler rod and flux, will provide a very sound fusion weld, as will the use of a cast iron filler rod and the carbon-arc process. The Thermit process can be employed for large fractures. All of these methods are actually 'casting' similar material into the fracture and provide the nearest match to the parent material. Pre-heating furnaces can be either permanent gas-fired or 'temporary' charcoal (Figure 17.4). The charcoal type can provide more equal heating for complicated shapes. If possible, the casting should be welded in the furnace, the furnace lid replaced and the casting left to cool down with the furnace.

With the arc-welding procedure, using stick electrodes, only complicated castings should require pre-heating, unless the weather is very cold, in which case they should be pre-heated to around 300°C (572°F) to minimize the risk of cracking.

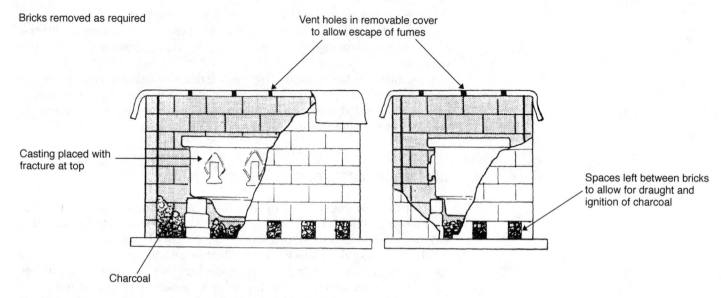

Bricks removed as required

Vent holes in removable cover to allow escape of fumes

Casting placed with fracture at top

Spaces left between bricks to allow for draught and ignition of charcoal

Charcoal

Fig. 17.4 *A temporary charcoal pre-heating furnace; bricks partially removed to show inside*

As with other materials, the prepared edges for cast iron welding must be clean, for thick sections vee-out both sides if possible, leaving a central portion of original fracture for alignment. Chipping or melting out is preferable to grinding, as grinding can smear graphite over the surfaces.

For gas welding, use cast iron rods of the high silicon type and a cast iron flux, as the surface oxide has a higher melting point than that of cast iron. The flux helps to break this down, as does stirring the molten pool with the filler rod and skimming any oxide out of the weld with the end of the rod. When adding filler metal always keep the end of the rod in the molten pool. Always use a strictly neutral flame.

For manual metal arc welding there are four main categories of electrode:

1. Those with a core wire of monel metal.
2. Those having a core wire of 55 per cent nickel/45 per cent iron.

Both these types will deposit a machinable weld metal and can be used for the repair of castings with or without pre-heating.

3. Those with a core wire of low-carbon mild steel. This type is sometimes used for the welding of thick-section repairs and will require full pre-heating if machining is required.
4. Those having a bronze core wire. These can be used for the repair of thin section castings and malleable iron where the colour of the deposit does not matter too much. Pre-heating is not required.

If castings are arc welded without pre-heat, the main problem will be to keep the contraction stresses to a minimum and avoid hard zones. This can be achieved to a large extent by keeping the casting as cool as possible and employing a filler metal giving a deposit that is capable of absorbing contraction stresses. Skip welding should be employed, depositing short runs (40 mm length) and, immediately, lightly peening the deposit with a ball peen hammer, in order to expand the shrinking weld deposit and, in so doing, reduce the contraction stresses (see Figure 17.5). The depositing of short weld beads helps to reduce the heat input, minimizing the chance of hard zones forming. As a test, you should be able to place your hand on the casting, about 80 mm away from any welding area before recommencing welding; if it is not possible to do this, the casting is too hot and welding should not be recommenced until it has cooled.

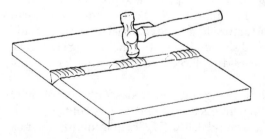

Fig. 17.5 *Peening and skip-welding method*

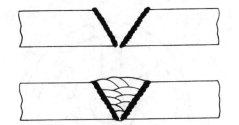

Fig. 17.6 *The buttering technique for welding cast iron*

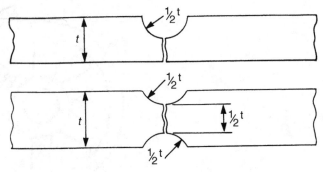

Fig. 17.7 *Examples of single-sided and double-sided preparations for welding cast iron fractures*

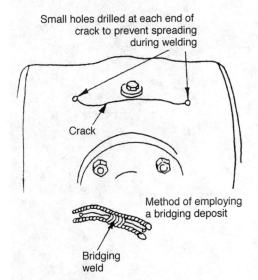

Small holes drilled at each end of crack to prevent spreading during welding

Crack

Method of employing a bridging deposit

Bridging weld

Fig. 17.8 *Methods to prevent a crack from spreading*

Fig. 17.9 *Studding as preparation for the welded repair of a cast iron gear tooth*

For full thickness welds (Figure 17.6), a buttering layer can be deposited on each face, using nickel alloy electrodes; this will ensure that the fusion zone will be soft, ductile and free from hard carbides. The bulk of the weld may then be completed by using less expensive low carbon steel electrodes, which, as a result of the buttering layer, will no longer be affected by the carbon in the casting.

When access to the repair is limited to one side, a grooved preparation can be employed (Figure 17.7) although preparation from both sides is preferable. As the deposited metal is stronger than the grey cast iron, the weld meet need not be made through the full section thickness and, as stated earlier, retaining an area of fracture face can assist in aligning the components.

Mild steel studs can be screwed into the fracture surfaces in order to give basic reinforcement for welded repairs. On fracture surfaces, studs should be staggered and should project from the fracture face for a distance of 3 mm (1/8 inch) less than the amount of build-up required.

Procedure for welding pre-heated castings with buttered faces

Deposit up to half the electrode in one run, as long runs can cause transverse cracking. Allow the welding area to cool down to the pre-heat temperature before depositing the next run. Use the skip welding technique to balance out heat input. Deposit on one side of the joint and then the other, joining together the metal deposited on both sides with the final runs.

Procedure for welding cold castings with prepared or buttered faces

If pre-heat is not possible, allow the heat to disperse after depositing each run – down to hand heat. Restrict the length of each run to a maximum of 75 mm (3 inches). Do not weave the electrode.

A tip when welding a crack in a casting is to drill a small hole at each end of the crack to prevent it spreading during welding (Figure 17.8).

A grey cast iron that has been subjected to repeated high temperatures will become unweldable owing to internal oxidation. Items such as fire-grates that have been subjected to high temperatures are therefore unweldable. If an oxy-acetylene flame is applied to such material, it will refuse to melt and give off a dazzling light similar to when a flame is directed on to a fire brick.

Figure 17.9 shows how studs can be screwed into a broken face (here the fracture is on a gear tooth). Welding round the studs will give extra strength to the repair. Figures 17.10 and 17.11 show the welding sequence and local

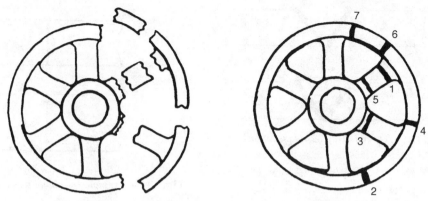

Fig. 17.10 *Welding sequence for repairing a badly broken cast iron gear wheel (see also Figure 17.12)*

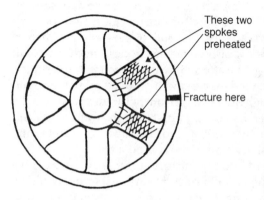

These two spokes preheated

Fracture here

Fig. 17.11 *Showing local pre-heat to spokes in order to weld fracture in rim*

Fig. 17.12 *Broken gear wheel teeth repaired by welding without pre-heat using nickel/iron electrodes*

(Photograph courtesy of UTP, UK/UTC, Nigeria and Ghana)

pre-heat required for two types of repair on a cast iron wheel. Figure 17.12 shows a typical cast iron repair weld. Such repairs can save a lot of money, for they not only save the cost of the original component and the cost of a new component, but much time in waiting for a new component to arrive, should one be available.

18

The welding of stainless steels, aluminium, copper and zinc based die castings

The welding of stainless steel

Stainless steel covers a group of corrosion and heat-resisting steels containing larger percentages of chromium or chromium and nickel than contained in the high tensile steels.

There are many different compositions but all the stainless steels fall into four main groups.

Martensitic stainless steel

This group contains from 11.5 to 14 per cent chromium and from 0.2 to 0.4 per cent carbon. Such steels are difficult to weld because they can form the very hard martensitic structure, regardless of the cooling rate, and are known as air hardening steels. They can be welded by pre-heating to around 350°C (662°F) and the use of a 25 per cent chromium and 20 per cent nickel flux-coated electrode. A post-heat of 750°C (1382°F) usually ensures acceptable ductility.

Ferritic stainless steel

These contain between 16 and 30 per cent chromium with a maximum of 0.1 per cent carbon, the base material being iron. These materials can be welded with a pre-heat of 150°C (302°F) and the use of the 25 per cent chromium and 20 per cent nickel core wire electrode. Sometimes combination welds are made, with the final layers being completed with electrodes of even higher chromium content (up to 30 per cent), in order to give the surface an extremely high resistance to corrosion. A post-heat at 730°C (1346°F) should be carried out immediately after welding to prevent brittleness.

Austenitic stainless steel

This third group contains chromium and nickel in amounts that give a predominantly austenitic structure. The composition of austenitic stainless steels can be varied to suit application, with chromium present from between 7 to 30 per cent and a range of nickel content from 6 to 36 per cent, with carbon content below 0.25 per cent, the base metal being iron. One of the most important steels in this group contains approximately 18 per cent chromium and 8 per cent nickel, with a carbon content less than 0.12 per cent. This steel is readily welded by many processes if the addition of small amounts of titanium or niobium is made in the manufacturing process (see section on 'weld decay' later in the chapter).

For special use at high temperatures, steels with 25 per cent chromium and 20 per cent nickel can be used.

In general, austenitic stainless steels are much more easily welded than martensitic and ferritic. Although they work-harden more easily than mild steel, they can be readily formed by the usual fabrication processes. Although oxy-acetylene cutting will not work on stainless steel, as it cannot form oxides, powder injection cutting and the plasma arc will cut through it easily, the best results being obtained with plasma.

Stainless steels have a higher coefficient of expansion and a much lower heat conductivity than does mild steel. This means that distortion can be more of a problem and, because of the lower conductivity, shorter electrodes are made to avoid overheating when welding with SMAW (stick electrodes).

Duplex stainless steel

These steels possess a duplex structure of roughly 50 per cent ferrite and 50 per cent austenite. The recommended welding approach is to use filler material that will match the base metal as closely as possible, but will also provide an as-welded structure near to 50 per cent austenite/50 per cent ferrite. This is achieved by increasing the nickel content.

The specific considerations regarding the welding of duplex are looked at later in this chapter.

Hot cracking (see Chapter 16) can be a problem in fully austenitic welds, because impurities in the form of low-melting-point compounds can form at the grain boundaries, weakening them.

This problem can be overcome when welding most austenitic materials by adjusting the filler rod/electrode material to produce around 4 to 5 per cent ferrite (in some cases up to 10 per cent ferrite but higher ferrite content can cause loss of ductility when 18Cr/10Ni and 25Cr/20Ni types are heated in the sigma-forming temperature range). This is because it is considered that harmful impurities are more soluble in ferrite than austenite.

As ferrite is magnetic, it can be detected in a non-magnetic fabrication and measured by a magnetic instrument (**ferritescope**) calibrated to AWS A4.2. Such an instrument will give readings in **ferrite number**.

The amount of ferrite can also be estimated from the composition of the base material and filler material by using a constitution diagram.

Various alloying elements may be classed as either austenite-formers, such as nickel, carbon, nitrogen and manganese, and ferrite-formers, such as chromium, molybdenum, silicon, niobium and aluminium.

The oldest constitution diagram, from 1948, is the Schaeffler (Figure 18.1), which has a chromium equivalent plotted along the horizontal axis and a nickel equivalent on the vertical axis.

When determining the deposited filler metal composition, dilution (Figure 18.2) must be taken into account. (This may be defined as the reduced alloy content of a weld deposit due to melting and incorporating melted base metal of **lower** alloy content.)

As a percentage, dilution may be defined as:

$$D = \frac{\text{weight of parent metal melted}}{\text{total weight of fused metal}} \times 100 \text{ (see also Figure 18.2)}$$

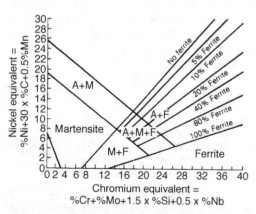

Fig. 18.1 *The Schaeffler constitution diagram*

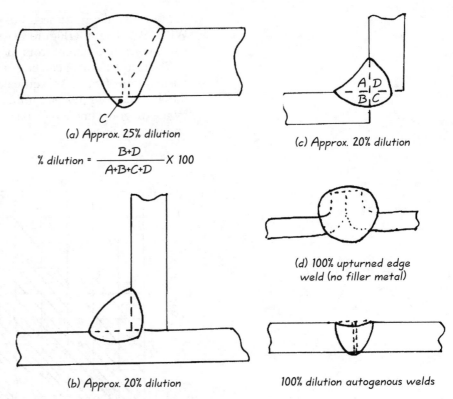

(a) Approx. 25% dilution

$$\% \text{ dilution} = \frac{B+D}{A+B+C+D} \times 100$$

(c) Approx. 20% dilution

(d) 100% upturned edge weld (no filler metal)

(b) Approx. 20% dilution

100% dilution autogenous welds

Fig. 18.2 *Determining the percentage of dilution*

The Schaeffler diagram is now considered to be outdated, one reason for this being that it does not consider the effects of nitrogen. An improvement, which includes nitrogen (N) and can be used to estimate the level of ferrite, is the WRC DeLong diagram (Figure 18.3).

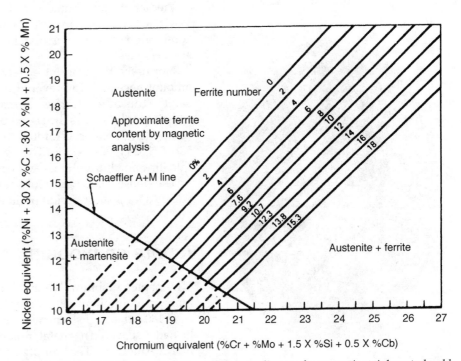

Fig. 18.3 *The Welding Research Council DeLong diagram for austenitic stainless steel weld metal*

However, the most accurate predicting diagram is considered to be the WRC 1988 diagram (Figure 18.4). On this diagram the Ni equivalent and Cr equivalent are different from those on the Shaeffler and WRC DeLong.

The ferrite number can be estimated by drawing a line horizontally across the diagram of the nickel equivalent number and a vertical line upwards from the chromium equivalent number. The ferrite number will then be indicated by the diagonal line which passes through the intersection of the horizontal and vertical lines.

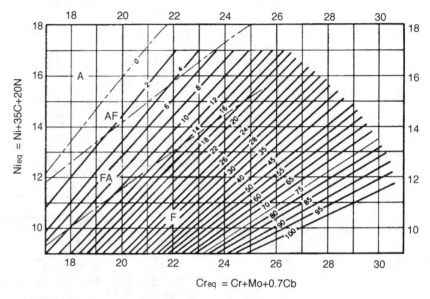

Fig. 18.4 *WRC diagram including solidification mode boundaries*[1]

Predictions for common grades are similar for both the WRC 1988 and the WRC DeLong diagrams, but the WRC 1988 is generally most accurate for less common grades such as manganese austenitic or duplex austenitic–ferritic stainless steels.

Normally, the amount of ferrite present should not be greater than an amount necessary to prevent hot cracking with some reasonable margin of safety. Too much ferrite can cause the formation of a crystallographic constituent known as the sigma phase. This can form slowly, in service if the component is exposed for long periods to high temperatures in the range of 538–871°C (1000–1600°F). The sigma phase increases hardness, but decreases ductility, notch-toughness and corrosion resistance of stainless steels.

The phase can be redissolved by heating the component to above 899°C (1650°F) for a short period of time.

A note on niobium, columbium and tantalum

(The names niobium, Nb and columbium, Cb are both used for the same material, columbium being more generally used in the USA).

In 1801, the English chemist Charles Hatchett (Figure 18.5) announced that he had discovered a "metal hitherto unknown" contained in "a mineral from North America". The sample of the black rock, now known as columbite, which was examined by Hatchett, had been discovered many years earlier by Governor John Winthrop the Younger, the first Governor of Connecticut.

Fig. 18.5 *Drawing of Charles Hatchett, 1765–1847 (from a picture at the Institution of Metallurgists)*

Some time later, his grandson had sent the sample to London, UK, where it had eventually been placed in the British Museum, and it was here that it attracted the interest of Hatchett.

Charles Hatchett was born in 1765 and gave up chemistry quite early in life in order to succeed his father in running the family's coach-making business. He died in Chelsea at the age of 82 on 10 March 1847.

Tantalum was discovered a year later than columbium, in 1802, by the Swedish chemist Anders Gustaf Ekeberg in a specimen of tantalite from Kimito, Finland, and also in yttrotantalite from Ytterby. Because it had been a tantalizing task to track the element down, he named it tantalum. Ekeberg was partially deaf and had lost the sight of one eye in 1801, when a flask had exploded in his hand.

These two discoveries of columbium and tantalum were not immediately accepted, with some scientists of the day claiming that they were both the same metal.

In 1846, however, the German analytical chemist Heinrich Rose established that they were, in fact, two elements. He proposed the names tantalum and niobium, the latter after the goddess Niobe, who in Greek mythology was the daughter of Tantalus. There has never been any doubt, however, that the niobium found by Rose in 1846 was the same element as the columbium found by Hatchett some forty five years earlier. There is, therefore, a good case that the name columbium should be retained for this element.

However, the International Union of Chemistry decided to recommend the name niobium from 1936, and this has tended to increase the use of this name in European countries.

It is for these reasons that information from the USA will tend to use the chemical symbol Cb, while Nb can be found on charts and information from European countries.

Weld decay

An 'unstabilized' stainless steel, that is one without additions of titanium or niobium, is likely to suffer from what is known as intercrystalline corrosion, or 'weld decay' if welding takes place. This condition is due to precipitation of chromium carbides at the grain boundaries in those areas where the steel has reached temperatures within the range of 450–850°C (840–1560°F), as a result of the heat of welding. This condition can occur at a slight distance from the actual weld, on either side in the plate material (see Figure 18.6).

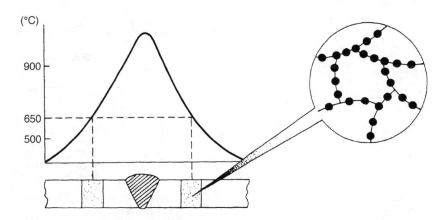

Fig. 18.6 *Weld decay. Precipitation of chromium carbides at grain boundaries leaves areas of chrome depletion open to attack*

The removal of chromium from the edges of the grains makes the material much less resistant to corrosion and can cause failure of the material.

Weld decay can be prevented by heating the work to 1100°C (2010°F) and then cooling very rapidly, or by using material with a very low carbon content (less than 0.04 per cent). The best solution, and the one most widely employed, is to use materials containing added elements that have a greater affinity for carbon than chromium has. These elements are called stabilizing elements and include titanium, niobium (columbium) and tantalum (see note on niobium and columbium above). It is fairly common practice for titanium to be added to a plate material and niobium to be added to electrodes.

There is a very wide range of stainless steels, and matching electrodes and filler rods are available, designed to give a deposit that will match the different parent metals as closely as possible.

These stabilized stainless steels can be welded by all the main processes: oxy-acetylene, manual metal-arc, tungsten arc gas shielded and metal-arc gas shielded will all produce sound welds when the following considerations are taken into account.

Because of the higher coefficient of expansion, great care must be taken in the prevention of distortion. For example, tacks must be made closer than for mild steel and they should be ground out before completing the joint. In order to minimize the chance of distortion taking place, careful consideration should be given to the sequence in which welds are deposited. Weaving should be kept to a minimum, or not used at all, and a short arc should be maintained. Weld preparations are usually the same as those employed for mild steel, although when welding thin sheet by the manual metal arc process (SMAW) up to and including 3 mm, it is good practice to deposit a run of weld on both sides (see Figure 18.7) to ensure that there is sound metal on both faces. When using a backing bar, this will not be necessary, as good penetration can be produced by welding from one side, using a larger-gauge electrode.

For oxy-acetylene welding, the joint area should be cleaned on both sides. Any oil or dirt should be removed using a solvent (read the section on health and safety in Appendix 5) and then wire-brushed using a stainless steel brush or stainless wire wool. Brush stainless steel flux paste on the underside of the joint; fluxing of the filler rod is not normally necessary. Use a strictly neutral flame and the leftward technique. On fabrications, plan the deposition of runs

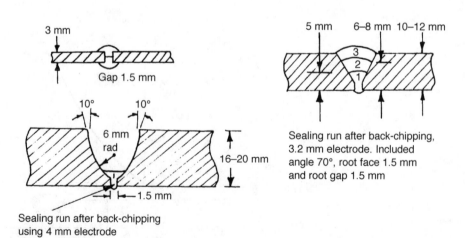

Sealing run after back-chipping
using 4 mm electrode

Sealing run after back-chipping,
3.2 mm electrode. Included
angle 70°, root face 1.5 mm
and root gap 1.5 mm

Fig. 18.7 *Example procedures for butt welding of austenitic stainless steels by the manual metal-arc process in the flat position (SMAW)*

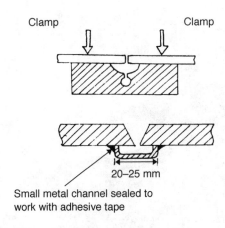

Small metal channel sealed to
work with adhesive tape

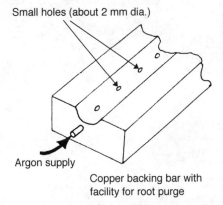

Small holes (about 2 mm dia.)

Argon supply

Copper backing bar with
facility for root purge

Fig. 18.8 *Methods of back-purging (protecting root with shielding gas) applicable to GTAW and GMAW welding*

to spread out heat input and to avoid an end crater, and withdraw the flame slowly, in order to allow the weld metal to solidify under its protection.

The tungsten arc gas shielded (GTAW) and metal-arc gas shielded (GMAW) processes are well suited to welding this material. Tungsten arc gas shielded is capable of producing welds of high quality, although it is of course much slower than manual metal-arc (SMAW) or metal-arc gas shielded (GMAW). Backing bars with the facility for back purging (see Figure 18.8) can be employed with both GTAW and GMAW in order to control and protect the penetration bead.

For GTAW welding, the plates should be cleaned as for gas welding, although of course no flux will be required. A direct current supply, having the tungsten electrode negative, is used. It is best to use a thoriated tungsten if possible, as this will produce a more stable arc and can be used with higher currents than pure tungsten electrodes. The shielding gas is usually argon, however mixtures of argon and helium or argon and hydrogen (up to 5 per cent) can be employed on thicker plate. These mixtures can help to give a more fluid weld pool and, because of the higher temperature of the arc, faster deposition rates. All draughts should be avoided with gas shielded methods and tents should be placed over welding points when welding out-of-doors in order that the protective shield is not blown away. The GTAW welding torch should be held almost vertical to the line of travel, feeding the filler rod into the leading edge of the molten pool and never removing the heated end of the rod from the protection of the gas shield while welding. To prevent excessive dilution (see Figure 18.2), maintain weaving to a minimum. The electrode extension (beyond the nozzle) should be as little as possible, at 3.0–5.0 mm for butt welds and 6.0–12 mm (max.) for fillet welds.

The arc length should be maintained at around 2 mm when welding without filler and between 3 and 4 mm when filler rod is employed. Gas shielding should be plentiful, using a gas lens (see Figure 7.4) if required, to prevent turbulence.

For GMAW welding, as with the other processes, the material should be cleaned around the joint area and the choice of filler material wire composition should match the material being welded and the properties required from the welded joint. Generally, all austenitic wires will fall into the following designations: 18/8 stainless steel, 18/8 stabilized stainless steel, 25/12 heat-resisting steel, 25/20 heat-resisting steel and 18/8/3 molybdenum stainless steel. The shielding gas most used for austenitic steels is argon with 2 per cent oxygen at flow rates of 1.15 m³/h for spray transfer and 0.80 m³/h for pulsed transfer.

For welding stainless steel to mild steel it is best to use a stainless steel electrode (or wire). This is even more necessary when making welds between stainless steel and low alloy or high carbon steels to prevent weld embrittlement and cracking. Dilution must also be considered (see Figure 18.2). Figures 18.9 and 18.10 show examples of SMAW welding of stainless steel and GTAW.

Welding duplex stainless steels

Duplex stainless steels are alloys of chromium and nickel with iron which can have twice the yield strength of 18Cr–18Ni stainless steels, together with low-temperature strength and high corrosion resistance.

Duplex material is used in natural gas processing within high-pressure systems that carry the wet natural gas for treatment. The material used has

Fig. 18.9 *Small stainless steel vessel fabricated using SMAW (manual metal-arc process) and an all-positional niobium-stabilized stainless steel electrode*

(Photograph courtesy of UTP, UK/UTC, Nigeria and Ghana)

Fig. 18.10 *Welded stainless steel obelisk at rear of AWS offices, Miami, USA (GTAW)*

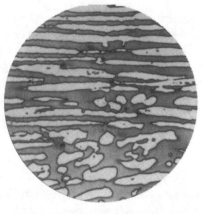

Fig. 18.11 *Photomicrograph showing ratio of austenite to ferrite in a duplex stainless steel. The light phase is austenite and the dark phase is ferrite*

(Photograph courtesy of AWS, Miami, USA and Lincoln Smitweld)

to be extremely corrosion resistant, as the water and carbon dioxide in the gas can combine to form carbonic acid. A further problem is hydrogen disulphide which is also highly corrosive and known as 'sour gas'.

Use in such conditions requires limits on the amount of nickel, with control of heat input to maintain corrosion resistance and control of weld hardness in order to prevent cracking.

In order to give the best compromise between corrosion resistance and toughness, the ferrite/austenite ratio should be as close as possible to 50/50 (Figure 18.11). The steels are called 'duplex' because the metallurgical structure consists of this mixture of both ferrite and austenite.

Duplex stainless steel is being used increasingly in the oil and chemical processing industries. There are many different types to suit specific working environments but the chemical composition generally falls within the following range: carbon – less than 0.3 per cent; chromium – 18.0 to 27 per cent (principal contributor to corrosion resistance); nickel – 5 to 10 per cent (improves toughness and assists austenite formation); molybdenum – 0 to 3.5 per cent (improves corrosion resistance); nitrogen – 0.1 to 0.35 per cent (helps to maintain a balanced ferrite/austenite ratio); and copper – 0 to 1.5 per cent.

Duplex stainless steels can be readily welded by GTAW (TIG), GMAW (MIG/MAG), SMAW (MMA), SAW, FCAW and PAW.

Weld-metal ferrite content should be controlled, as high ferrite levels above about 80 per cent will reduce the material's toughness and increase the possibility of hydrogen embrittlement.

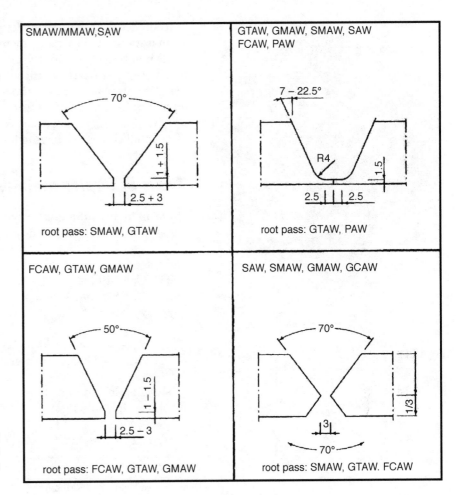

Fig. 18.12 *Typical joint preparations for welding duplex stainless steel by various processes (after Hilkes and Bekkers, Lincoln International and Lincoln Smitweld)*

Semi-automatic or automatic welding processes can quite accurately control the amount of filler-metal addition. When manual processes are employed, care must be taken with filler-metal control, as insufficient filler metal can cause a major increase in ferrite content, resulting in possible embrittlement.

For this reason, 'V' preparations with root gap should be employed, to ensure that the root run contains the correct amount of filler, with minimum dilution. Gas backing with argon is generally recommended.

The physical properties of duplex can require special welding procedures. For example, when welding heavy sections, the slower cooling rates of base metal and the heat affected zone can prolong the amount of time spent within the temperature range at which embrittlement can occur. The weld metal and localized base metal can also retain heat longer during welding, making weld-pool control more difficult. Differences in the electrical conductivity can require a reduced welding current, in order to avoid overheating of the electrode.

Figure 18.12 shows typical joint preparations for welding duplex stainless steel by various processes.

Duplex stainless steels

Introduction

The term **duplex stainless steel** covers a group of high-strength, corrosion-resistant steels that contain higher chromium percentages and lower nickel

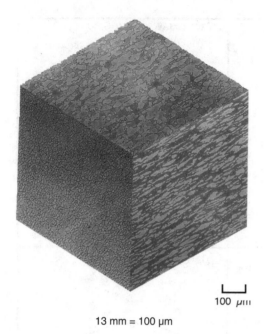

Fig. 18.13 *The X–Y–Z microstructure of Zeron 100 pipe parent material,[2] produced by the powder technology route*

100 μm

13 mm = 100 μm

percentages than their equivalent austenitic stainless steel, and have nitrogen as an important alloying element.

As mentioned previously, duplex stainless steels receive their name from their two-phase microstructure, the parent material comprising approximately equal volumes of ferrite and austenite, as shown in Figure 18.13. This microstructure is obtained from a solution-anneal heat treatment from about 1050°C (1920°F), and consists of islands of austenite in a ferrite matrix, the directional properties having been induced by hot working.

Types of duplex stainless steel

There are many different compositions, but duplex stainless steels generally fall into four main grades, as follows:

Grade		Pre_N*
23% Cr	Mo-free	~25
22% Cr	Standard	30–36
25% Cr	0–2.5% Cu	32–40
25% Cr	Superduplex	>40

* Pitting resistance equivalent (PRE_N) = %Cr + 3.3 × %Mo + 16 × %N

The chemical compositions (per cent) of selected popular **European** grades are shown in Table 18.1.

Table 18.1 *Commercial grades of duplex stainless steel*

Manufacturer	Grade	Cr	Ni	Mo	N	Cu	W
Avesta-Sheffield	SAF 2304	23	4	—	0.1	—	—
Sandvik	SAF 2205	22	5	3	0.14	—	—
Creusot Ind.	UR 52N	25	7	3	0.16	0.2	—
Weir Materials	Zeron 100	25	6.5	3.7	0.25	0.7	0.7

Technical welding considerations

All the general rules applied to austenitic stainless steels with regard to handling, segregation of materials, plasma or mechanical cutting, weld preparation, cleanliness etc. also apply to duplex stainless steels. The parent material and welding consumable manufacturers make readily available a wealth of technical information on their products and application.

The weldability of duplex stainless steel is similar to that of the austenitic grades and achieving sound welds is not a problem. Under normal circumstances they are virtually free from the risk of hydrogen cracking and solidification cracking. However, as the alloying-element content is increased, the need to control heat input becomes more important. Many complex deleterious precipitations can occur if the heat input is too high and cooling rate is too low, as shown in Figure 18.14.

The weld-metal microstructure is quite distinct from that of the parent material and is controlled by composition and cooling rate. On solidification, the structure is fully ferritic, as shown in Figure 18.15. As the temperature falls below the ferrite solvus line, a partial transformation from ferrite to austenite occurs by diffusion-controlled reactions. Too rapid cooling rates, from low heat input welding processes, result in high ferrite levels. Too low cooling rates, from high heat input welding processes and uncontrolled inter-

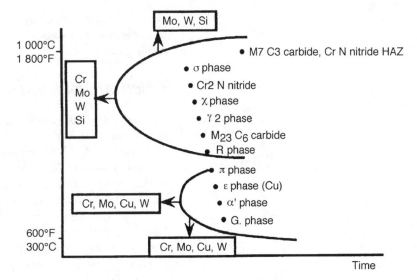

Fig. 18.14 *Time, temperature, transformation (T.T.T.) diagram showing the effects of deleterious precipitations that can cause embrittlement, and loss of mechanical and corrosion-resistance properties (after J. Charles[3])*

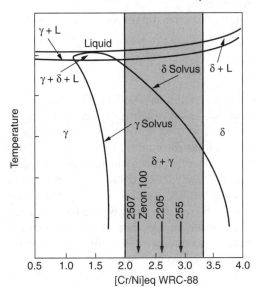

Fig. 18.15 *Modified ternary section of the FE–Gr–Ni phase diagram, showing a number of commercial grades of duplex stainless steel[4]*

pass temperatures, result in the deleterious precipitations previously mentioned. Thus, care has to be taken with these materials.

To assist this austenite formation, welding consumables have a higher nickel content than the parent material (usually about 9 per cent nickel).

Unfortunately, the high-temperature heat affected zone (HTHAZ) is also heated to above the ferrite solvus and, not having a subsequent solution-anneal heat treatment, nor the benefit of the higher nickel content, can become highly ferritic, as shown in Figure 18.16. This can result in reduced corrosion resistance. Weld-metal austenite contents can also vary through a multi-pass weld. Typical weld-metal microstructures are also shown, depicting increased austenite formation in the cap, as a result of dilution in the previous run with its increased nickel content, the root pass being diluted with the lower nickel parent material.

It can be clearly understood that duplex stainless steels are fairly complex materials and, as such, additional care must be taken with them. Of particular importance is the training of welders and inspection personnel, and to have

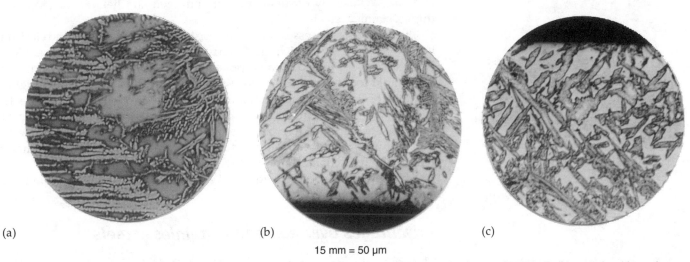

15 mm = 50 µm

Fig. 18.16 *The weld microstructures from Sandvik SAF 2507 pipe parent material showing: (a) ferritized HTHAZ; (b) typical weld metal at root; (c) typical weld metal at cap[2]*

(Photographs courtesy of Roy Bradshow)

clearly defined welding-procedure specification parameters, inter-pass temperature and deposition sequences in order to control heat input and cooling rates.

Practical welding considerations

Duplex stainless steels can be readily welded to oil and gas industry requirements using the commonly used welding processes SMAW, GTAW, GMAW, FCAW, SAW and also PAW.[5]

Autogenous welding is not recommended unless a nitrogen addition to the argon shielding gases can be made, nitrogen being a strong austenite-former. If a solution-anneal heat treatment can be applied, then autogenous welding and the use of matching welding consumables can be performed. However, in the vast majority of cases, welding consumables are overalloyed with nickel to give the required phase balance.

Heat inputs are generally within the following ranges for manual welding processes:

2205	0.5–2.5 kJ/mm
Superduplex	0.5–1.5 kJ/mm

Inter-pass temperatures are usually restricted to 100°C (212°F) max.

Open joints are preferred for manual one-sided welding. This encourages the welder to use considerable amounts of filler and, thus, makes autogenous welding or very highly diluted welds and their high ferrite content less likely. GTAW is the most commonly used process for critical work, producing the highest quality with control over root penetration. Unfortunately, it is also the process most open to abuse by the welder with regard to heat input. Welders must be aware that techniques that ensure good fusion with infrequent filler wire additions may exceed the heat-input limitations. Welding techniques that 'just' give adequate fusion are usually preferred. Thus, bead techniques with high filler wire feeding rates are usually specified and welders trained accordingly.

SMAW basic electrodes are often used to fill and cap pipe joints rooted with GTAW, in order to increase deposition rates.

GMAW and FCAW consumables are also available, however consistency in wire quality, especially for wire of less than 1 mm diameter, can be problematic. Many proprietary gas mixes are becoming available for GTAW and GMAW, usually containing argon, nitrogen and helium for usage on duplex stainless steels.

Other welding processes, including resistance (RW), laser (LW) and electron beam welding (EBW), which have been routinely applied to austenitic stainless steels, can cause problems when applied to duplex stainless steels. This is due to the rapid cooling rates associated with these processes, which lead to highly ferritized HTHAZs. As such, they are not yet applied to duplex stainless steels.

Gas (GW) and electro-slag welding (ESW) are also unsuitable for duplex stainless steel application, because of the high heat inputs giving rise to deleterious precipitation and resulting in unacceptable and dramatic reductions in mechanical properties and corrosion resistance.

Applications of friction welding (FW) to date have been limited, but have shown some promise.

Advantages over austenitic stainless steels

The main advantage of duplex stainless steel grades over their equivalent austenitic stainless steel grades comes from their outstanding mechanical and corrosion-resistance properties. The yield strength is almost double that of the austenitic grades, with higher ultimate tensile strengths, while maintaining good elongation.

Duplex stainless steels also have good resistance to general corrosion and chloride-induced (that is, marine-environmental) pitting corrosion, with outstanding resistance to both chloride- and hydrogen-induced stress corrosion cracking.

In addition, their price is reasonable, because of the relatively low nickel content.

Disadvantages

All grades display only moderate toughness and they are prone to thermal ageing. Thus, they cannot be used for cryogenic purposes or as a heat-resistant material. Their normal safe usage is restricted to about −20° to +250°C.

Practical applications

The first serious usage of duplex stainless steels was in the oil and gas industry. The materials' corrosion-resistance properties and high strength have been utilized in pipelines, process pipework, vessels, pumps and heat exchangers.

On offshore platforms, their weight-saving capability (resulting from their high yield strength) topside was quickly exploited. Their use onshore in other industries is continuing to grow.

The welding of aluminium

Aluminium and aluminium alloys are the lightest commercial metals in use in large quantities. Correct alloying gives the alloys good tensile strength. The manual metal-arc and oxy-acetylene processes require corrosive fluxes made from chlorides and fluorides in order to dissolve the layer of aluminium oxide (alumina) from the metal's surface and also prevent further oxidation during welding. These fluxes require that joint design must be such that flux will not become trapped and allowed to 'eat' away at the joint during service. These fluxes and flux residues must also be thoroughly removed after welding. This is carried out by washing and brushing with hot water after the component has been cooled, or by immersion in a 5 per cent solution of nitric acid in water, followed by washing and rinsing with hot water.

If manual metal-arc and oxy-acetylene processes are the only ones available, then lap and fillet joints must be avoided, as they will tend to trap residues of this corrosive flux in areas that cannot be reached by post-weld cleaning.

If TIG (GTAW) or MIG (GMAW) welding is available, it will allow these types of joints to be readily welded, as, of course, no flux is required with these two processes (the oxide film being removed by the electrical cathodic cleaning action and further contamination during welding being prevented by the inert gas shield).

For all processes, care should be taken that the area to be welded is free from contamination (paint, oxides, oil etc.); it should therefore be cleaned immediately before welding takes place. (Read the safety notes at the end of Appendix 5; if solvents are used, they must be completely evaporated from the area before welding takes place.)

Aluminium has a low melting point, around 660°C (1220°F) and a high thermal conductivity. These properties, combined with the fact that there is virtually no colour change before melting takes place, require practice to get used to and it is recommended that, if a welder has never welded aluminium before, some time should be spent in practice on scrap materials to master the technique fully, before attempting a real job.

For oxy-acetylene welding, a good flux having a melting point of around 570°C (1058°F) will be necessary to dissolve the oxide film, which has a higher melting point than the aluminium. The filler rod should be dipped in the flux and then made to wash down the rod using the torch flame. On thin plate, the upturned edge preparation can be used, as for mild steel, without the use of a filler rod. For thicker sections, over 3 mm, the edges should have a 90 degree 'V'. For plates over 6 mm in thickness, a double 90 degree 'V' may be used. The flame should be adjusted to neutral and the leftward technique employed. Post-weld cleaning should then be carried out after welding.

For manual arc-welding, the electrode should be connected to the positive terminal of a direct current welding supply. It will be found that a layer of flux can form on the end of the electrode in use and that it has therefore to be struck very hard in order to re-ignite the arc.

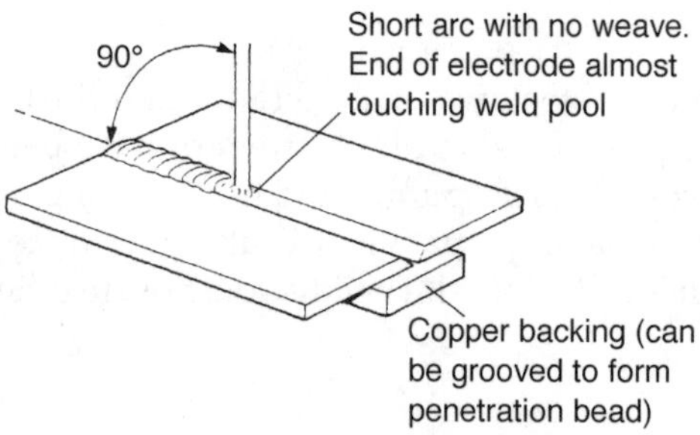

The electrode should be maintained at right-angles to the weld and a short arc with no weaving employed (see Figure 18.17). Too long an arc length can allow oxidation and the formation of a brittle weld. Castings should be pre-heated to 200°C (392°F) [for pure aluminium and non-heat-treatable alloys 350°C (660°F)] prior to welding; this reduces the tendency to cracking and speeds up welding.

Tungsten arc gas shielded welding is usually employed on aluminium using an alternating current supply, as the oxide film is removed when the tungsten is positive on one half-cycle, allowing the tungsten to cool when it is negative on the next half-cycle. However, direct current pulsed arc welding can be employed on very thin sheet of high purity.

Argon or helium is employed as the shielding gas and electrodes for use on AC should be of 0.8 per cent zirconiated tungsten. The electrode tip should be ground as shown in Figure 18.18.

For best results, filler rods should be used as soon as they are removed from the packet.

When metal-arc gas shielded welding is employed, argon is the recommended shielding gas. Helium can be used for welding very heavy sections as it tends to provide deeper penetration welds, but the arc is more erratic in helium and the resulting weld bead appearance is usually not as good as for a weld produced with argon shielding. Mixtures of argon and helium can be employed to give advantages not obtained with either gas on its own.

Some aluminium wires, such as the pure aluminiums (99.8 or 99 per cent aluminium), are too soft to feed through a long conduit from a spool based on the welding machine and a spool-on-gun arrangement (Figure 18.19) can be employed in such cases. The addition of magnesium and other alloying elements can help to stiffen the wire, allowing the use of large machine-mounted reels. Again, cleaning of the joint area will be required and over-handling of the electrode wire should be avoided. Wear clean cotton gloves when loading the spool and replace the plastic cover over the spool electrode wire as soon as possible to prevent possible contamination. The gun should point in the direction of welding at an angle of 15 degrees to the vertical. Ensure that the welding area is free from draughts and that fume extraction equipment is not over-powerful, as this can disturb the protective gas shield. As with the other processes, temporary or permanent backing can be employed. Pipes can be filled with argon from a separate supply, using inserted dams to prevent wastage, in order to give a protective back purge.

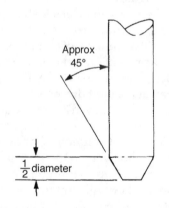

90°

Short arc with no weave. End of electrode almost touching weld pool

Copper backing (can be grooved to form penetration bead)

Fig. 18.17 *Manual metal-arc welding of aluminium (SMAW)*

Approx 45°

$\frac{1}{2}$ diameter

Fig. 18.18 *Electrode shape for AC TIG (GTAW) welding. It will round up once the arc is struck*

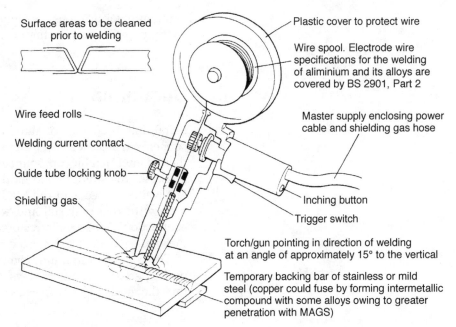

Surface areas to be cleaned prior to welding

Plastic cover to protect wire

Wire spool. Electrode wire specifications for the welding of aluminium and its alloys are covered by BS 2901, Part 2

Wire feed rolls

Welding current contact

Guide tube locking knob

Shielding gas

Master supply enclosing power cable and shielding gas hose

Inching button

Trigger switch

Torch/gun pointing in direction of welding at an angle of approximately 15° to the vertical

Temporary backing bar of stainless or mild steel (copper could fuse by forming intermetallic compound with some alloys owing to greater penetration with MAGS)

Fig. 18.19 *Spool-on-gun arrangement for MIG/MAGS (GMAW) welding using 'soft' aluminium wire electrodes*

Welding heat has an adverse affect on the mechanical properties of all aluminium alloys and the strength of the weld area will be less than that of the unaffected parent material. Satisfactory results can be obtained by upgrading the filler metal with suitable alloys, increasing the welding speed and ensuring adequate reinforcement.

The problem of hot cracking (Figure 18.20) can be caused by contraction stresses as the weld cools. They usually form in the weld, although they can sometimes occur in the heat affected zone adjacent to the weld. Certain types of alloys are 'hot-short', which means that they are weak at high temperatures and can crack. Pure aluminium is not usually prone to hot cracking, however cracking is likely to occur if the alloy content of the weld is within the following composition ranges:

Silicon 0.5 to 1.2 per cent   Copper 2.0 to 4.0 per cent
Magnesium 2.0 to 5.0 per cent   Manganese 1.5 to 2.5 per cent
Zinc 4.0 to 5.0 per cent

The composition of the filler material has therefore to be considered with great care, as, when dilution is taken into account, the resultant weld deposit must have a composition of individual alloy content either above or below the hot-short ranges.

Fast welding speeds help to prevent hot-shortness, as the weld metal will be in the hot-short range for less time and the faster cooling rates will produce a finer grain structure that is less prone to cracking. The use of pre-heat and an unrestrained joint design will also reduce stresses. Joint designs in

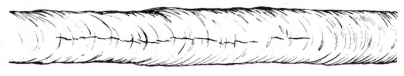

Fig. 18.20 *A typical hot-short crack in an aluminium weld*

aluminium fabrications are often best done by an evaluation of sample weld tests if no previous information is available.

Duralumin

Duralumin is an alloy of aluminium that contains approximately 4 per cent copper. When added to the aluminium, the copper first forms a solution and then an intermetallic compound ($CuAl_2$). In the correct composition, this alloy possesses the strength and hardness of mild steel, but is much lighter. The fast welding speeds of the MIG (GMAW) process reduce the heat affected zone with this material and will reduce the risk of fracture caused by contractional stresses. To restore the strength of this material after welding, the whole job must be reheated to 500°C (932°F) and then quenched.

The copper–aluminium compound will then precipitate out of solution over a period of about six days if the alloy is kept at room temperature, and thus the mechanical properties will be restored.

The welding of copper

Good ventilation (see Health and Safety section in *Basic Welding*, pages 16 and 17) must always be employed when welding copper and its alloys, including brasses. For prolonged welding, either carry out work out-of-doors or wear a respirator, as the fumes are dangerous. Tough pitch copper, that is copper containing copper oxide, is very difficult to weld and welds can crack. Copper that is being purchased for welding should therefore be specified as the deoxidized type. Because of the high heat conductivity of copper it will be necessary to pre-heat, between 500°C and 600°C (932°F and 1112°F), material thicker than 4.5 mm, in order to ensure the correct amount of fusion. Thin plates can be prepared with an upturned edge, and plates of 1.5 mm will require a gap of half the sheet thickness. Over 1.5 mm, a single 'V' with a 90 degree included angle is recommended. For even thicker plates, the double 'V' preparation can be employed, or for arc welding methods the single or double 'U' with a 6 mm radius at the bottom.

For oxy-acetylene welding of deoxidized copper, the plate surfaces and edges must be cleaned and, because of the high coefficient of expansion, it will be necessary to set the plate edges with a widening gap, tapering at the rate of 3–4 mm for every 100 mm of weld run. Because copper is weak at high temperatures (hot-short), it should be supported by firebricks or other material, such as synthetic asbestos placed on a steel backing. For welding long seams, as tacks are not recommended because they can fail when they become hot, clamps and / or jigs should be used to maintain the taper spacing. A nozzle of a size larger than that required for mild steel welding will be needed and a neutral or **very slightly** carburizing flame employed. For a butt weld, it is good practice to clamp in the middle and then start welding one-third from the end of the joint towards the far end of the seam. When the far end is reached, come back to the starting point and complete the weld to the other end in the opposite direction. Copper can be welded by the oxy-acetylene process without a flux, or borax may be used. The technique can be either leftward or rightward and, for very thick plates, it can be welded from both sides at once, in the vertical position using the 'double operator technique'. The only difference from the technique used on mild steel (apart from the larger nozzle) is that the blowpipe angle should be steeper, at 60–80 degrees to the plate, in order to put in as much heat as possible. Maintain the rod in the molten pool for the duration of the weld. After welding, the strength of the weld can be improved by light peening, while the work is still hot. Annealing can be carried out, if required, by heating to 600–650°C (1112–1202°F).

Electrodes for manual metal-arc welding are usually of the silicon–bronze or tin–bronze type, as copper electrodes tend to give a porous weld with this process. Therefore, if the weld has to have characteristics that match the parent material, for electrical conductivity or other reasons, the tungsten arc gas shielded (GTAW) or metal-arc gas shielded (GMAW) processes should be used. If using the manual metal-arc processes, the electrode should be connected to the positive terminal of a DC power source. After pre-heat, the weld should be made with a short arc and, holding the electrode almost vertical, perform a crescent-shaped weave, pausing slightly at each fusion face.

For TIG (GTAW) welding, direct current with the electrode negative should be employed using a pure argon shielding gas (helium can be used for very heavy sections). Mild or stainless steel backing can be employed but these should be coated with anti-spatter spray to stop them fusing.

With MIG (GMAW) welding, there is a choice of shielding, and the following gases can be employed: argon, nitrogen, helium, argon–nitrogen mixtures, or argon–helium mixtures. Nitrogen and helium will give off large amounts of spatter but give a high heat input, so they will reduce the pre-heat temperatures required. Argon–helium mixtures produce welds having the best appearance.

The pre-heating will oxidize the area around the joint to be welded. This can be reduced by mixing borax with alcohol and painting the joints prior to pre-heating. All flux residue must be removed after welding to prevent corrosion. Figure 18.21 shows an etched cross-section of a 'T' fillet weld in copper (GMAW) using a nitrogen gas shield. Note the lack of fusion on the left-hand weld, second pass, which is caused by the inter-run temperature being allowed to fall below the required level.

Fig. 18.21 *Macro-etched cross-section of copper fillet weld. Welded using nitrogen gas shield*

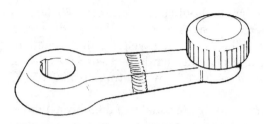

Fig. 18.22 *Welded repair of zinc die-case door handles*

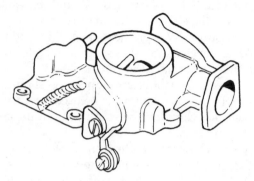

Fig. 18.23 *Typical repair of zinc die-cast carburettor*

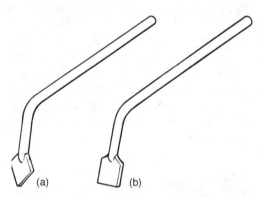

Fig. 18.24 *(a) Diamond-shaped spatula for forming 'V' preparations. (b) Square-shaped spatula for pushing molten metal into place*

Zinc based die castings

Although various items that used to be made from this material are now made from different types of plastic, many small components, such as door handles (Figure 18.22), carburettors (Figure 18.23), sewing machine parts etc. are still made from zinc based materials. These can be rather difficult to weld but success is often achieved by use of either the oxy-acetylene process or the TAGS process. The alloy content very often varies, but if the component is scrap if it cannot be repaired, then it is sometimes worth attempting a repair.

The preparation will be determined by the size, shape and thickness of the casting and the actual location of the fracture. If required, a 'V' preparation of 60–80 degrees can be made on the fracture area, leaving small areas of the original fracture faces as 'location' spots for alignment of the components. The preparation can be made with a spatula (see Figure 18.24) and two types of spatula should be available, or made prior to carrying out this type of repair.

About 6 mm on both sides of the upper fracture surfaces should be filed to remove oxide, that is until bright is exposed. The weld area should be supported with a suitable moulding material or metal inserts and any holes near to the weld area should be plugged. In some cases, it may be best to support the whole component and there is a 'carbon putty paste' available for this purpose.

Use a filler rod of zinc base alloy and file off the oxide on the rod, again until bright metal is exposed, before commencing welding. A slightly carburizing flame should be employed and no flux is required.

Pre-heating will not be necessary on small castings or thin sections but may be applied, with the blowpipe flame only, on slightly larger or more complex shapes (larger than, say, an engine carburettor). The welding technique should be leftward. Melt at the starting point and use a spatula to remove oxide from each face. Add the filler material without agitating the rod and allow the metal to 'flow' into the weld pool underneath the oxide skin. A mould may also be used to contain the weld metal as well as for support. After completing the weld, remove any surplus metal with a spatula. Always keep the weld area free of excessive oxide, by use of the spatula, and keep the spatula cool, that is, avoid it being heated by the oxy-acetylene flame, as if it becomes overheated it will vaporize zinc oxide, making the weld area dirty.

Small aluminium and magnesium castings can be welded in a similar manner but, of course, a flux will have to be employed and aluminium or magnesium filler rods respectively. Pre-heating may also be necessary.

If a magnesium casting catches fire, extinguish the blowpipe and any pre-heat burners, and if possible smother the flames with flux powder. If the casting is in a furnace, close the furnace door. Cover the casting with dry sand – **ON NO ACCOUNT USE WATER**.

The TIG (GTAW) process using argon shielding, if available, will often be the best choice for repairing small aluminium or magnesium castings.

General note on casting repairs

When repairing **any** casting, it is important to be able to reposition all broken pieces exactly before welding. This is often achieved by only partially preparing the mating faces with a 'V', leaving the original fracture faces at the root.

The seemingly impossible repair of the differential housing shown in Figures 18.25 and 18.26 was carried out in this way. The broken pieces were tacked together and then SMAW (manual metal-arc welded) using a nickel deposit.

Fig. 18.25 *Broken differential house*
(*Photograph courtesy of UTP, UK/UTC, Nigeria and Ghana*)

Fig. 18.26 *Differential house pieced together and manual metal-arc welded using nickel electrode*
(*Photograph courtesy of UTP, UK/UTC, Nigeria and Ghana*)

References

1. T.A. Stewert, C.N. McCowan and D.L. Olson, *AWS Welding Journal*, December 1988
2. R. Bradshaw, The Influence of Nitrogen on the Weldability and Corrosion Resistance of Duplex Stainless Steels. PhD Thesis, Corrosion and Protection Centre, UMIST, UK
3. J. Charles, Superduplex stainless steels: structure and properties. In *Proceedings Duplex Stainless Steels '91 Conference*, Beaune, France, 1991
4. J.C. Lippold *et al.*, Microstructural evolution in duplex stainless steel weldments. In *Proceedings Duplex Stainless Steels '91 Conference*, Beaune, France, 1991
5. D.J. Kotecki and J.L.P. Hilkes, Welding processes for duplex stainless steels. In *Proceedings Duplex Stainless Steels '94 Conference*, Glasgow, Scotland, 1994

19

Rebuilding worn surfaces, hard surfacing and Stelliting

Oxy-acetylene, manual metal-arc, TIG/TAG (GTAW) or MIG/MAG (GMAW) welding processes can all be used for building-up operations on worn parts by using a deposit similar to that of the parent metal. This is an economical method for building up worn shafts, gear wheels etc. When depositing a layer or layers on a worn shaft, consideration should be made to balancing the amount of heat input (see *Basic Welding*, pages 63–65, distortion and stresses in welding).

To form a hard surface on components both worn and new, rods that contain carbon, chromium, manganese and silicon can be used to deposit surfaces to give the required amount of hardness or resistance to wear and corrosion. However, these surfaces are deposited by a 'sweating' method in order that they do not become diluted with the softer base material. The oxy-acetylene flame is adjusted to have an excess of acetylene with a white plume from 2 to $2\frac{1}{2}$ times the length of the inner cone. The work surface will absorb carbon from such a flame and its melting point will be reduced, which causes the surface to sweat. The rod is then melted on to this sweating surface, producing a sound bond between the hard deposit and the softer parent material but with a minimum amount of dilution taking place.

An extremely hard surface can be deposited using special tubular rods that contain granules of tungsten carbide. These rods can be obtained with granules of various sizes.

One of the most well-known hardsurfacing materials is Stellite. It was originally developed in the United States in 1900, being an alloy of cobalt, chromium and tungsten with carbon. There is a range of Stellite materials available for different applications. Stellite tips are used for machine cutting tools and can be brazed on to steel shanks. Stellite can also be deposited directly on to surfaces that will have to stand up to extreme wear and corrosion conditions in service. With oxy-acetylene, the flame is adjusted with an excess of acetylene, again 2 to $2\frac{1}{2}$ times the length of the inner cone, and when the steel surface begins to sweat, the Stellite rod is brought into the flame and a drop of Stellite melted on to the surface. The Stellite will spread out, making a good bond. This method is repeated, without the inner cone touching the work, until the deposit is complete. The surface should be clean prior to Stelliting and it is common to use a pre-heat and controlled cooling to prevent cracking. When building up deposits on hardened parts such as camshafts, it is a common practice to direct small jets of water on each side of the deposit, in order to limit the heat flow and thus reduce distortion.

With MAGS and manual metal-arc welding, it is important to select the electrode that will give the best deposit to match the required service conditions. (This is of course also true for the filler rod when oxy-acetylene or TAGS welding.)

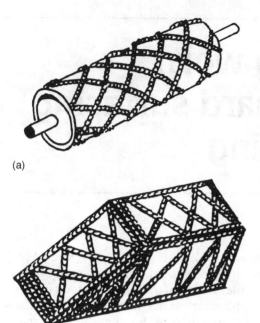

(a)

(b)

Fig. 19.1 *Various patterns of chequered reinforcement on (a) a crushing roller and (b) a digger tooth*

The approximate hardness of the deposit will not necessarily be the only criterion for suitability. The ability to withstand impact and/or corrosion may be just as important.

The following gives a rough guide to the electrodes available for both GMAW (MIG) and manual metal-arc (SMAW) resurfacing applications:

Service condition required	Type of deposit	Approximate Vickers hardness (VPN)
High-impact resistance coupled with medium abrasion resistance	Low alloy steel	350 VPN
Medium impact with higher abrasion resistance	Medium alloy steel	500 VPN
Excellent resistance and work hardening under impact conditions to resist abrasion	13 per cent manganese steel or austenitic steel	250 VPN (500 VPN when work hardened)
Maximum abrasion resistance with medium impact resistance	Tungsten carbide	1800 VPN
Maximum resistance to both impact and abrasion	Chromium carbide	700 VPN

General Hints

When hardfacing new parts, some of the surface can be cut away, or prepared to better accommodate the hardsurface metal. Build up areas where wear is likely to occur. On worn parts, ensure first that you have identified the parent material correctly. Remove any old hardsurfacing and prepare by cutting away if necessary. If using a tungsten carbide electrode, restrict the deposit thickness to a maximum of 6 mm, to avoid the deposit breaking off. Where damage is severe on manganese steel components, sections of metal can be welded on to the part using an austenitic electrode, thus reducing the total amount of welding required. Always plan heat input to avoid distortion. The amount of welding can often be minimized by using a chequered or rib pattern of weld runs when complete resurfacing is not required (see Figure 19.1). Figures 19.2 to 19.8 illustrate a variety of rebuilding and surfacing techniques.

What is hardness?

The **hardness** of a material is commonly defined as the resistance of the material to plastic deformation, or the ability to resist penetration.

A number of load–penetration tests are employed to measure the hardness of metals. Some of the most commonly used methods are described below.

Brinell hardness test

This test was devised by the Swedish metallurgist, Dr Johan August Brinell. In the test a hardened steel ball is forced into the surface of the metal under test by applying a suitable standard load (typically, 3000 kg for 30 seconds on a hardened steel ball 10 mm in diameter). The diameter of the impression is then measured and the Brinell hardness number (H) found from

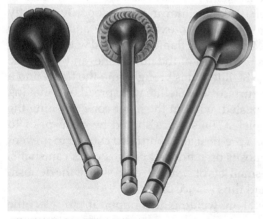

Fig. 19.2 *Engine exhaust valves built up using oxy-acetylene process and Stellite rod*

(Photograph courtesy of UTP, UK/UTC, Nigeria and Ghana)

Fig. 19.3 *Excavator-bucket teeth rebuilt using high chrome/carbon hard surfacing electrodes*
(Photograph courtesy of UTP, UK/UTC, Nigeria and Ghana)

Fig. 19.4 *Manganese steel digger tooth repaired using austenitic Cr–Ni-Mn electrode*
(Photograph courtesy of UTP, UK/UTC, Nigeria and Ghana)

$$H = \frac{\text{Load } (P)}{\text{Surface area of impression}}$$

The surface area of the impression is

$$\pi \frac{D}{2} \left(D - \sqrt{D^2 - d^2} \right.$$

where D is the diameter of the ball and d the diameter of the impression, so that

$$H = \frac{P}{\pi \dfrac{D}{2} \left(D - \sqrt{D^2 - d^2} \right.}$$

Fig. 19.5 *Roll welding in operation*
(Photograph courtesy of UTP, UK/UTC, Nigeria and Ghana)

Fig. 19.6 *Roll built back up to size with layers of weld*
(Photograph courtesy of UTP, UK/UTC, Nigeria and Ghana)

255

Fig. 19.7 *Rock crushing machine. Working surfaces built up with work-hardening manganese deposit*

Fig. 19.8 *Hardfacing a caster roll using a Lincoln Automatic submerged-arc welding machine*

(Photograph courtesy of AWS, Miami, USA)

Table 19.1 *Comparison between various hardness test numbers*

Brinell number	Vickers number	Rockwell number		
		Scale A	*Scale B*	*Scale C*
95	100	43	54	—
115	120	47	65	—
155	160	53	83	—
195	200	59	94	—
235	240	61	100	20
275	280	64	—	27
295	300	66	—	30
360	380	69	—	36
415	440	73	—	44
460	500	75	—	48
535	600	77	—	54
595	700	80	—	58
630	750	81	—	61
—	800	82	—	62
—	900	83	—	65
—	1000	84	—	68
—	1100	85	—	69
—	1400	91	—	71

To avoid tedious calculations, H is generally obtained by reference to tables drawn up for different loads and ball diameters.

Vickers pyramid hardness test

This type of test is similar to the Brinell, except that a diamond square-based pyramid (which gives geometrically similar impressions under different loads) is used instead of a hardened ball. The size of the indentation is read through a calibrated measuring microscope fitted to the machine. The Vickers hardness number is then given by reference to a table. Brinell and Vickers hardness numbers are very similar up to 500 (see Table 19.1).

Rockwell hardness test

This test was devised in the USA and is particularly useful for rapid, routine testing of finished items, as the final hardness reading is given direct on a dial. There are three scales on the dial:

Scale A, which is used in conjunction with a diamond cone and a 60 kg load.
Scale B, which uses a 1.5 mm (1/16 inch) -diameter ball and a 100 kg load.
Scale C, which uses a diamond cone and a 150 kg load.

The welding of pipes

Fig. 20.1 *The 'roll' welding of pipe*

The welding of pipe is mentioned in other chapters, because most processes used to join plates together can also be used to join pipework.

In recent years much work has been undertaken in developing automatic arc welding systems and 'one-shot' pipewelding methods such as 'flash-butt' welding, for joining pipe.

However, regardless of these advances, manual and semi-automatic welding methods still play a large part in many pipeline constructions. When these methods are used, high degrees of skill are required by the welding operator in order to weld in the overhead, vertical and flat (downhand) positions if the pipe cannot be rotated.

Also, pipelines play a significant role in the economic and environmental considerations of countries. Some carry water to help irrigate desert areas, others deliver gas over vast distances, and those that carry liquid fuels often unseen as as they are buried underground, save thousands of road tanker journeys, worldwide, per day.

For these special reasons, as well as being mentioned in other chapters regarding specific processes, the welding of pipes has been given its own chapter.

Before attempting to weld pipe by any process it is usual to master the welding of plate first. The easiest method of welding is if the pipe can be rotated, and this method is known as **'roll welding'** (Figure 20.1). With SMAW (manual metal-arc 'stick welding'), the arc is struck at the top of the pipe and held in this position to deposit the weld while the pipe is steadily rotated under the electrode. This method can be adopted for other welding processes, particularly if the rate of pipe rotation can be varied.

The real skill of manual and semi-automatic welding comes when the pipe cannot be rotated, as this will necessitate welds being made in the flat, vertical and overhead positions and also the ability to change from one position to the next without having to stop welding.

Welding techniques

Figure 20.2 shows the various welding test positions and their designations for groove welds in pipe. Figure 20.3 illustrates one type of automatic pipewelding machine employing the gas metal-arc welding process (MIG); some other pipewelding machines are illustrated in Chapter 7. Although automatic machines are employed on certain pipewelding applications, by far the largest amount of pipewelding is done by one of the manual or semi-automatic processes.

Pipe in the horizontal position (5G) for use in industrial applications is usually welded vertically upwards from the 6 o'clock position to 12 o'clock

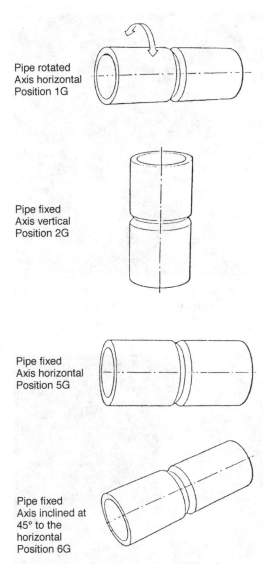

Pipe rotated
Axis horizontal
Position 1G

Pipe fixed
Axis vertical
Position 2G

Pipe fixed
Axis horizontal
Position 5G

Pipe fixed
Axis inclined at
45° to the
horizontal
Position 6G

Fig. 20.2 *The 'G' positions for welding pipe*

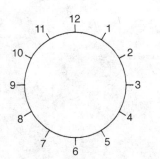

Fig. 20.4 *The clock positions used as reference when pipewelding*

Fig. 20.3 *Automatic GMAW (MIG) welding of large-diameter pipe*
(Photograph courtesy of TWI, Cambridge, UK)

(Figure 20.4). The preparation is usually a 60 degree or 70 degree included angle with a root gap of up to 3 mm (depending on the wall thickness), to ensure full penetration.

The technique employed for vertical-up pipewelding is the same as for the vertical welding of plate, in that a **keyhole** or **onion** is maintained in the root. For very-high-quality work the root run can be deposited by the GTAW (TIG/TAGS) process and then the remaining runs can be deposited by SMAW (manual metal-arc process). Another method of pipewelding often employed for transmission pipelines across country is the '**downhill**' or '**stovepipe**' method.

Figure 20.5 shows a student practising this technique, while Figure 20.6 shows a student practising repair welds on pipe.

With SMAW, stovepipe welding is undertaken with cellulose or cellulose/iron powder electrodes, welding downwards from the 12 o'clock position to 6 o'clock in multiple runs. The GMAW (MAGS) process can also be used for stovepipe welding, or for a combination technique, where the root run is deposited vertically down and then the hot pass and capping run are deposited vertically upwards. Figure 20.7 shows this technique being carried out (note also the air supply to the welding helmet in this picture; this helps to eliminate fumes from entering around the edges).

With stovepipe welding, the root gap generally need not be as wide as for vertical-up welding – a 1.5 mm (1.16 inch) root face and root gap usually being adequate.

Fig. 20.5 *Student practising 'downhill' or 'stovepipe' method of depositing first-pass root run in pipe (SMAW)*
(Photograph courtesy of AWS, Miami, USA)

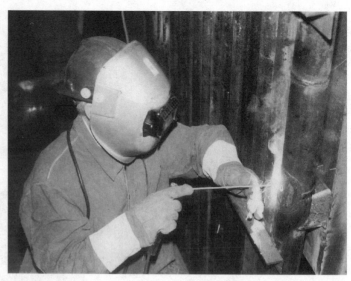

Fig. 20.6 *Practising the making of repair welds to pipes in awkward positions by using 'mock-up' in training school (SMAW)*
(Photograph courtesy of AWS, Miami, USA)

Fig. 20.7 *GMAW (MAGS) welding of a pipe butt joint*
(Photograph courtesy of TWI, Cambridge, UK)

Figure 20.8 shows the various points relating to GMAW (MIG/MAGS) butt welding of pipes. The root (penetration) run (a) when deposited vertically downwards is completed without weaving but, in order to obtain uniform penetration, the small keyhole or onion must be maintained at the leading edge.

It is common on transmission pipelines, when using the GMAW (MIG/MAGS) process, to use three runs of weld; the amount deposited is increased as the pipe wall thickness increases. The root run is deposited vertically downwards, without weaving, and then the hot pass and capping run are deposited vertically upwards, using the type of weave patterns shown in Figure 20.8d.

The penetration bead (Figure 20.8c) should be uniform and not excessive, as too much penetration can interfere with the flow within the pipe (such dimensions are usually specified in the welding procedure).

The various inspection methods used on the pipe welds and the criteria for weld acceptance are determined by the welding code that is being worked to. This, in turn, is determined by the intended use of the pipeline.

Torch angles are shown in Figure 20.8, (e) and (f). If the pipe can be rotated, then the angle can be maintained, turning the pipe one segment at a time, or the pipe can be slowly rotated (at the correct welding speed), with the torch in a fixed position. Figure 20.9 shows one test used in the training of pipewelders.

When depositing the root run vertically upwards, which is more common with SMAW (manual metal-arc), GTAW (TIG/TAGS) and OAW (oxy-acetylene process), the keyhole must be maintained in order to obtain the required penetration (Figure 20.8, (a) and (b)), and it usually takes many practice welds to perfect this (see Figures 20.8c and 20.9).

For very-high-quality welds a combination of processes can be employed. For example, the root run can be deposited using the GTAW (TIG/TAGS) process and then the weld can be filled and capped using SMAW (manual metal-arc welding).

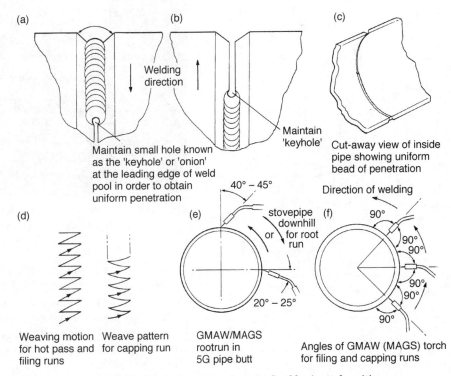

Fig. 20.8 *GMAW (MAGS) welding of pipe butts in fixed horizontal position*

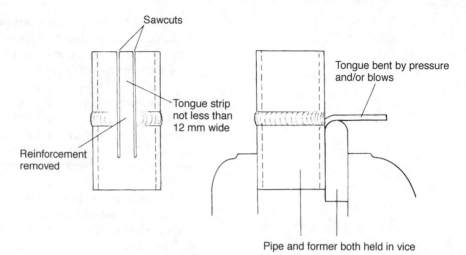

Fig. 20.9 *The tongue bend test is a workshop test used in the training of pipewelders. The diameter D of the former should be equal to 4 ×t, where t is the thickness of the pipe wall. The weld is usually considered satisfactory if the angle of the bend reaches 90 degrees without fracture*

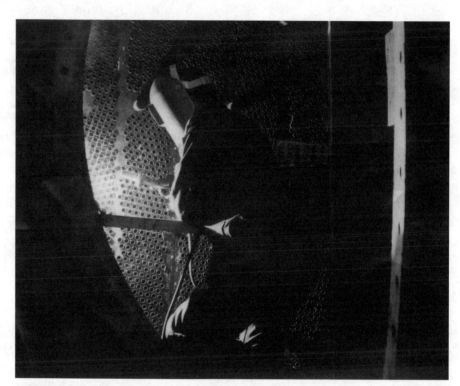

Fig. 20.10 *Welding tubes to tube-plate by the manual metal-arc process* (Photograph courtesy of UTP, UK/UTC, Nigeria and Ghana)

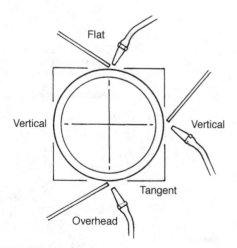

Fig. 20.11 *End view of a fixed pipe, showing positions of blowpipe and filler rod. If the pipe can be rotated, then the welding position can remain fixed*

Tubes can be joined to tube plates for boiler applications by a number of methods. Automatic tungsten arc gas shielded welding employs a small torch that rotates at a pre-determined speed, welding the end face of the tube to the plate, while explosive welding will join several tubes to the tube plate at once, by inserting small internal charges into the ends of each pipe to be joined and then detonating them from a safe distance; the force of the internal explosions welds the tubes to the plate. Figure 20.10 shows the welding of tubes to a tube plate by SMAW (manual metal-arc welding).

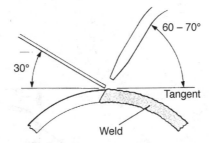

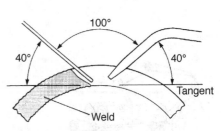

Fig. 20.12 *The leftward technique of pipewelding*

Fig. 20.13 *The rightward technique of pipewelding*

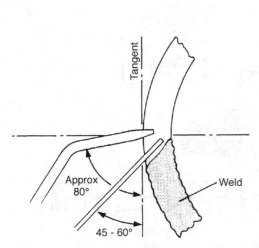

Fig. 20.14 *The all-position rightward technique of pipewelding*

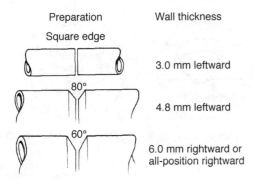

Fig. 20.15 *Appropriate pipewelding techniques for different wall thicknesses*

Oxy-acetylene welding is still widely used for welding pipes and can, of course, be performed in areas where there is no electrical supply. Figure 20.11 illustrates the various welding positions in relation to the plane of the tangent. There are three oxy-acetylene techniques that can be employed: leftward (Figure 20.12), rightward (Figure 20.13) and all-position rightward (Figure 20.14). Which technique is chosen will depend on the pipe-wall thickness, the welding position, and whether or not the pipe can be rotated.

The leftward method is usually used on pipe with wall thickness up to 6 mm ($\frac{1}{4}$ inch). Above this thickness, pipe is welded more satisfactorily by the rightward or all-position rightward method (Figure 20.15). Just as with the welding of plate, edge preparations are required as the wall thickness increases. This can be achieved by either oxy-acetylene bevelling (pipe rotated under a fixed cutting torch set at the required angle, or the use of small machines that will travel round a fixed pipe) or by using a large lathe.

Figures 20.16 and 20.17 show two scenes during the arc welding (SMAW) of a long-distance gas transmission pipeline, and Figure 20.18 illustrates the three-run method of cross-country pipe welding. Figure 20.19 shows how the completed section of pipe is lowered into the trench ready for covering over.

Pipe branches

In pipework there is often a need to branch one pipe off from another. This work is made easier by having a set of **templates** for different sizes of pipe and types of branches (Figure 20.20).

Fig. 20.16 *Mobile arc welding station for use on long-distance transmission pipelines. Note the tent to give all-weather protection while each weld is being made*

Fig. 20.17 *Gas-transmission pipe-line welding completed above surface*

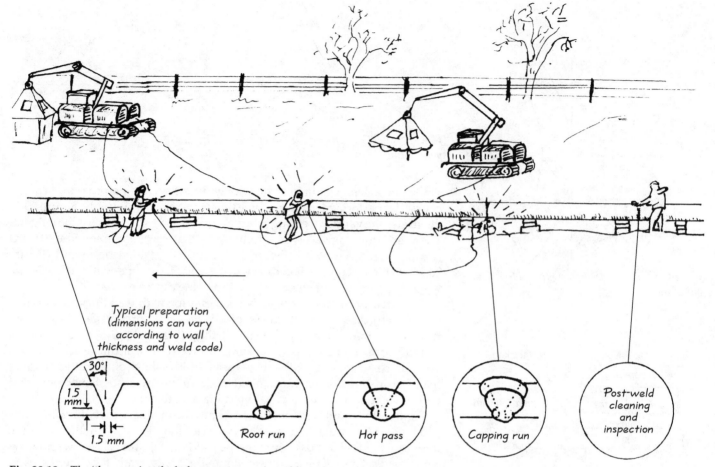

Typical preparation
(dimensions can vary
according to wall
thickness and weld code)

30°

1.5 mm

1.5 mm

Root run

Hot pass

Capping run

Post-weld
cleaning
and
inspection

Fig. 20.18 *The 'three-run' method of cross-country pipewelding. On completion of each run, the team advances to the left. One welder completes the root run, the next the hot pass, and the third finishes the weld with the 'capping run'*

To develop cutting templates for the hole in the main pipe and the shape of the branch pipe end, draw out the side and end views of the main pipe (or header) and the branch as shown in Figure 20.21, (a) and (b). Then divide the lower half of the circumference of the branch into equal parts, and draw lines from these points, parallel to the centre line to intersect the outside and inside diameter of the main pipe.

Draw horizontal lines from intersecting points a, b, c, d, e (Figure 20.21a) to intersecting points J, K, L, M, N (Figure 20.21b). Set out the circumference of the branch pipe as a baseline A–B–A (Figure 20.21c) and mark off the equal

Fig. 20.19 *Method of lowering completed pipe into trench using tracked mobile cranes*

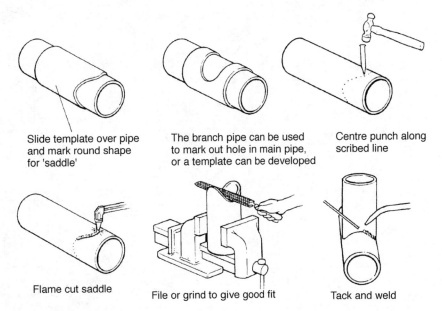

Slide template over pipe and mark round shape for 'saddle'

The branch pipe can be used to mark out hole in main pipe, or a template can be developed

Centre punch along scribed line

Flame cut saddle

File or grind to give good fit

Tack and weld

Fig. 20.20 *Pipe branches: using templates for marking out cutting lines for 'saddle' on branch pipe and hole in main pipe*

distances 1–2, 2–3, 3–4, etc. Plot from this baseline the distances from A–B to the points of intersection with the main pipe wall a–a, b–b, c–c, etc. Join these points with a smooth curve, which will give the required shape of the template for preparing the end of the branch pipe.

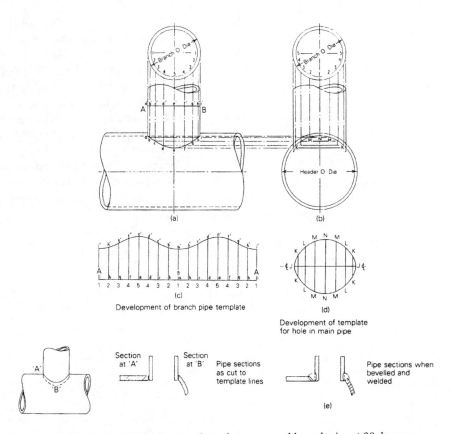

Fig. 20.21 *Method of developing templates for an unequal branch pipe at 90 degrees*

265

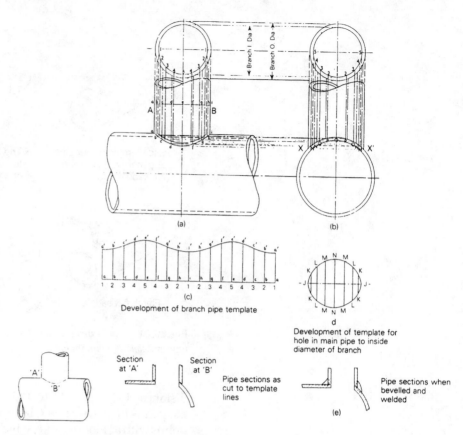

Development of branch pipe template

d

Development of template for
hole in main pipe to inside
diameter of branch

Section
at 'A'

Section
at 'B'

Pipe sections as
cut to template
lines

Pipe sections when
bevelled and
welded

(e)

Fig. 20.22 *Method of developing templates for a set-on branch joint*

To develop a template for the hole in the header, lay out horizontal centre line JJ and vertical centre line NN, as shown in Figure 20.21d. On centre line JJ, plot the distances NM, ML, LK, KJ from Figure 20.21b, these distances being measured from intersecting points on the main pipe curve. This determines the length of the template to allow for the curve.

To obtain the required width, which will be equal to the outside diameter of the branch pipe, plot the distances from Figure 20.21a, measured in a straight line, using e as the centre line, ei, eh, eg, ef. This will then give NN, MM, etc. Connecting the intersecting points with a smooth curve will give the template for cutting the hole. When cut, the hole in the main pipe can be prepared with the required bevel ready for welding (Figure 20.21e).

Figure 20.22 shows the development of template patterns for a set-on branch pipe.

Underwater welding

Regardless of welding process, there are two main types of underwater welding. These are: **hyperbaric**, where a dry diving chamber is used, so that the actual welding can be done in the dry; and **wet welding**, where there is no protection given to the weld from the water. The biggest problem to be overcome with wet welding arises from the quenching effect of the water on the weld, as fast as it is deposited.

Although wet welding is primarily used for the repair welding of underwater structures, advances in electrode design have allowed it to be used for other applications and, with correct training, various types of high-quality weld can be made. Figures 21.1, 21.2 and 21.3 show examples of welds produced by the underwater 'wet' welding method.

Typical applications include the attachment of sacrificial anodes to marine structures and the attachment of lifting lugs to submerged structures during salvage operations.

Fig. 21.1 *Examples of welds produced by wet-welding techniques*
(Photograph courtesy of Hydromech Technical Services Ltd)

Fig. 21.2 *Underwater (wet) fillet weld*
(Photograph courtesy of Hydromech Technical Services Ltd)

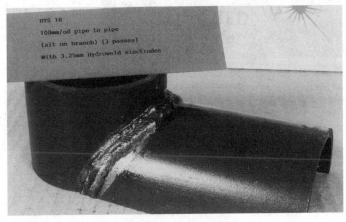

Fig. 21.3 *Pipe-to-pipe underwater (wet) weld*
(Photograph courtesy of Hydromech Technical Services Ltd)

Fig. 21.4 *David J. Keats, Managing Director of Hydromech Technical Services Ltd, demonstrating underwater wet-welding techniques*
(Photograph courtesy of Hydromech Technical Services Ltd)

Fig. 21.5 *While the diver/welder works below, two other divers watch over the control panel and keep in constant contact with the working diver*
(Photograph courtesy of Hydromech Technical Services Ltd)

Fig. 21.6 *Underwater welder training taking place in the training tank at The Welding Institute*
(Photograph courtesy of TWI, Cambridge, UK)

Electrodes are usually coated with a varnish to protect the flux. Quality control procedures and welder-qualification tests are required for underwater welding, just as for surface welding, and comprehensive welder/diver training is essential.

Figures 21.4 and 21.5 show a demonstration of wet welding at the premises of Hydromech, and Figure 21.6 shows a welder/diver under training in the training tank at the Welding Institute, Cambridge, UK.

With hyperbaric welding, special **underwater habitats** (UWH) are used (Figure 21.7). These can encase, or fasten around the work to be welded and, in the case of underwater pipeline welding, for example, the divers fasten seals at the intersection of the pipes with the habitat walls. Once the seals are installed, the water can be displaced from the habitat with a gas, usually helium. Because the habitat is filled with a large percentage of helium gas and also contains an environmental control system, the welder/divers can spend prolonged periods inside, before having to come back to the surface.

Some hyperbaric systems use robotic welding methods, controlled from a ship on the surface, as television cameras relay close-up pictures of the welding operation to an on-board television screen, enabling welding engineers to

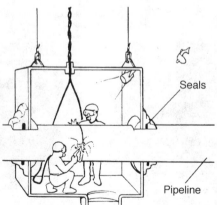

Fig. 21.7 *Welders working in a dry habitat*

Fig. 21.8 *Diver/welder attaching a stud by friction welding*

make any adjustments required. Hyperbaric welding is very expensive and is therefore only employed when no other system will suffice.

As in other areas, development work in underwater welding is constantly being carried out. One fairly recent innovation is the use of underwater friction welding for the attachment of studs (Figure 21.8). This has proved to be a very effective method of welding underwater.

When 'wet-welding' with coated electrodes, it is recommended that the diver's head is insulated from the helmet by a skull cap and also by the use of a piece of rubber tape or other insulation on the button of the exhaust valve. The current should be switched off at a switch on the surface while changing electrodes. The diver should then position the new electrode over the restart position before calling for the current to be switched back on.

The protection of the diver from electric shock is catered for by use of insulated clothing, special equipment and the use of this surface switch on the verbal command of the diver.

There is, however, a further less well known hazard, which can arise when 'wet-welding'. The bubbles produced from the welding operation tend to contain around 70 per cent hydrogen, which cannot be burnt owing to the lack of oxygen present. If welding is taking place in a closed or semi-closed compartment, this hydrogen can build-up above the surface of the water. In this location there is sufficient oxygen to allow the mixture to explode if a spark or flame were to come in contact with it.

It is therefore of vital importance that any such pockets, corners or compartments should be fitted with adequate positive ventilation before the welding operation commences.

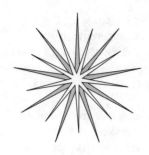

Weld testing and quality control standards

In order to include as much welding process information as possible within viable size limitations, coverage of this vast area has been limited to simple testing methods within chapters of this book and *Basic Welding*, while many specialized tests, such as those for corrosion, fatigue and brittle fracture, have not been dealt with.

In the interests of world trade, international standards are being developed. One of the major bodies involved is the International Organization for Standardization or ISO (ISO being derived from the Greek 'iso', meaning equal).

Several countries have welding institutes and societies belonging to the International Institute of Welding (IIW) and provide input to ISO, therefore, if there is any doubt as to the correct standard to work to (as many standards relating to welding are under review), it is always best to consult ISO/IIW or the individual welding organization for a particular country. These organizations are listed in Appendix 6.

Appendix 1: Temperature measurement

Temperature measurement in the welding workshop

The pre-heat temperatures used when welding certain types of alloy steel and casting must be measured accurately, You can estimate the temperature of low alloy, high tensile steels, medium and high carbon steels by the colours of the oxides on the surface. These are known as **temper colours**. If the surface has been freshly filed, these temper colours will show as listed in Table A1.1.

Table A1.1 *Temper colours of surface oxides*

Surface colour	Temperature (approximate) in degrees Celsius
Pale yellow	200
Straw	230
Brown	245
Purple	270
Dark purple	280
Blue	340

In the same way, you can estimate the temperature of iron and steels from the colour changes on their surfaces as you heat them with the preheating flame. The colour changes are as listed in Table A1.2.

When certain types of soap are rubbed on to the surface of a hot material they will change colour at definite temperatures. This idea was developed into **temperature-indicating paints** or **crayons**, which can now be purchased to measure temperatures accurate to one degree in stages up to 1370°C (2523°F) (Figure A1.1).

Fig. A1.1 *A temperature-indicating crayon*

Table A1.2 *Surface colours for iron and steel at various temperatures*

Colour	Temperature (approximate) in degrees Celsius
Faint red	500
Blood red	650
Cherry red	750
Bright red	850
Salmon	900
Orange	950
Yellow	1050
White	1200

Seger cones are another workshop method of temperature measurement. They are small pyramid-shaped cones of clay and oxide mixtures, made to collapse at specific temperatures from 600°C (1126°F) upwards. When you are

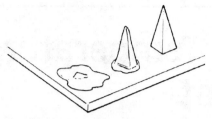

Cone below temperature melted.
Required-temperature cone just beginning to melt.
Higher-temperature cone unmelted

Fig. A1.2 *Seger cones*

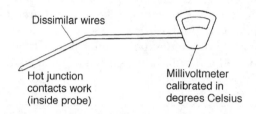

Dissimilar wires

Hot junction
contacts work
(inside probe)

Millivoltmeter
calibrated in
degrees Celsius

Fig. A1.3 *Probe thermocouple pyrometer*

using Seger cones to determine a pre-heat temperature, it is best to use three: one that will melt below the required temperature, one that will melt at the required temperature and one just above the required temperature. In this way, when the first cone melts, you know that the right temperature is approaching, and you can be ready to start welding as soon as the second cone begins to start to bend. You should hold the temperature so that the third cone does not melt (Figure A1.2).

The probe thermocouple pyrometer

This instrument can measure temperatures up to around 1700°C (3117°F) and is a very useful workshop tool. It consists of a probe which is touched to the work. Inside the end of the probe is a joint between two wires made of different metals, known as the hot junction or **thermocouple**. The other ends of these wires are connected to a millivoltmeter to complete the circuit (Figure A1.3).

The electrical resistance of a metallic conductor varies with temperature, and the amount of this variation is different for different metals. So when the hot junction is brought into contact with a hot surface, an electrical pressure (or voltage) is set up, because there are dissimilar metals at the junction. This very small electrical voltage is measured on the millivoltmeter. The greater the temperature, the greater the voltage will be, and the voltage increase is proportional to the temperature rise.

The millivoltmeter can be calibrated in degrees Celsius, and the temperature of the surface can be read off directly.

Appendix 2: Assessing different metals

Being able to recognize different metals and alloys is obviously important, particularly in repair work. If the component can be readily identified, the welder with a good mechanical background will be aware that certain machine parts are always made from specific materials. The use to which the component is put is, therefore, a good starting point in the identification process.

Some metals can be readily recognized by their colour and weight – copper, aluminium and brass, for example.

For the various types of steel and iron, spark-testing the metal on a grinding wheel or with a hand grinder (Figure A2.1) is a good method of identification. (Figure A2.2 shows some of the spark patterns, however many welding repair shops have labelled samples of known steels so that the spark patterns can be compared with the sample being identified.

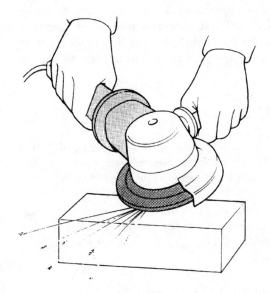

Fig. A2.1 *Sparks given off when using a hand grinder*

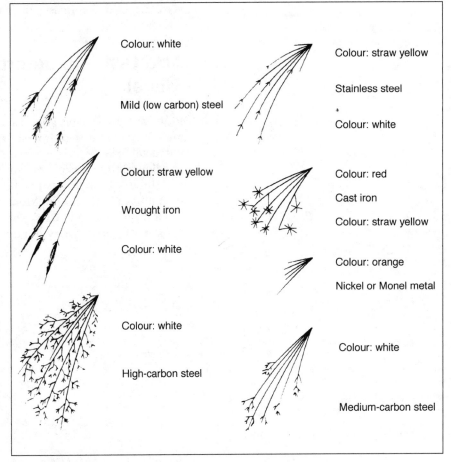

Fig. A2.2 *Using the spark test to identify metals*

273

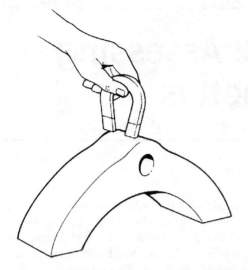

Fig. A2.3 *Checking the material with a magnet*

Testing with a magnet

It is useful to have a magnet or a piece of magnetized metal in the workshop as you will then be able to test if the material that requires welding is magnetic or not (Figure A2.3). The majority of steels and cast irons are very magnetic; only manganese steel and austenitic stainless steel are non-magnetic.

Most of the non-ferrous metals are non-magnetic, the exceptions being nickel, which is very magnetic, and Monel, which can be slightly magnetic. Some nickel alloys, like Inconel and certain types of Monel, are non-magnetic.

Sometimes, components are coated, or 'clad' with nickel or stainless steel. In such cases, it will be necessary to clean back the surface coating on either side of the joint to be welded and to use two welding techniques / procedures, one for the base material and one for the coating. By welding in such a way, the original properties of the repaired component should be restored.

Acoustic or sound test

With experience, metals can be identified by the sound they give off when struck by a hammer (Figure A2.4). For example, steel will give off a higher-pitched tone than cast iron. The sound test can also be used to indicate if a component has a flaw or crack inside it. This method is still employed to check railway wheels. For example, if the wheel is free from defect when it is struck by a hammer, it will give out a high-pitched ringing sound; however, if there is a crack or defect inside, the sound will be a dull 'thud', with no ringing.

Acid test – to determine stainless steel and Monel

Protective gloves and goggles must always be worn when carrying out this test. Drop one or two drops of concentrated nitric acid on a clean surface of the metal to be tested (Figure A2.5); no reaction will indicate stainless steel or an alloy high in nickel chromium content. If there appears to be a reaction, add three or four drops of water, one at a time; if the area turns greenish-blue then it is an indication of Monel.

Only use acids if you have been trained in the safety procedures, otherwise get a laboratory to carry out this test.

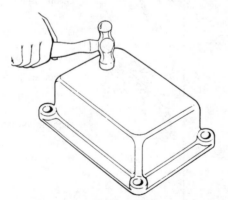

Fig. A2.4 *Acoustic test (sound test)*

Fig. A2.5 *Acid test on a sample plate. Always wear gloves and goggles*

Measuring the properties of metals

There is a close relationship between a metal's various properties. If one property is changed, it will cause a change in the others. For example, if a piece of steel is treated so that its hardness is increased, the tensile strength will also be increased, but the ductility will be reduced. Likewise, if a metal is heated and it becomes soft, with an increase in ductility, its tensile strength will be lowered.

Properties are therefore very important to welders. Generally, we can take it that the harder a material is, the more brittle it is likely to be, and special precautions such as pre-heating and post-heating may well have to be used in order to carry out a satisfactory weld without cracking.

If a piece of scrap material, the same as the metal to be welded, is available, you can perform a bend test to measure the ductility. The file hardness test provides a workshop method of estimating hardness and tensile strength.

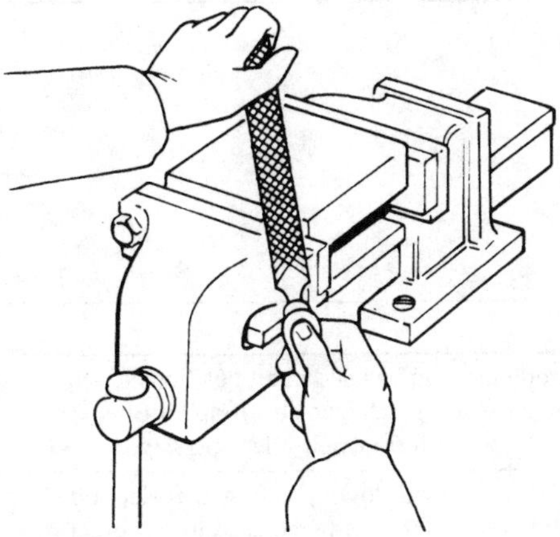

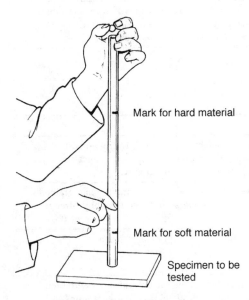

Fig. A2.6 *The file hardness test*

File hardness test

This test can give a rough estimate of a material's Brinell hardness number, by determining the material's resistance to the cutting action of the file (Figure A2.6). The test is best performed with a new file of possible. Table A2.1 explains how the method works.

Table A2.1 *The file hardness test*

Approximate Brinell hardness number	Ease with which the steel surface can be filed with a new file
100	File cuts into surface very easily and metal is very soft
200	File still removes metal but a bit more pressure is required. Metal still quite soft.
300	Metal begins to exhibit resistance to file
400	File will still remove metal but with difficulty: metal is quite hard
500	File barely removes metal, showing the metal is only slightly softer than file
600	File will slide over surface without removing metal. The file teeth are dulled

Bouncing ball test

This is another workshop hardness test that can be developed into quite an accurate method. For this method you need a steel ball bearing and a glass tube. The ball bearing is allowed to fall down the glass tube on to the surface being tested (Figure A2.7) and the height of bounce is measured. A steel ball bearing is manufactured with great accuracy as to its hardness, so that it provides a cheap tool that will give reasonably accurate results. The glass tube will guide the fall and the bounce of the steel ball. If the glass has no markings on it, you can make these carefully with a file at regular intervals in order to measure the bounce.

The test can be made quite accurate by calibrating the glass tube after comparative tests of known hardness. The heights and bounces can be marked on the glass, one low and one high. The spaces between these two marks can then be divided and marked evenly. To obtain a very hard specimen and a very soft specimen the file hardness test can be used.

Mark for hard material

Mark for soft material

Specimen to be tested

Fig. A2.7 *The bouncing ball test. The ball should run freely in a glass tube. The higher it bounces off the surface, the harder the material*

Appendix 3: Electrode and filler metal charts

Electrode chart 1

Mild steel electrodes

Designation	Standard Classification	Coating/Current Data	Application
MS-110	BS639:E5121R21 AWS A5.1:E6013	Rutile AC 50 V min. DC +/–ve	General-purpose electrode that can be used in all positions (Ex.vd), it can also be used as a contact-type electrode for fillet welds where weld finish is the criterion. Self-releasing slag.
MS-120	BS636:E4322R12 AWS A5.1:E6013	Rutile Cellulose AC 50 V min. DC –ve	Designed for welding in the vertical-down position, this electrode exhibits excellent welding characteristics when used in any other position with easy slag removal and low spatter loss.
MS-125	BS639:E5132RR31 AWS A5.1:E7024	Rutile Iron Powder AC 50 V min. DC +/–ve	This electrode combines high welding speeds with 140 per cent metal recovery for superb weld finish in downhand butt and horizontal-vertical fillet welds.
MS-130	BS639:E5153B24(H) AWS A5.1:E7016	Basic AC 70 V min. DC +/–ve	A positional electrode used on joints where high-notch toughness is required on structures operating at temperatures down to –20°C.
MS-135	BS639:E5154B24(H) AWS A5.1:E7018	Rutile Basic AC 70 V min. DC –/+ve	General fabrication of heavy platework to radiographic standard and highly restrained joints on structural steelwork for use at low temperatures down to –20°C.
MS-136	BS639:E5154B24(H) AWS A5.1:E7018	Rutile Basic AC 70 V min. DC –/+ve	Ease of use and notch-free welds are achieved with this electrode in all positions (Ex.vd) when operated on either AC or DC welding current. The weld appearance is excellent.

Low alloy steel electrodes

Designation	Standard Classification	Coating/Current Data	Application
MS-145	BS2493:MoBH AWS A5.5:E7018-A1	Basic AC 70 V min. DC +/–ve	This electrode is used for the fabrication of 0.5Mo creep-resistant steels in all positions except vertical-down. Typical applications are to be found in the power generation industry.
MS-150	BS2493:2NiBH AWS A5.5:E8018-C1	Basic AC 70 V min. DC +/–ve	For welding low-temperature, low-alloy 2.5Ni and the fine-grained C–Mn steel when sub-zero toughness combined with high strength is the criterion.
MS-155	BS2493:3NiBH AWS A5.5:E8018-C2	Basic AC 70 V min. DC +/–ve	Designed for the fabrication of 3.5Ni type steels for use at temperatures down to –80°C.
MS-160	AWS A5.5:E8018-G	Basic AC 70 V min. DC +/–ve	For welding high-strength fine-grained structural steels with high tensile strength requirements, typical values 550–600 N/mm². The electrode can be used in all positions except vertical-down.
MS-165	AWS A5.5:E9018-M	Basic AC 70 V min. DC +/–ve	For welding high-strength fine-grained structural steels with high tensile strength requirements, typical values 600–700 N/mm². All positions except vertical-down on HY 80 and Naxtra 70.

Reproduced by courtesy of Westbrook Welding Alloys.

Low alloy steel electrodes (continued)

Designation	Standard Classification	Coating/Current Data	Application
MS-170	BS2493:(NiMoBH) AWS A5.5:E10018-M	Basic AC 70 V min. DC +/−ve	For welding high-strength fine-grained structural steels with high tensile strength requirements, typical values 700–800 N/mm². All positions except vertical-down on HY 80 and RQT 600.
MS-175	BS2493:(2MnMoBH) AWS A5.5:E10018-D2	Basic AC 70 V min. DC +/−ve	For general fabrication of high-strength manganese–moly type steels that require sub-zero toughness after PWHT is carried out.
MS-180	BS2493:(2NiMoBH) AWS A5.5:E11018-M	Basic AC 70 V min. DC +/−ve	Manufactured with Ni and Mo, this electrode is suitable for use on high-strength, notch tough Q&T steels when used in the as-welded condition.

Electrode chart 2

Stainless steel electrodes

Designation	Standard Classification	Coating/Current Data	Application
SS-600	BS2926:13.4 Mo RMP AWS A5.4:E410NiMo-26	Rutile AC 70 V min. DC +ve	A high-strength martensitic-type stainless steel giving improved corrosion and sub-zero toughness when compared with the plain 12 per cent Cr steels when Ni/Mo are acceptable.
SS/601	BS2926:13.4 R AWS A5.4:E410-26	Rutile AC 70 V min. DC +ve	A martensitic 12 per cent Cr stainless steel for overlaying steels and joining the many variations of these alloys in wrought and cast form, such as 403, 405, 409, 410.
SS-605	 AWS A5.4:E630-16~	Rutile AC 70 V min. DC +ve	For welding Cu containing precipitation-hardening stainless steel ASTM type 630, cast alloys ASTM A747 CB-7Cu-1/Cu-2 and proprietary alloys such as 17-4PH (Armco Steel).
SS-610	BS2926:19.9.R AWS A5.4:E308-16	Rutile AC 70 V min. DC +ve	With a C content of 0.04–0.08 per cent, this electrode is for use on type 308H stainless steel when used at elevated temperatures in all positions except vertical-down.
SS-620P	BS2926:19.9.L.R AWS A5.4:E308L-16	Rutile AC 50 V min. DC +ve	A low-carbon stainless steel electrode for general fabrication of the corrosion-resistant steels AISI 302, 304, 305 and 304L in their many forms, wrought, cast and pipework.
SS-630P	BS2926:19.12.3.L.R AWS A5.4:E316L-16	Rutile AC 50 V min. DC +ve	With a good resistance to pitting, this electrode is recommended for 1.5–3.0 per cent Mo-bearing austenitic stainless steels of the 316/316L types.
SS-635	BS2926:19.12.3.L.RMP AWS A5.4:E316L-17	Rutile AC 45 V min. DC +ve	The analysis of this electrode is similar to that above but it has a coating containing a high percentage of alloys and iron powder to ensure a superb weld finish and a metal recovery of 160 per cent.
SS-640	BS2926:19.9.Nb.R AWS A5.4:E347-16	Rutile AC 50 V min. DC +ve	A low-carbon niobium-stabilized electrode for the fabrication of type AISI 347 niobium and AISI 321 titanium-stabilized steels also AISI 302, 304, 304L.
SS-650	BS2926:19.12.3.Nb.R AWS A5.4:E318-16	Rutile AC 50 V min. DC +ve	This molybdenum-bearing electrode is stabilized with niobium for use on austenitic stainless steels, types AISI 318 and 316Ti and joints between stabilized and low-carbon types.
SS-660P	BS2926:23.12.L.R AWS A5.4:E309L-16	Rutile AC 50 V min. DC +ve	Suitable for overlaying C–Mn steel as a buffer layer prior to depositing 308 or 308L for corrosion resistance. The electrode is used for dissimilar joints between stainless and C–Mn steels.
SS-670P	BS2926:23.12.2.L.R AWS A5.4:E309MoL-16	Rutile AC 50 V min. DC +ve	Suitable for overlaying C–Mn steel as a buffer layer prior to depositing 316 or 316L for corrosion resistance. The electrode is used for dissimilar joints between stainless and C–Mn steels.
SS-680	BS2926:25.20.R AWS A5.4:E310-16	Rutile AC 70 V min. DC +ve	A fully austenitic electrode for wrought or cast heat-resistant AISI 310 base materials, the weld material has low magnetic permeability. Also recommended for dissimilar joints.

Reproduced by courtesy of Westbrook Welding Alloys.

(continued)

Stainless steel electrodes (continued)

Designation	Standard Classification	Coating/Current Data	Application
SS-690	AWS A5.4:E2209-16	Rutile AC 50 V min. DC +ve	Designed for use on Duplex stainless steels of types such as UNS S31803 and DIN No. 1.4462. The weld metal has a PREN number of approximately 35.
SS-695	BS2926: 20.25.5.L.CuNb.R AWS A5.4:385-16~	Rutile AC 50 V min. DC +ve	The weld metal from this electrode is low carbon with Mo and Cu, and is recommended to resist corrosion when used with the acids sulphuric, phosphoric and other organic and inorganic acids.

Chart 3

Electrodes for cast iron

Electrode	Standard Classification	Coating/Current Data	Application
CI-200	AWS A5.15:ENi–CI	Graphite AC/DC –ve	For use on all types of cast iron giving a fully-machinable deposit, easy to use even in position.
CI-210	AWS A5.15:ENiCu-B	Graphite AC/DC –ve	With a Ni–Cu core wire, this electrode can be used on all types of cast iron where a good colour match is required.
CI-220	AWS A5.15:ENiFe-CI	Graphite AC/DC –ve	A Ni–Fe electrode for joining and surfacing the higher-strength cast irons and joining cast iron to mild steel.
CI-230	AWS A5.15:ENiFe/CI	Graphite AC/DC –ve	A Ni–Fe electrode with 130 per cent recovery for surfacing most types of cast iron and buttering prior to joining.

Chart 4

Chrome–moly electrodes

Electrode	Standard Classification	Coating/Current Data	Application
CM-900	BS 2493:1CrMoLBH AWS A5.5:E8015-B2L	Basic DC +ve/AC	This electrode is for use on 1.25Cr/0.5Mo type steels for use at elevated service temperatures.
CM-910	BS 2493:1CrMoLBH AWS A5.5:E8018-B2L	Basic DC +ve/AC	As above but with a metal recovery of up to 120 per cent.
CM-920	BS 2493:2CrMoLBH AWS A5.5:E9018-B3L	Basic DC +ve/AC	This electrode is for use on 2.25Cr/1.0Mo type steels for use in high-temperature/corrosive environments.

Aluminium electrodes

AL-1000	DIN 1732:EL-AlSi12	Halide Type DC +ve	For fabrication/repair of aluminium and aluminium–silicon alloys and castings requiring good corrosion resistance.
AL-1010	DIN 1732:EL-AlMn1	Halide Type DC +ve	For welding aluminium–manganese and aluminium–magnesium alloys in wrought and cast form.

Cutting and gouging electrodes

CG-1100	No Classification	Acid Type AC/DC –ve	A gouging/cutting electrode used for chamfering prior to welding and for removing weld or casting defects.
CG-1110	No Classification	Acid Type AC/DC –ve	A cutting/piercing electrode producing a forceful arc for hole piercing and cutting ferrous and non-ferrous metals.

Reproduced by courtesy of Westbrook Welding Alloys.

Brazing alloys

Electrode	Standard Classification	Coating/Current Data	Application
BA-1200 BA-1200Fc BA-1200Fl	BS 1453:C2 BS 1453:C2 BS 1453:C2	Bare wire Flux coated Flux impreg.	A general-purpose brazing alloy for use on steel, copper, cast iron to themselves or any combination. This alloy is a good colour match for brass.
BA1210 BA-1210Fc	BS 1453:C5 BS 1453:C5	Bare wire Flux coated	A brazing alloy with 9 per cent Ni for fabrications that require a superior finish combined with high mechanical values.

Chart 5

Alloy steel electrodes

Electrode	Classification	Coating Data	Applications
AS-300	AWS A5.4:E307-16	Rutile Basic AC/DC +ve	High-recovery electrode for difficult-to-weld steels and also buffer layers prior to hardsurfacing. Recovery 180 per cent approx.
AS-310	AWS A5.4:E312-16	Rutile Basic AC/DC +ve	Designed for maintenance and repair of difficult-to-weld steels and dissimilar joints between them.
AS-320	AWS A5.4:E312-16	Rutile Basic AC/DC +ve	As above but the welding characteristics offer improved operability when positional welding.
AS-330	AWS A5.4:E312-16	Rutile Basic AC/DC +ve	A high-recovery, 180 per cent, for applications similar to those of AS-310 above.
AS-340	BS 2926:20.9.3.R	Rutile Basic AC/DC +ve	A balanced analysis of Cr–Ni–Mo makes this electrode suitable for use on high tensile steels and armour plate.

Hardfacing/tool steel electrodes

Electrode	Classification	Coating Data	Applications
HF-400		Rutile Basic AC/DC +ve	30–38 RC for overlaying or building up surfaces to resist heavy impact and yet still be fully machinable.
HF-415		Rutile Basic AC/DC +ve	57 RC. A Cr–Mo–V martensitic-type electrode for surfacing areas of high impact and limited abrasion.
HF-425	AWS A5.13:EFeCr-A1	Rutile AC/DC +ve	58–60 RC. The deposit is a C–Cr–Fe weld metal with 140 per cent recovery to resist high abrasion up to 200°C.
HF-430		Rutile Basic AC/DC +ve	25 RC work-hardening 45 RC for application of severe impact and low abrasion
HF-440		Tubular Type AC/DC +ve	The weld deposit contains complex carbides of Cr–Mo–V in a matrix of hard alloy steel. Deposit 68 RC approx.
HF-450		Tubular Type AC/DC +ve	A tungsten carbide deposit of approximately 1800 HV to resist very severe abrasion.
HF-460		Cast Tungsten	A very smooth, dense tungsten deposit to resist extremely severe fine particle abrasion. Hardness 68–72 RC.
TS-500		Rutile Basic AC/DC +ve	A high-recovery Mo–Cr–W–V high-speed type tool steel resistant to abrasion, erosion and impact. Hardness 62 RC.
TS-510	AWS A5.13:ECrCo-A	Basic AC/DC +ve	A Co–Cr–W alloy with excellent resistance to impact and corrosion at elevated temperatures. Hardness 40–45 RC.
TS-520	AWS A5.13:ECrCo-B	Basic AC/DC +ve	A Co–Cr–W alloy with excellent resistance to abrasion and corrosion at elevated temperatures. Hardness 48–52 RC.
TS-530		Basic AC/DC +ve	For the reclamation of C–Cr–Mo–V hot working die steels. Hardness of a typical air-cooled deposit 50–55 RC.
TS-532		Rutile Basic AC/DC +ve	For fabrication/repair of hot and cold working tools and dies to resist abrasion, compression and heat up to 500°C.

Reproduced by courtesy of Westbrook Welding Alloys.

(continued)

Hardfacing/tool steel electrodes (continued)

TS-533		Rutile Basic AC/DC +ve	For fabrication/repair of hot and cold working tools and dies to resist abrasion, impact and heat up to 550°C.
TS-534		Rutile Basic AC/DC +ve	For fabrication/repair of hot and cold working tools and dies to resist impact, abrasion and heat up to 550°C.

Chart 6

Silver brazing alloys

Alloy Desig.	Classification	Temp. Range (°C)	Applications
SB-20		690–810	A complete range of silver brazing alloys for joining steel, copper, copper alloys, nickel and nickel alloys as used in many industrial applications.
SB-30	BS 1845:Ag21*	650–750	
SB-30Cd	BS 1845:Ag 12	600–690	
SB-40	BS 1845:Ag20*	650–710	
SB-40Cd	BS 1845:Ag3*	595–630	
SB-55	BS 1845:Ag14	630–660	

*Denotes nearest equivalent

Copper-phosphorus brazing alloys

CP-2	BS 1845:CP2	645–810	These alloys are self-fluxing on copper to copper. When used on or in combination with brass or bronze, a silver brazing flux is required. Do **NOT** use on ferrous or nickel alloys. Minimum brazing temp. for these alloys is 700–740°C.
CP-5	BS 1845:CP4	650–815	
CP-15	BS 1845:CP1	645–800	

Reproduced by courtesy of Westbrook Welding Alloys.

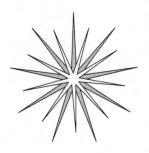

Appendix 4: Tables, notation, and useful measures and information

The Greek alphabet

(Greek characters are often used as symbols in scientific notation)

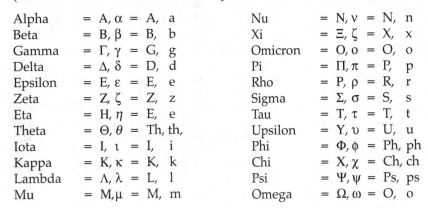

Alpha	= A, α = A, a	Nu	= N, ν = N, n
Beta	= B, β = B, b	Xi	= Ξ, ζ = X, x
Gamma	= Γ, γ = G, g	Omicron	= O, o = O, o
Delta	= Δ, δ = D, d	Pi	= Π, π = P, p
Epsilon	= E, ε = E, e	Rho	= P, ρ = R, r
Zeta	= Z, ζ = Z, z	Sigma	= Σ, σ = S, s
Eta	= H, η = E, e	Tau	= T, τ = T, t
Theta	= Θ, θ = Th, th,	Upsilon	= Y, υ = U, u
Iota	= I, ι = I, i	Phi	= Φ, φ = Ph, ph
Kappa	= K, κ = K, k	Chi	= X, χ = Ch, ch
Lambda	= Λ, λ = L, l	Psi	= Ψ, ψ = Ps, ps
Mu	= M, μ = M, m	Omega	= Ω, ω = O, o

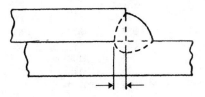

Effective throat thickness

Fig. A4.1 *Throat thickness of concave and convex fillet welds*

The strength of fillet welds

The strength of a fillet weld is dependent upon:

1. the effective length of the fillet weld;
2. the effective throat thickness of the weld.

Throat thickness

The **throat thickness** of a fillet weld is the minimum thickness of weld metal measured along a line passing through the root of the weld. Figure A4.1 shows the dimension representing the effective throat thickness for convex and concave fillets, the effective throat thickness being that dimension which is taken for normal stress calculations. From this it can be seen that the optimum economical shape of a fillet weld section is triangular, that is, neither concave nor convex.

The thickness of a fillet weld is calculated from its nominal size. BS 499 defines the size of a normal fillet weld (Figure A4.2) as the minimum leg length, and the size of a deep-penetration fillet weld (Figure A4.2) as the minimum nominal leg length plus the depth of penetration beyond the root.

For a right-angle fillet weld the effective throat thickness is $0.7 \times$ the minimum leg length or specified weld size. Where the angle between the fusion faces is not 90 degrees, however, factors other than 0.7 must be used. For instance, if the angle between the fusion face is 120 degrees, the throat thickness is $0.5 \times$ the fillet size.

Normal penetration

(a) Normal fillet weld

Increased penetration

(b) Deep penetration fillet weld

Fig. A4.2

Conversions for Fahrenheit–Celsius temperature scales

Find the number to be converted in the centre (boldface) column. If converting Fahrenheit degrees, read the Celsius equivalent in the column headed '°C'. If converting Celsius degrees, read the Fahrenheit equivalent in the column headed '°F'.

°C		°F	°C		°F	°C		°F	°C		°F
−273	−459		−13.3	8	46.4	60	140	284	332	630	1166
−268	−450		−12.2	10	50.0	66	150	302	338	640	1184
−262	−440		−11.1	12	53.6	71	160	320	343	650	1202
−257	−430		−10.0	14	57.2	77	170	338	349	660	1220
−251	−420		− 8.9	16	60.8	82	180	356	354	670	1238
−246	−410		− 7.8	18	64.4	88	190	374	360	680	1256
−240	−400		− 6.7	20	68.0	93	200	392	366	690	1274
−234	−390		− 5.6	22	71.6	99	210	410	371	700	1292
−229	−380		− 4.4	24	75.2	100	212	414	377	710	1310
−223	−370		− 3.3	26	78.8	104	220	428	382	720	1328
−218	−360		− 2.2	28	82.4	110	230	446	388	730	1346
−212	−350		− 1.1	30	86.0	116	240	464	393	740	1364
−207	−340		0.0	32	89.6	121	250	482	399	750	1382
−201	−330		1.1	34	93.2	127	260	500	404	760	1400
−196	−320		2.2	36	96.8	132	270	518	410	770	1418
−190	−310		3.3	38	100.4	138	280	536	416	780	1436
−184	−300		4.4	40	104.0	143	290	554	421	790	1454
−179	−290		5.6	42	107.6	149	300	572	427	800	1472
−173	−280		6.7	44	111.2	154	310	590	432	810	1490
−168	−270	−454	7.8	46	114.8	160	320	608	438	820	1508
−162	−260	−436	8.9	48	118.4	166	330	626	443	830	1526
−157	−250	−418	10.0	50	122.0	171	340	644	449	840	1544
−151	−240	−400	11.1	52	125.6	177	350	662	454	850	1562
−146	−230	−382	12.2	54	129.2	182	360	680	460	860	1580
−140	−220	−364	13.3	56	132.8	188	370	698	466	870	1598
−134	−210	−346	14.4	58	136.4	193	380	716	471	880	1616
−129	−200	−328	15.6	60	140.0	199	390	734	477	890	1634
−123	−190	−310	16.7	62	143.6	204	400	752	482	900	1652
−118	−180	−292	17.8	64	147.2	210	410	770	488	910	1670
−112	−170	−274	18.9	66	150.8	216	420	788	493	920	1688
−107	−160	−256	20.0	68	154.4	221	430	806	499	930	1706
−101	−150	−238	21.1	70	158.0	227	440	824	504	940	1724
− 96	−140	−220	22.2	72	161.6	232	450	842	510	950	1742
− 90	−130	−202	23.3	74	165.2	238	460	860	516	960	1760
− 84	−120	−184	24.4	76	168.8	243	470	878	521	970	1778
− 79	−110	−166	25.6	78	172.4	249	480	896	527	980	1796
− 73	−100	−148	26.7	80	176.0	254	490	914	532	990	1814
− 68	− 90	−130	27.8	82	179.6	260	500	932	538	1000	1832
− 62	− 80	−112	28.9	84	183.2	266	510	950	543	1010	1850
− 57	− 70	− 94	30.0	86	186.8	271	520	968	549	1020	1868
− 51	− 60	− 76	31.1	88	190.4	277	530	986	554	1030	1886
− 46	− 50	− 58	32.2	90	194.0	282	540	1004	560	1040	1904
− 40	− 40	− 40	33.3	92	197.6	288	550	1022	566	1050	1922
− 34	− 30	− 22	34.4	94	201.2	293	560	1040	571	1060	1940
− 29	− 20	− 4	35.6	96	204.8	299	570	1058	577	1070	1958
− 23	− 10	14	36.7	98	208.4	304	580	1076	582	1080	1976
− 17.8	0	32	37.8	100	212.0	310	590	1094	588	1090	1994
− 16.7	2	35.6	43	110	230	316	600	1112	593	1100	2012
− 15.6	4	39.2	49	120	248	321	610	1130	599	1110	2030
− 14.4	6	42.8	54	130	266	327	620	1148	604	1120	2048

Conversions for Fahrenheit–Celsius temperature scales (continued)

°C		°F	°C		°F	°C		°F	°C		°F
610	1130	2066	888	1630	2966	1166	2130	3866	1443	2630	4766
616	1140	2084	893	1640	2984	1171	2140	3884	1449	2640	4784
621	1150	2102	899	1650	3002	1177	2150	3902	1454	2650	4802
627	1160	2120	904	1660	3020	1182	2160	3920	1460	2660	4820
632	1170	2138	910	1670	3038	1188	2170	3938	1466	2670	4838
638	1180	2156	916	1680	3056	1193	2180	3956	1471	2680	4856
643	1190	2174	921	1690	3074	1199	2190	3974	1477	2690	4874
649	1200	2192	927	1700	3092	1204	2200	3992	1482	2700	4892
654	1210	2210	932	1710	3110	1210	2210	4010	1488	2710	4910
660	1220	2228	938	1720	3128	1216	2220	4028	1493	2720	4928
666	1230	2246	943	1730	3146	1221	2230	4046	1499	2730	4946
671	1240	2264	949	1740	3164	1227	2240	4064	1504	2740	4964
677	1250	2282	954	1750	3182	1232	2250	4082	1510	2750	4982
682	1260	2300	960	1760	3200	1238	2260	4100	1516	2760	5000
688	1270	2318	966	1770	3218	1243	2270	4118	1521	2770	5018
693	1280	2336	971	1780	3236	1249	2280	4136	1527	2780	5036
699	1290	2354	977	1790	3254	1254	2290	4154	1532	2790	5054
704	1300	2372	982	1800	3272	1260	2300	4172	1538	2800	5072
710	1310	2390	988	1810	3290	1266	2310	4190	1543	2810	5090
716	1320	2408	993	1820	3308	1271	2320	4208	1549	2820	5108
721	1330	2426	999	1830	3326	1277	2330	4226	1554	2830	5126
727	1340	2444	1004	1840	3344	1282	2340	4244	1560	2840	5144
732	1350	2462	1010	1850	3362	1288	2350	4262	1566	2850	5162
738	1360	2480	1016	1860	3380	1293	2360	4280	1571	2860	5180
743	1370	2498	1021	1870	3398	1299	2370	4298	1577	2870	5198
749	1380	2516	1027	1880	3416	1304	2380	4316	1582	2880	5216
754	1390	2534	1032	1890	3434	1310	2390	4334	1588	2890	5234
760	1400	2552	1038	1900	3452	1316	2400	4352	1593	2900	5252
766	1410	2570	1043	1910	3470	1321	2410	4370	1599	2910	5270
771	1420	2588	1049	1920	3488	1327	2420	4388	1604	2920	5288
777	1430	2606	1054	1930	3506	1332	2430	4406	1610	2930	5306
782	1440	2624	1060	1940	3524	1338	2440	4424	1616	2940	5324
788	1450	2642	1066	1950	3542	1343	2450	4442	1621	2950	5342
793	1460	2660	1071	1960	3560	1349	2460	4460	1627	2960	5360
799	1470	2678	1077	1970	3578	1354	2470	4478	1632	2970	5378
804	1480	2696	1082	1980	3596	1360	2480	4496	1638	2980	5396
810	1490	2714	1088	1990	3614	1366	2490	4514	1643	2990	5414
816	1500	2732	1093	2000	3632	1371	2500	4532	1649	3000	5432
821	1510	2750	1099	2010	3650	1377	2510	4550			
827	1520	2768	1104	2020	3668	1382	2520	4568			
832	1530	2786	1110	2030	3686	1388	2530	4586			
838	1540	2804	1116	2040	3704	1393	2540	4604			
843	1550	2822	1121	2050	3722	1399	2550	4622			
849	1560	2840	1127	2060	3740	1404	2560	4640			
854	1570	2858	1132	2070	3758	1410	2570	4658			
860	1580	2876	1138	2080	3776	1416	2580	4676			
866	1590	2894	1143	2090	3794	1421	2590	4694			
871	1600	2912	1149	2100	3812	1427	2600	4712			
877	1610	2930	1154	2110	3830	1432	2610	4730			
882	1620	2948	1160	2120	3848	1438	2620	4748			

$$C = \tfrac{5}{9}(F - 32) \qquad F = \tfrac{9}{5}C + 32$$

Tables of elements

Element	Symbol	Atomic weight	Melting point (°C)
Actinium	Ac	227	—
Aluminium	Al	26.97	658.7
Americium	Am	241	—
Antimony	Sb	121.77	630
Argon	Ar	39.94	–188
Arsenic	As	74.96	850
Astatine	At	211	—
Barium	Ba	137.36	850
Berkelium	Bk	245	—
Beryllium	Be	9.02	1280
Bismuth	Bi	209.00	271
Boron	B	10.82	2200–2500
Bromine	Br	79.91	–7.3
Cadmium	Cd	112.41	320.9
Caesium	Cs	132.81	26
Calcium	Ca	40.07	810.0
Californium	Cf	246	—
Carbon	C	12.00	3600
Cerium	Ce	140.13	635
Chlorine	Cl	35.45	–101.5
Chromium	Cr	52.01	1615
Cobalt	Co	58.94	1480
Copper	Cu	63.57	1083
Curium	Cm	242	—
Dysprosium	Dy	162.5	—
Erbium	Er	167.64	—
Europium	Eu	152	—
Fluorine	F	19.0	–223
Francium	Fa	223	—
Gadolinium	Gd	157.26	—
Gallium	Ga	69.72	30.1
Germanium	Ge	72.60	958
Gold	Au	197.2	1063
Hafnium	Hf	179	2200
Helium	He	4.00	–272
Holmium	Ho	165	—
Hydrogen	H	1.0078	–259
Indium	In	114.8	155
Iodine	I	126.932	113.5
Iridium	Ir	193.1	2350
Iron	Fe	55.84	1530
Krypton	Kr	83.7	–169
Lanthanum	La	138.90	810
Lead	Pb	207.22	327.4
Lithium	Li	6.94	186
Lutecium	Lu	175	—
Magnesium	Mg	24.32	651
Manganese	Mn	54.93	1230
Mercury	Hg	200.61	–38.87
Molybdenum	Mo	96	2620

Tables of elements (continued)

Element	Symbol	Atomic weight	Melting point (°C)
Neodymium	Nd	144.27	840
Neon	Ne	20.18	–253
Neptunium	Np	237	—
Nickel	Ni	58.69	1452
Niobium (Columbium)	Nb (Cb)	92.9	1950
Nitrogen	N	14.008	–210
Osmium	Os	190.8	2700
Oxygen	O	16.000	–218
Palladium	Pd	106.7	1549
Phosphorus	P	30.98	44
Platinum	Pt	195.23	1755
Plutonium	Pn	239	—
Polonium	Po	210	—
Potassium	K	39.1	62.3
Praseodymium	Pr	140.92	940
Promethium	Pm	147	—
Protactinium	Pa	231	—
Radon	Rn	222	–71
Radium	Ra	226.1	700
Rhenium	Re	186	3167
Rhodium	Rh	102.91	1950
Rubidium	Rb	85.44	38
Ruthenium	Ru	101.7	2450
Samarium	Sm	150.43	1300–1400
Scandium	Sc	45.10	1200
Selenium	Se	78.96	217–220
Silicon	Si	28.06	1420
Silver	Ag	107.88	960.5
Sodium	Na	22.997	97.5
Strontium	Sr	87.63	800
Sulphur	S	32.06	112.8
Tantalum	Ta	181.5	2900
Technetium	Tc	99	—
Tellurium	Te	127.5	452
Terbium	Tb	159.2	—
Thallium	Tl	204.39	302
Thorium	Th	232.12	1700
Tin	Sn	118.70	231.9
Thulium	Tm	169.4	—
Titanium	Ti	47.9	1800
Tungsten	W	184.0	3400
Uranium	U	238.14	1850
Vanadium	V	50.96	1720
Xenon	Xe	131.3	–140
Ytterbium	Yb	173.6	1800
Yttrium	Y	88.92	1490
Zinc	Zn	65.38	419.4
Zirconium	Zr	91.22	1700

Conventional symbols and methods of indicating butt welds on drawings (after British Standards)

DESCRIPTION OF WELD	SYMBOL	SECTIONAL REPRESENTATION	SYMBOLIC REPRESENTATION
SQUARE BUTT			
SINGLE–V BUTT			
DOUBLE –V BUTT			
SINGLE–U BUTT			
DOUBLE–U BUTT			
SINGLE–BEVEL BUTT			
DOUBLE–BEVEL BUTT			
SINGLE– J BUTT			
DOUBLE– J BUTT			

See also *Basic Welding*, pages 127–129.

Welding symbols (after British Standards)

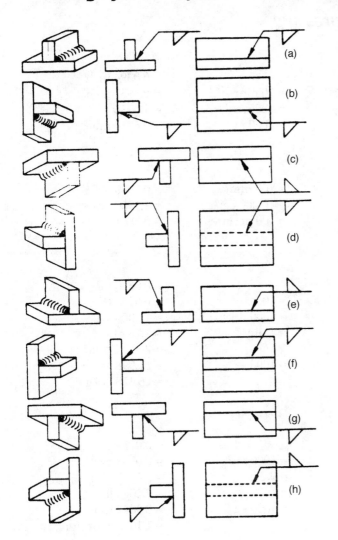

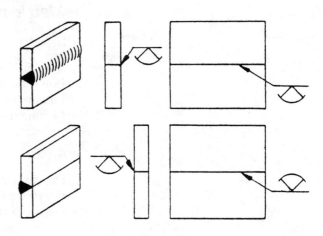

Single-V butt welds without a sealing run.

(left)

Examples showing the significance of the arrow and position of the weld symbol in relation to the reference line in the case of fillet welds on one side of a tee.

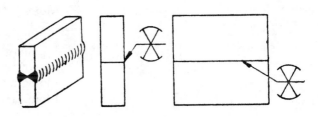

Single–V butt weld with a sealing run.

See also *Basic Welding*, pages 127–129.

Metric and Imperial units

Metric measures

Length

1 millimetre (mm)		= 0.0394 inch
1 centimetre (cm)	= 10 mm	= 0.3937 inch
1 metre (m)	= 100 cm	= 1.0936 yd
1 kilometre (km)	= 1000 m	= 0.6214 mile

Area

1 sq cm (cm²)	= 100 mm²	= 0.1550 in²
1 sq metre (m²)	= 10 000 cm²	= 1.1960 yd²
1 sq km (km²)	= 100 hectares	= 0.3861 mile²

Volume/capacity

1 cu cm (cm³)		= 0.0610 in³
1 cu decimetre (dm³)	= 100 cm³	= 0.0353 ft³
1 cu metre (m³)	= 1000 dm³	= 1.3080 yd³
1 litre (l)	= 1 dm³	= 1.76 pt
		= 2.113 US liq pt
1 hectolitre (hl)	= 100 l	= 21.998 gal
		= 26.418 US gal

Mass (weight)

1 milligram (mg)		= 0.0154 grain
1 gram (g)	= 1000 mg	= 0.0353 oz
1 metric carat	= 0.2 g	= 3.0865 grains
1 kilogram (kg)	= 1000 g	= 2.2046 lb
1 tonne (t)	= 1000 kg	= 0.9842 ton
		= 1.1023 short ton

Pressure, stress

1 hectobar	= 0.6475 tonf/in²	
1 megapascal (MPa)	= 0.0647 tonf/in²	= 145.038 lbf/in²
1 atm	= 14.696 lbf/in²	
1 kgf/cm²	= 14.223 lbf/in²	
1 bar	= 14.504 lbf/in²	
1 kPa (kN/m²)	= 20.885 lbf/ft²	
1 pascal (Pa = N/m²)	= 0.000 145 lbf/in²	

Energy (work, heat)

1 megajoule (MJ)	= 0.2778 kWh
1 joule (J)	= 0.7376 ft lbf
1 calorie	= 0.003 97 Btu

Temperature conversion

$$C = \tfrac{5}{9}(F - 32) \qquad\qquad F = \tfrac{9}{5}C + 32$$

Imperial and US measures

Length

1 inch (in)		= 2.54 cm
1 foot (ft)	= 12 in	= 0.3048 m
1 yard (yd)	= 3 ft	= 0.9144 m
1 mile	= 1760 yd	= 1.6093 km
1 int nautical mile	= 2025.4 yd	= 1.852 km

Area

1 acre	= 4046.86 m^2	= 0.4047 ha
1 sq yard (yd^2)	= 0.8361 m^2	
1 sq foot (ft^2)	= 0.0929 m^2	
1 sq inch (in^2)	= 645.16 mm^2	

Volume/capacity

1 cu yard (yd^3)	= 0.7645 m^3	
1 cu foot (ft^3)	= 0.0283 m^3	= 28.3168 dm^3
1 cu inch (in^3)	= 16.3871 cm^3	
1 UK gallon	= 4.5461 dm^3	
1 US gallon	= 3.7854 dm^3	
1 pint (pt)	= 0.5683 dm^3	

Mass (weight)

1 ton	= 1016.05 kg	= 1.016 t
1 pound (lb)	= 0.4536 kg	
1 ounce (oz)	= 28.3495 g	
1 grain	= 64.7989 mg	= 0.324 metric carats

Pressure, stress

1 tonf/ft^2	= 107.252 kPa	
1 tonf/in^2	= 15.4443 MPa	
1 lbf/ft^2	= 47.8803 Pa	
1 lbf/in^2	= 6.8948 kPa	= 68.9476 mbar

Energy (work, heat)

1 therm	= 105.506 MJ
1 hp h (horsepower × hour)	= 2.6845 MJ
1 kWh	= 3.6 MJ
1 British thermal unit (Btu)	= 1.0551 kJ

Power

1 horsepower (hp)	= 745.700 W	= 0.7457 kW

American Welding Society

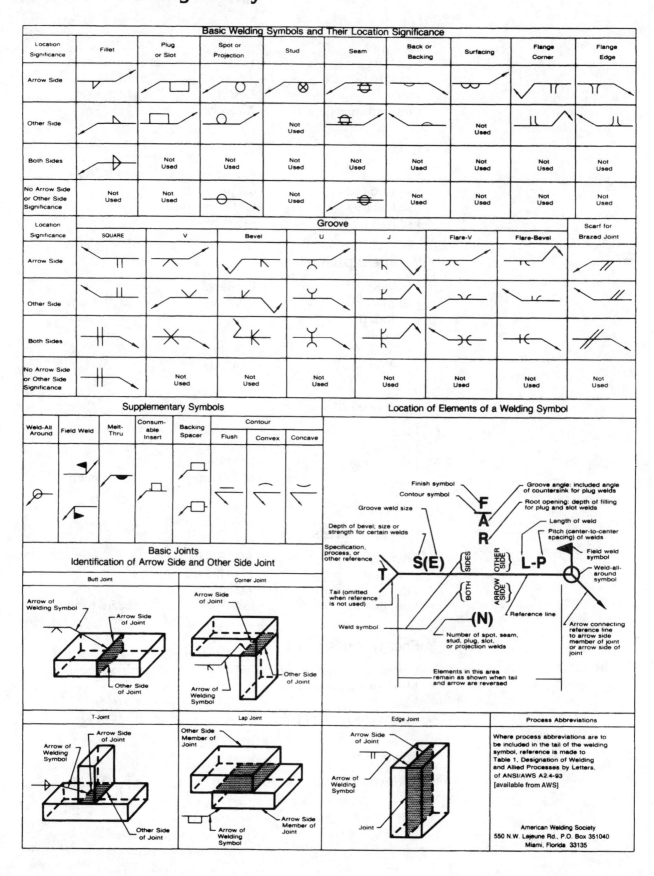

American Welding Society

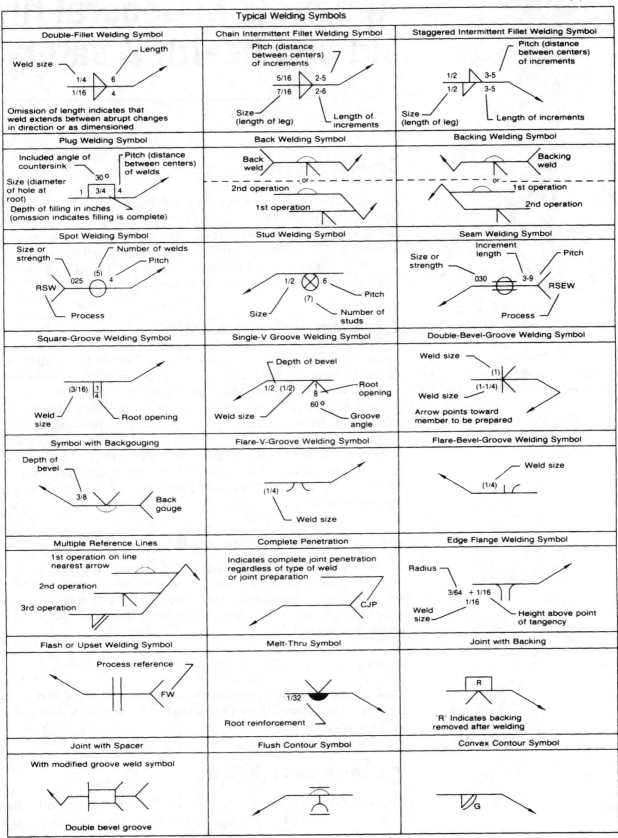

Typical Welding Symbols

* It should be understood that these charts are intended only as shop aids. The only complete and official presentation of the standard welding symbols is in A2.4. [available from AWS]

Appendix 5: General first aid and health & safety

General first aid

For any injuries other than very minor ones, you should call for expert medical attention, but first aid can help and comfort the patient until the doctor or ambulance arrives. First aid can save lives and there are many organizations worldwide that run short courses to train qualified first-aiders.

The following notes contain some of the basic points of first aid that anyone working in the welding and fabrication industries should be aware of.

Treatment for burns or scalds

Burns cause damage to body tissue by heat, chemicals or radiation. Burns caused by steam or hot liquids are called **scalds**. With welding processes, burns are the most common injury and are mostly of a minor nature. Welders' gloves and other protective equipment give protection most of the time, but minor burns can be caused by sparks or by accidentally touching hot metal.

If possible, immerse the burn or scald in cold water or under cold running water for at least 10 minutes, until pain reduces (Figure A5.1). Then cover with dry gauze or clean fabric and bandage. Do not remove clothing or apply ointments. Do not break blisters. If a burn is more than 3 cm in diameter, seek medical advice.

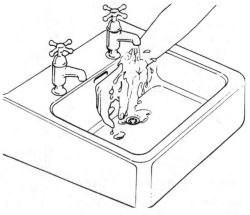

Fig. A5.1 *Immerse a small burn or scald in cold water or under cold running water for at least 10 minutes*

Arc burn: similar to severe sunburn

If skin is exposed to the rays from an electric arc, it can become tender and swollen, and may possibly blister. The affected skin will also feel hot. This is why you should ensure that all skin is covered with suitable protective clothing when electric arc welding, in addition to the protection provided against sparks and hot metal.

Treatment

Cool the skin gently by sponging it with cold water. If the skin is not broken, apply sunburn cream. Do not break blisters. If there is extensive blistering, seek qualified medical aid.

Arc eye or welder's flash

The front of the eye, called the **cornea**, can be injured if exposed to the ultra-violet light produced by the electric welding arc. This can sometimes happen when you walk past welding operations that are not screened off properly, or if your welding helmet is not positioned correctly or has a cracked filter

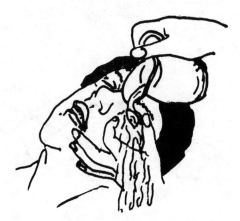

Fig. A5.2 *Bathing eyes with clean, cold water*

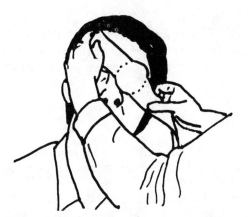

Fig. A5.3 *Applying clean pads*

lens. The symptoms usually appear up to 6 hours after exposure to the welding arc. They consist of intense pain in the affected eye(s) and a feeling as if they are full of sand. Eyes affected will be sensitive to light, red in colour, and may water in severe cases.

Treatment

Bathe the eye(s) with cold water and then apply pads of clean, non-fluffy material (Figures A5.2 and A5.3). Arc eye can last for up to 48 hours. In severe cases, seek medical attention; special welder's eye drops are available from a doctor or hospital which will help to ease the pain. Wearing dark glasses can help to ease discomfort in the later stages of arc eye.

Electrical injuries

If an electric current passes through the body it can cause severe and sometimes fatal injuries. It can affect the heart muscle, causing the heart to stop beating and breathing to stop.

Never touch the casualty with bare hands until you are sure that the current has been turned off or the casualty is no longer in contact with the electrical source. With high-voltage electricity, such as electrical transmission lines, do not approach the accident area until police or authority in charge says that it is safe to do so. High-voltage electricity can 'arc' over considerable distances and insulating materials will not provide any protection in these circumstances.

Treatment

If the injured person's breathing and heart have stopped, start resuscitation by using mouth-to-mouth ventilation and applying external chest compression (Figures A5.4 and A5.5). If the injured person is breathing normally but unconscious, place them in the 'recovery position' on their side, unless you suspect a fracture of the spine.

Tend any obvious injuries and treat the casualty for shock while you are waiting for medical assistance to arrive. Tell medical authorities how long the casualty was in contact with the electrical source.

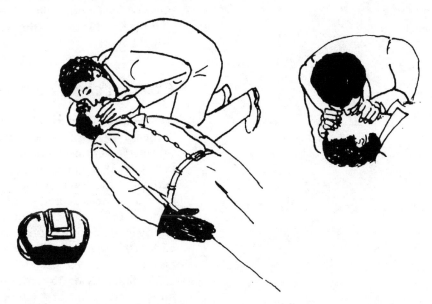

Fig. A5.4 *Mouth-to-mouth ventilation*

Fig. A5.5 *Method of external chest compression*

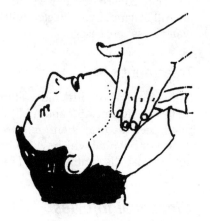

Fig. A5.6 *Position of the carotid pulse*

Fig. A5.7 *The recovery position. The arm should be carefully moved from under the casualty and placed parallel (as shown), to prevent the casualty from rolling over on to his or her back*

Mouth-to-mouth ventilation

Remove any obvious obstructions from the casualty's face and loosen their collar. Open the airway by placing one hand under casualty's neck, the other hand on the forehead, and tilt their head backwards.

Transfer your hand from the neck and push the casualty's chin upwards. This will lift their tongue forwards, clearing the airway. Remove any debris that might be in the mouth or throat.

Open your mouth wide and take a deep breath. Pinch the casualty's nostrils together with your fingers. Sealing your lips around the mouth, blow into the casualty's lungs until you can see the chest rise. Then remove your mouth well away from the casualty's and breath out any excess air. Watch the chest fall, then take a fresh breath and repeat the procedure. Check the casualty's pulse to ensure that the heart is beating. Figure A5.6 shows the position of the pulse on the carotid artery.

If the heart is not beating you must carry out external chest compression straight away.

External chest compression

Lay the casualty on their back and kneel alongside. Place the heel of one hand in the centre of the lower region of the breastbone, keeping your fingers off the ribs. Cover this hand with your other hand and lock fingers together (see Figure A5.5). Keeping your arms straight, move forwards until your arms are vertical. Press down on the lower part of the breastbone about 4 to 5 cm ($1\frac{1}{2}$ to 2 inches) for an average adult. Then move backwards, releasing the pressure. Perform 15 compressions at the rate of about 80 per minute. (To judge the approximate timing count one and two and three, and repeat.) Then move back to the casualty's head, re-opening the airway and giving two mouth-to-mouth ventilations.

Continue giving 15 compressions followed by two ventilations, checking for heartbeat after one minute. Continue, checking heartbeat every three minutes. As soon as the heartbeat returns, stop compressions immediately

Fig. A5.8 *One method of supporting an injured arm in a padded sling*

Fig. A5.9 *Securing an injured arm if the elbow cannot be moved*

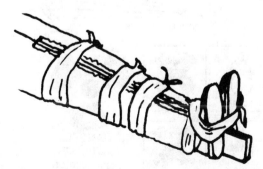

Fig. A5.10 *Supporting a fractured leg if a splint is available. Pad the splint and bandage, avoiding the site of the fracture*

and continue mouth-to-mouth ventilation until natural breathing returns. Place the casualty in the recovery position (Figure A5.7). When resuscitation is successful, the carotid pulse will return.

Shock

If a person has been injured they often become quite weak because they are in a **state of shock**. You should immediately reassure and comfort the casualty, loosen any tight clothing and keep the casualty warm with extra clothing or a blanket. If the casualty is thirsty, moisten their lips with water but do not give them anything to drink.

Bleeding and wounds

Get the casualty to sit or lie down and elevate the bleeding limb if no fracture is suspected. Apply a clean pad and bandage. If blood seeps through, apply a further pad and then pressure. Continue applying pressure with the second pad until bleeding stops or medical help arrives.

Treatment of a fracture

The prime first-aid treatment is to prevent movement in the area of the fracture until qualified medical attention arrives. If you must move a casualty, then support the part with padding or by hand. For a short journey to hospital, you may be able to immobilize the injured part by securing it to a sound part of the body using padding and bandages. Bandages must be firm, but not so tight that they affect the blood circulation. For a long journey over rough ground, you may need extra bandages and splints. You can use virtually any suitable strong stick or piece of metal as a splint – even rolled-up newspaper. Keep checking to ensure that the bandages are not too right. See Figures A5.8, A5.9 and A5.10.

> It is very important to remember that, if you suspect a fracture of the neck or spine, do not move the patient unless it is absolutely necessary because their life is in danger from some other cause. It is always best to comfort the patient and carefully tend to obvious injuries, restricting movement until qualified medical help arrives. Moving a patient with a fracture of the spine can cause permanent injury – even paralysis – if done incorrectly.

Summary

- For any injuries other than very minor ones, get expert medical attention as soon as possible.
- For burns of less than 3 cm diameter, immerse the area under cold or cold running water for at least 10 minutes.
- If skin is exposed to the rays of an electric arc, the area can become tender and swollen. This condition is known as 'arc burn'. It should be treated by gently sponging with cold water. If the skin is not broken, you can apply sunburn cream.

- The condition called 'arc eye' or welder's flash is fairly common. Eyes can be affected either by direct or by reflected exposure to the ultra-violet light produced by the electric welding arc. The symptoms – pain and 'sand in the eye' feeling – can last up to 48 hours. Bathe the eyes with cold water and cover with pads of clean, non-fluffy material. Special welder's eye drops which can help to ease the pain, are available from a doctor or hospital.
- If someone has received an electric shock, **never** touch them until you are sure that the current has been turned off.
- If breathing and heart have stopped, carry out mouth-to-mouth ventilation and apply external chest compression.
- If a person has been injured they often become quite weak because they are in a state of shock. Reassure and comfort them, loosen any tight clothing, but keep the casualty warm with extra clothing or a blanket. If the casualty is thirsty, moisten their lips with water but do not give them anything to drink.
- If a casualty is bleeding, get them to sit or lie down, elevate the bleeding limb if possible, and apply a clean pad and bandage.
- If you suspect a fracture, prevent movement in the area of the fracture until qualified medical attention arrives.
- When bandages are applied, keep checking to ensure that they are not too tight.

First aid is the method of treating a casualty until expert medical attention arrives, and the measures stated briefly here have been used many times to bring comfort to the casualty, ease pain and save lives. A first-aid course will ensure that you are properly trained and able to help should an accident occur.

Health and safety

Introduction

Safety is everyone's responsibility. We are responsible both for our own safety and for that of other people.

Fig. A5.11 *Some common hazard signs*

Because of this, we must understand correct working methods and be able to notice if something is not as it should be. It is our duty to report unsafe equipment and to warn others about unsafe working practices.

See the Health and Safety section in *Practical Welding* (ISBN 0–333–60957–3), pages 10 to 20 and *Basic Welding* (ISBN 0–333–57853–8), pages 5 to 17. (Both books published by Macmillan Press, Basingstoke and written by the author of *Advanced Welding*.)

This covers safety points that are in most cases common to all welding processes and welding workshops. The names of regulations may vary from one country to another, but the content is concerned with safety in welding and therefore applies to any areas or workplaces where welding takes place.

The Health and Safety at Work etc. Act 1974

This Act is very important. It covers the legal duties of employers, employees and self-employed persons. Health and safety are everyone's responsibility. One of the important aims of the Act is to encourage employers and employees to work together to make a safer workplace.

Appendix 6: International Institute of Welding List of Member Societies/Liste des Sociétés Membres (with addresses), 1995

Fig. A6.1 *Reception building at The Welding Institute, Cambridge, UK*

Fig. A6.2 *The American Welding Society's Precision Joining Center, Denver, USA*

International Institute of Welding
Institut International de la Soudure

List of Member Societies/Liste des Sociétés Membres 1994/95

ARGENTINA/ARGENTINE

Fundación Latinoamericana de
  Soldadura
Calle 18, No 4113
1672 Villa Lynch,
Buenos Aires
Argentina

Tel: 54 1 753 4039
Fax: 54 1 755 1268

AUSTRALIA/AUSTRALIE

Welding Technology Institute of
  Australia
PO Box 6165
Silverwater
NSW 2128
Australia

Tel: 61 2 748 4443
Fax: 61 2 748 2858

AUSTRIA/AUTRICHE

Osterreichische Gesellschaft für
  Schweisstechnik
Arsenal, Objekt 12
A-1030 Wien
Austria

Tel: 43 1 798 2168
Fax: 43 1 798 2168-15

Sweisstechnische Zentralanstalt
Arsenal, Objekt 207
A-1030 Wien
Austria

BELGIUM/BELGIQUE

Institit Belge de la Soudure
21 Rue des Drapiers
1050 Brussels
Belgium

Tel: 32 3 512 2892
Fax: 32 3 512 7457

BRAZIL/BRESIL

Brazilian Committee of IIW
Rua Sao Francisco Xavier 601
Maracana
20550-011 Rio de Janeiro - RJ
Brazil

Tel: 55 21 254 0203
Fax: 55 21 284 2191

BULGARIA/BULGARIE

National Welding Society
c/o Union of Mechanical
  Engineering of Bulgaria
Rakovski Street 108, POB 431
Sofia 1000
Bulgaria

Tel: 359 2 877290
Fax: 359 2 879360/802365

CANADA

Welding Institute of Canada
Canadian Council of the IIW
391 Burnhamthorpe Road East
Oakville
Ontario
L6J 6C9
Canada

Tel: 1 905 257 9881
Fax: 1 905 257 9886

CHILE/CHILI

Centro Tecnico Chileno de la
  Soldadura
Camino a Melipilla 7060
P O Box 13850
Santiago
Chile

Tel: 56 2 571 777
Fax: 56 2 557 3471

CHINA/CHINE

Chinese Welding Society
111 Hexing Lu
Harbin 150080
Peoples Rep of China

Tel: 86 451 6322012
Fax: 86 451 6325871

CROATIA/CROATIE

Hrvatsko Drustvo Za Tehniku
  Zavarivanja
Ivana Lucica 1
41000 Zagreb
Croatia

Tel: 385 41 512 689
Fax: 385 41 512 689

CZECH REPUBLIC/
REPUBLIQUE TCHEQUE

Czech Welding Society
Novotného lávka 5
110 01 Praha 1
Czech Republic

Tel: 42 4 2310 124
Fax: 42 4 261 897

DENMARK/DANEMARK

Danish Welding Society
Park Alle 345
DK-2605 Broendby
Denmark
Tel: 45 43 96 8800
Fax: 45 43 96 2636

Force Institute
Park Alle 345
DK-2605 Broendby
Denmark

EGYPT/EGYPTE

CMRDI & ESEMI
P O Box 87 Helwan
Cairo
Egypt

Tel: 20 790775 / 790003
Fax: 20 790898 / 794108

FINLAND/FINLANDE

Suomen Hitsausteknillinen
Yhdistys r.y.
Makelankatu 36 A 2
FIN-00510 Helsinki
Finland

Tel: 358 0 773 2199
Fax: 358 0 773 2661

FRANCE

Institut de Soudure
Zi Paris Nord II
BP 50362
95942 Roissy CDG Cedex
France

Tel: 33 1 49 90 36 10
Fax: 33 1 49 90 36 50

Sociétés des Ingénieurs Soudeurs
ZI Paris Nord II
BP 50362
95942 Roissy CDG Cedex
France

GERMANY/ALLEMAGNE

Deutscher Verband für
  Schweisstechnik e.V.
Postfach 10 19 65
40010 Düsseldorf
Germany

Tel: 49 211 59 10
Fax: 49 211 1 59 1200

GREECE/GRECE

Hellenic Institute of Welding
  Technology
P O Box 64070
157 10 Zografos
Greece

Tel: 30 1 77 00 671
Fax: 30 1 77 59 213

HUNGARY/HONGRIE

Gépipari Tudományos Egyesület
Fo utca 68
POB 433
1371 Budapest
Hungary

Tel: 36 1 202 0582 / 0656
Fax: 36 1 202 0252

IRAN

Iranian Institute of Welding and
  NDT
P O Box 14155-4686
Tehran
Iran

Tel: 98 21 653 096
Fax: 98 21 651 809

ISRAEL

The Israeli Metallurgical Society
Israel Institute of Metals
Technion City
P O Box 4910, 32 000 Haifa
Israel

Tel: 972 4 23 5104
Fax: 972 4 22 1581

ITALY/ITALIE

Istituto Italiano della Saldatura
Lungobisagno Istria 15
16141 Genova
Italy

Tel: 39 10 83 411
Fax: 39 10 836 7790

Registro Italiano Navale
Via Corsica 12
16129 Genova GE
Italy

Tel: +39 10 53851
Fax: +39 10 591877

JAPAN/JAPON

Japan Institute of Welding
1-11 Kanda, Sakuma-Cho
Chiyoda-Ku
Tokyo 101, Japan

Tel: 81 33257 1521
Fax: 81 33255 5196

Library Section
Science Council of Japan
22-34 Roppongi 7-Chome
Minato-Ku
Tokyo 106
Japan

KOREA/COREE

The Korean Welding Society
Yusung P O Box 104
Tae-job, 305-343
Korea

Tel: 82 42 861 2696
Fax: 82 42 861 1172

MEXICO/MEXIQUE

Colegio de Ingenieros Mecanicos y
  Electricistas
Oklahoma 89
CP 03810
Mexico D.F.
Mexico

Tel: 52 5 5231123 / 5231254
Fax: 52 5 5437902

NETHERLANDS/PAYS-BAS

Nederlands Instituut voor
  Lastechniek
Krimkade 20
2251 KA Voorschoten
The Netherlands

Tel: 31 71 61 12 11
Fax: 31 71 61 14 26

**NEW ZEALAND/
  NOUVELLE-ZELANDE**

New Zealand Welding Committee
POB 76-134
Manukau City, Auckland
New Zealand

Tel: 64 9 262 2885
Fax: 64 9 262 2856

NORWAY/NORVEGE

Norsk Sveiseteknisk Forening
PO Box 7072 Homansbyen
N-0306 OSLO
Norway

Tel: 47 22 465 820
Fax: 47 22 461 838

POLAND/POLOGNE

Instytut Spawalnictwa
ul Bl Czeslawa 16 / 18
44-100 GLIWICE
Poland

Tel: 48 32 31 00 11
Fax: 48 32 31 4652

PORTUGAL

Instituto de Soldadura e Qualidade
Estrada Nacional 249 - km 3
Cabanas - Leiao (Taguspark)
Apartado 119
2781 Oeiras Codex
Portugal

Tel: 351 1 421 1307 / 1323
Fax: 351 1 421 1406 / 1518

ROMANIA/ROUMANIE

Institute of Welding & Material
 Testing
Bv Mihai Viteazul Nr 30
1900 Timisoara
Romania

Tel: 40 56 191831 / 191835
Fax: 40 56 192797

RUSSIA/RUSSIE

Russian Welding Society
Shelaputinksky per. 1
109004 Moscow
Russia

Tel: 07 95 915 0917
Fax: 07 95 915 0841

SERBIA/SERBIE

Savez Za Zavarivanje
Grcica Milenka 67
11000 Beograd
Serbia

Tel: 381 111 419 255
Fax: 381 11 402 849

SLOVAKIA/SLOVAQUIE

Výskumný ústav zváračský
Račianska 71
832 59 Bratislava
Slovakia

Tel: 42 7 255 966
Fax: 42 7 254 867

SLOVENIA/SLOVENIE

Zveza Drustev Za Varilno Tehniko
 Slovenije
c/o Institut Za Varilstvo
Ptujska 19
61000 Ljubljana
Slovenia

Tel: 386 61 182 533
Fax: 386 61 349 282

SOUTH AFRICA/ AFRIQUE DU SUD

South African Institute of Welding
POB 527
Crown Mines
2025 Johannesburg
South Africa

Tel: 27 11 836 4121
Fax: 27 11 836 4132

SPAIN/ESPAGNE

Dr. J. Fernandez-Ballesteros
c/o Centro Nacional de
 Investigaciones de Soldadura
Avda. Gregorio del Amo 8
28040 Madrid
Spain

Tel: 34 1 553 89 00
Fax: 34 1 534 74 25

SWEDEN/SUEDE

Svetskommissionen
Box 5073
S-10242 Stockholm
Sweden

Tel: 46 8 791 2900
Fax: 46 8 679 9404

Tryckkarlsstandardiseringen
Box 49126
S-100 28 Stockholm

Svetstekniska Foreningen
Box S 5073
S-19242 Stockholm